Studies in Computational Intelligence

Volume 756

Series editor

Janusz Kacprzyk, Polish Academy of Sciences, Warsaw, Poland
e-mail: kacprzyk@ibspan.waw.pl

The series "Studies in Computational Intelligence" (SCI) publishes new developments and advances in the various areas of computational intelligence—quickly and with a high quality. The intent is to cover the theory, applications, and design methods of computational intelligence, as embedded in the fields of engineering, computer science, physics and life sciences, as well as the methodologies behind them. The series contains monographs, lecture notes and edited volumes in computational intelligence spanning the areas of neural networks, connectionist systems, genetic algorithms, evolutionary computation, artificial intelligence, cellular automata, self-organizing systems, soft computing, fuzzy systems, and hybrid intelligent systems. Of particular value to both the contributors and the readership are the short publication timeframe and the world-wide distribution, which enable both wide and rapid dissemination of research output.

More information about this series at http://www.springer.com/series/7092

Vassil Sgurev · Vincenzo Piuri
Vladimir Jotsov
Editors

Learning Systems: From Theory to Practice

Editors
Vassil Sgurev
Bulgarian Academy of Sciences
Institute of Information
and Communication Technology
Sofia
Bulgaria

Vladimir Jotsov
University of Library Studies
and Information Technologies
Sofia
Bulgaria

Vincenzo Piuri
Department of Computer Science
Università degli Studi di Milano
Crema
Italy

ISSN 1860-949X ISSN 1860-9503 (electronic)
Studies in Computational Intelligence
ISBN 978-3-030-09164-4 ISBN 978-3-319-75181-8 (eBook)
https://doi.org/10.1007/978-3-319-75181-8

Printed on acid-free paper

This Springer imprint is published by the registered company Springer International Publishing AG part of Springer Nature
The registered company address is: Gewerbestrasse 11, 6330 Cham, Switzerland

Dedicated to the rise of ideas and memories of Lotfi Zadeh

Preface

Nowadays, fuzzy sets and computing with words systems are penetrating into all aspects of intelligent systems and applications. This volume presents recent advances in this area, which are increasingly important in the fuzzy-related theories. Computing with words is addressed from various points of view, from theory to applications, to provide a comprehensive vision and trends of the research and practical use of this technology.

Chapter “From von Neumann Architecture and Atanasoffs ABC to Neuro-Morphic Computation and Kasabov’s NeuCube: Principles and Implementations” is written by a large team of authors: Neelava Sengupta, Josafath Israel Espinosa Ramos (New Zealand), Enmei Tu (Singapore), Stefan Marks, Nathan Scott (New Zealand), Jakub Weclawski (Poland), Akshay Raj Gollahalli, Maryam Gholami Doborjeh, Zohreh Gholami Doborjeh, Kaushalya Kumarasinghe, Vivienne Breen, and Anne Abbott (New Zealand). Neuromorphic computing beyond von Neumann architecture had been discussed. As a result, the first neuromorphic spatiotemporal data machine NeuCube was introduced aiming at unsupervised learning. The neuromorphic computing aspires to move away from the bit-precise computing paradigm toward the probabilistic models. On the other hand, we can easily establish links from the described pattern recognition, classification, or regression problems to uncertain and fuzzy logic applications in forthcoming generations of neuro-fuzzy systems.

Chapter “Applications of Computational Intelligence in Industrial and Environmental Scenarios” written by Ruggero Donida Labati, Angelo Genovese, Enrique Muñoz, Vincenzo Piuri, and Fabio Scotti (Italy) is dedicated to novel Computational Intelligence (CI) techniques that are able to aggregate inputs from several heterogeneous sensors, adapt themselves to wide ranges of operational and environmental conditions, and cope within complete or noise-affected data. Fuzzy logic is quoted as one of the most commonly used methods. Its usage had been mentioned in data preprocessing, data fusion, and classification stages. The authors consider a combination of artificial neural networks, support vector machines, fuzzy logic, or evolutionary computation aiming to obtain flexible models that can cope with noise, data incompleteness, and varying operational conditions, while having a limited computational complexity.

Chapter "Real, Imaginary and Complex Sets" is written by Vassil Sgurev (Bulgaria). An extension of the classical set theory had been applied to the propositional logic environment. In this direction, fuzzy set applications are welcome to be discussed in this case aiming at wide range of contemporary intelligent applications.

Chapter "Intercriteria Analysis over Patterns" authored by Krassimir Atanassov is devoted to a new method for decision support, aimed at detection of dependencies between pairs of criteria in a multiobject, multicriteria problem, on the basis of pairwise comparisons of the evaluations or mesurements of the set of objects against the set of criteria. The presence of sufficiently high coefficients of dependence between some criteria can support the decision-making process, for instance, to skip measurements of some or all the objects against a criterion, if these measurements are rather expensive, slow, or in other way cost unfavorable. The method employs the concepts of intuitionistic fuzzy sets and index matrices, and renders an account of the inherent levels of uncertainty in real-life applications. The first attempt for application of the intercriteria analysis over patterns is described. Pattern recognition, scene analysis, computing with words, and other potential applications are discussed in the chapter.

Boris Stilman (USA) created Chapter "From Primary to Conventional Science". A primary language is introduced as the Language of Visual Streams (mental movies) that operate via multiple thought experiments. Based on all those results, the author suggested that the essence of the discoveries is in the visual streams. As in most previous chapters, nowhere the word "fuzzy" is mentioned in this chapter, but it is obvious that the proposed reasoning component (visual reasoning) would be more effective using fuzzy sets and applications. This reflects any of the introduced stream classes, namely, observation, construction, and validation.

Chapter "Intelligent Two-Level Optimization and Model Predictive Control of Degrading Plants" written by Mincho Hadjiski, Alexandra Grancharova, and Kosta Boshnakov (Bulgaria) is dedicated to intelligent control and plant optimization using case-based and other intelligent tools and technologies. Since more than a decade fuzzy model predictive control methods proved high efficiency in many analogical approaches. Contemporary researchers do not use both the mentioned directions in one system because of their high computational complexity but when speaking on perspective applications, this question will be resolved.

Chapter "Collaborative Clustering: New Perspective to Rank Factor Granules" by Shihu Liu, Xiaozhou Chen (China), and Patrick S. P. Wang (USA) and in this chapter fuzzy granules have been used for clustering purposes. Factor granules are considered in this chapter. They are ranked in terms of collaborative clustering. During the process of constructing the ranking algorithm, the privacy of each factor granule is well preserved.

Chapter "Learning Through Constraint Applications" is written by Vladimir Jotsov, Pepa Petrova, and Evtim Iliev (Bulgaria). Since many decades, constraint satisfaction approaches are widely used. New types of constraints are introduced in this chapter and they have been used in logical-based applications under control of synthetic methods. The synthetic puzzle method is discussed in this chapter. One of the important results of this research is that the logical-oriented constraints help agents or analogical software to concentrate the attention on most important events or objects. Computing with words is the motto of the proposed enhanced modeling and knowledge processing.

Chapter "Autonomous Flight Control and Precise Gestural Positioning of a Small Quadrotor" is written by Nikola G. Shakev, Sevil A. Ahmed, Andon V. Topalov, Vasil L. Popov, and Kostadin B. Shiev (Bulgaria). Intelligent autonomous flight control strategy has been considered in this chapter. Gestural control is also mentioned, and in this direction type-2 fuzzy topic models for human action recognition had been analyzed. It is shown by practical examples that the proposed gesture control is a promising approach.

Chapter "Introduction to the Theory of Randomized Machine Learning" is prepared by Yuri S. Popkov, Yuri A. Dubnov, and Alexey Y. Popkov (Russia). Probability density functions of random parameters had been introduced aiming at minimization of errors during the machine learning process. The authors apply this approach to text classification and dynamic regression problems. As in previous chapters, the word "fuzzy" is found nowhere in this material, but this chapter is the same example of possible applications to Big Data problems, where fuzzy classification or prediction methods are essential part of the learning cycle.

Chapter "Grid-Type Fuzzy Models for Performance Evaluation and Condition Monitoring of Photovoltaic Systems" represents the innovative research conducted by Gancho Vachkov and Valentin Stoyanov (Bulgaria). The main idea in the presented research is to construct special grid-type fuzzy models with a partial fuzzy rule base, where the number of the fuzzy rules and their locations depend on the amount of the available data and their distribution in the input space.

Chapter "Intelligent Control of Uncertain Switched Nonlinear Plants: NFG Optimum Control Synthesis via Switched Fuzzy Time-Delay Systems" is written by Georgi M. Dimirovski (FYROM, Turkey), Jinming Luo, and Huakuo Li (China). Intelligent control had been applied using the Lyapunov of Lurie-type functions. As a result, a closed-loop system had been designed that is asymptotically stable. A switched fuzzy system had been applied.

Chapter "Multidimensional Intuitionistic Fuzzy Quantifiers and Level Operators" is written by Krassimir Atanassov, Ivan Georgiev (Bulgaria), Eulalia Szmidt, and Janusz Kacprzyk (Poland). The paper is a continuation of previous authors' research, in which the concepts of multidimensional intuitionistic fuzzy sets and logic are introduced. Here, some groups of multidimensional intuitionistic fuzzy quantifiers and some different intuitionistic fuzzy-level operators are introduced. Their basic properties are studied and open problems are formulated.

Chapter "Data Processing and Harmonization for Intelligent Transportation Systems: An Application Scenario on Highway Traffic Flows" is written by Paulo Figueiras, Guilherme Guerreiro, Ricardo Silva, Ruben Costa, and Ricardo Jardim-Gonçalves (Portugal). The research concerns Intelligent Transportation Systems and ETL approaches realized in an ongoing work under the EU

H2020 OPTIMUM project. Same as in research represented in chapters "Collaborative Clustering: New Perspective to Rank Factor Granules" or "Introduction to the Theory of Randomized Machine Learning", the application here is tightly linked to Big Data applications and its extentsion will inevitably lead to fuzzy clustering and/or realizations of strategies style "Computing with Words".

Nowadays, a significant and increasing portion of research in intelligent systems leverages on fuzzy approaches in different ways. The rise of Lotfi Zadeh's ideas is as never before and we anticipate the rise of new types of fuzzy systems. Lotfi Zadeh's ideas went far beyond fuzzy sets and fuzzy logics. They concern-effective processing of invisible or informal relations between different parts of knowledge in a wide range of intelligent applications. The book aims at contributing to the dissemination of various innovative ideas among the recent research trends for advanced intelligent technologies. We hope that every reader finds something interesting, challenging, and stimulating: something useful for her own applications or for developing new research ideas.

Sofia, Bulgaria Vassil Sgurev
Crema, Italy Vincenzo Piuri
Sofia, Bulgaria Vladimir Jotsov

Contents

From von Neumann Architecture and Atanasoffs ABC to Neuro-Morphic Computation and Kasabov's NeuCube: Principles and Implementations 1
Neelava Sengupta, Josafath Israel Espinosa Ramos, Enmei Tu, Stefan Marks, Nathan Scott, Jakub Weclawski, Akshay Raj Gollahalli, Maryam Gholami Doborjeh, Zohreh Gholami Doborjeh, Kaushalya Kumarasinghe, Vivienne Breen and Anne Abbott

Applications of Computational Intelligence in Industrial and Environmental Scenarios 29
Ruggero Donida Labati, Angelo Genovese, Enrique Muñoz, Vincenzo Piuri and Fabio Scotti

Real, Imaginary and Complex Sets 47
Vassil Sgurev

Intercriteria Analysis over Patterns 61
Krassimir Atanassov

From Primary to Conventional Science 73
Boris Stilman

Intelligent Two-Level Optimization and Model Predictive Control of Degrading Plants 117
Mincho Hadjiski, Alexandra Grancharova and Kosta Boshnakov

Collaborative Clustering: New Perspective to Rank Factor Granules 135
Shihu Liu, Xiaozhou Chen and Patrick S. P. Wang

Learning Through Constraint Applications 149
Vladimir Jotsov, Pepa Petrova and Evtim Iliev

Autonomous Flight Control and Precise Gestural Positioning of a Small Quadrotor . . . 179
Nikola G. Shakev, Sevil A. Ahmed, Andon V. Topalov, Vasil L. Popov and Kostadin B. Shiev

Introduction to the Theory of Randomized Machine Learning . . . 199
Yuri S. Popkov, Yuri A. Dubnov and Alexey Y. Popkov

Grid-Type Fuzzy Models for Performance Evaluation and Condition Monitoring of Photovoltaic Systems . . . 221
Gancho Vachkov and Valentin Stoyanov

Intelligent Control of Uncertain Switched Nonlinear Plants: NFG Optimum Control Synthesis via Switched Fuzzy Time-Delay Systems . . . 251
Georgi M. Dimirovski, Jinming Luo and Huakuo Li

Multidimensional Intuitionistic Fuzzy Quantifiers and Level Operators . . . 267
Krassimir Atanassov, Ivan Georgiev, Eulalia Szmidt and Janusz Kacprzyk

Data Processing and Harmonization for Intelligent Transportation Systems: An Application Scenario on Highway Traffic Flows . . . 281
Paulo Figueiras, Guilherme Guerreiro, Ricardo Silva, Ruben Costa and Ricardo Jardim-Gonçalves

From von Neumann Architecture and Atanasoffs ABC to Neuro-Morphic Computation and Kasabov's NeuCube: Principles and Implementations

Neelava Sengupta, Josafath Israel Espinosa Ramos, Enmei Tu, Stefan Marks, Nathan Scott, Jakub Weclawski, Akshay Raj Gollahalli, Maryam Gholami Doborjeh, Zohreh Gholami Doborjeh, Kaushalya Kumarasinghe, Vivienne Breen and Anne Abbott

1 Introduction

The breakthrough work of Alan Turing in the 1940s, stating the possibility of using just 0s and 1s to simulate any process of formal reasoning [1] lead to massive development in the field of information theory and computer architecture. Simultaneously, significant progress was made by the neuroscientists in understanding the most efficient and intelligent machine known to man, the human brain. These parallel advancements in the middle of the last century had made man's imagination of creating 'intelligent' systems a possibility. These rational systems/agents were thought ideally to be able to perceive the external environment and take actions accordingly to maximise its goal, mimicking the human brain. The improvements in computer architecture with its advances in input/output, storage and processing power meant that the dream of artificial intelligence (AI) was now a reality and hugely reliant.

The field of AI and machine learning has grown strength to strength from the simple McColluch and Pitt's linear threshold based artificial neuron model [2] to the latest era of deep learning [3], which builds very complex models by performing a combination of linear and non-linear transformations. This is done using millions of neuron stacked in a layered fashion forming an interconnected mesh. The tremendous push of AI towards emulation of real intelligence has been sustained by the

N. Sengupta (✉) · J. I. E. Ramos · N. Scott · A. R. Gollahalli
M. G. Doborjeh · Z. G. Doborjeh · K. Kumarasinghe · V. Breen · A. Abbott
KEDRI, AUT, Auckland, New Zealand
e-mail: nsengupt@aut.ac.nz

E. Tu
Rolls Royce@NTU-corporate Lab, NTU, Singapore

S. Marks
Colab, AUT, Auckland, New Zealand

J. Weclawski
Warsaw University of Technology, Warsaw, Poland

V. Sgurev et al. (eds.), *Learning Systems: From Theory to Practice*, Studies in Computational Intelligence 756, https://doi.org/10.1007/978-3-319-75181-8_1

realisation of the Moore's law [4] which states that the processing power of the of central processing units (CPU) doubles in every couple of years. The scalable computer architecture proposed by John von Neumann in 1945 as part of the draft of EDVAC computer [5] had to play a substantial role in accomplishing the continuous miniaturisation of the CPU chips. In the more recent years, the CPU chip manufacturing companies have spent billions of dollars in CMOS technology to shrink the transistor size to a minuscule ($\approx$14 nm) and thus keep Moore's law alive. It is evident that this is non-sustainable and as per well-supported predictions will reach its boundary in the next five years [6].

The saturation in the scalability of the von Neumann architecture led to new developments in computer and computing architectures. Neuromorphic computing coined by Carver Mead in the 1980s [7] and further developed recently is one of the paradigms of computing which has come into prominence. As the name 'neuromorphic' suggests, this paradigm of computing is inspired heavily by the human brain. Moreover, as the existence of AI is complimented by computing architectures and paradigms, having a real neuromorphic computer architecture oriented processing unit is a step towards the development of highly neuromorphic AI.

1.1 Neuromorphic Computing Beyond von Neumann Architecture

Throughout the continuous evolution of the traditional computers, von Neumann or the stored program architecture has continued to be the standard architecture for computers. It is a multi-modular design based on rigid physically separate functional units. It specifically consists of three different entities:

- Processing unit: The processing unit can be broken down into a couple of subunits, the arithmetic and logical unit (ALU), the processing control unit and the program counter. The ALU compute the arithmetic logic needed to run programs. The control unit is used to control the flow of data through the processor.
- I/O unit: The I/O unit essentially encompasses all I/O the computer could possibly do (printing to a monitor, to paper, inputs from a mouse or keyboard, and others.).
- Storage unit: The storage unit stores anything the computer would need to store and retrieve. This includes both volatile and non-volatile memory.

These units are connected over different buses like data bus, address bus and control bus. The bus allows for the communication between the various logical units. Though very robust, as shown in Fig. 1a, this architecture inherently suffers from the bottleneck created due to the constant shuffling of the data between the memory unit and the central processing unit. This bottleneck leads to rigidity in the architecture as the data needs to pass through the bottleneck in a sequential order. An alternate solution of parallelising the computers has been proposed where millions of processors are interconnected. This solution, though, increases processing power, is still limited by the bottleneck in its core elements [9].

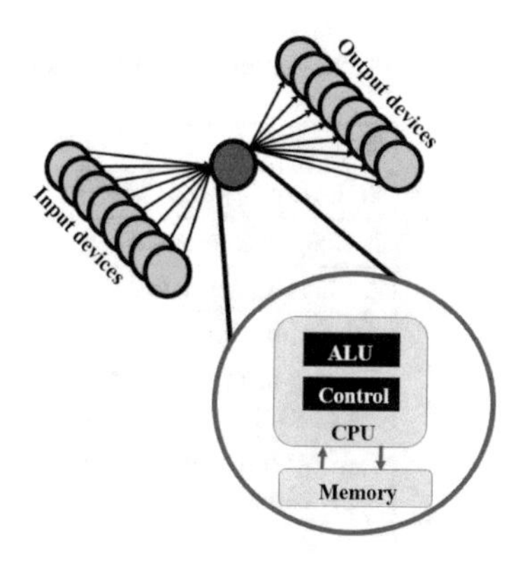

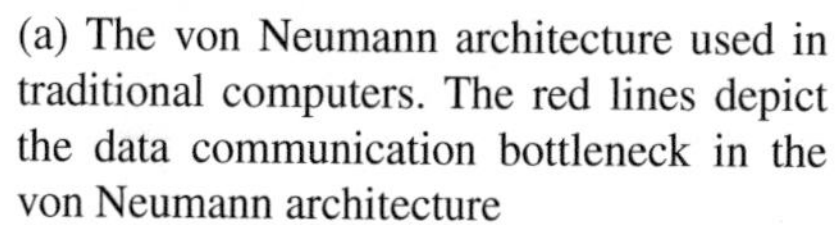
(a) The von Neumann architecture used in traditional computers. The red lines depict the data communication bottleneck in the von Neumann architecture

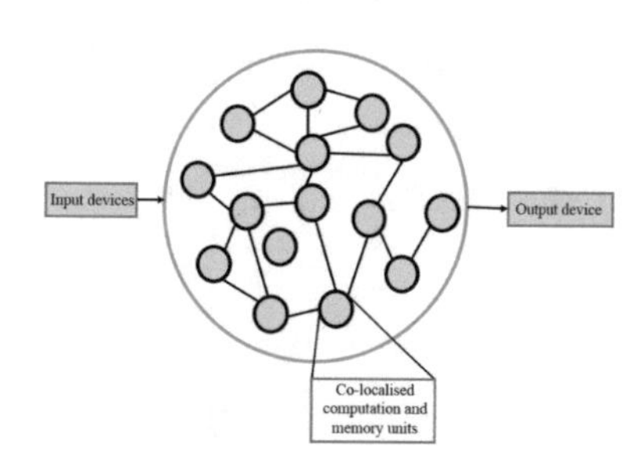

(b) A graphical representation of a general neuromorphic architecture. In this architecture, the processing and memory are decentralised across different neuronal units(the yellow nodes) and synapses(the black lines connecting the nodes), creating a naturally parallel computing environment via the mesh-like structure

Fig. 1 A graphical comparison of the von Neumann and Neuromorphic architecture [8]

The neuromorphic computing paradigm as shown graphically in Fig. 1b, on the contrary, draws great inspiration from our brain's ability to manage tens of billions of processing units connected by the hundreds of trillions of synapses using tens of watts of power on an average. The vast network of the processing units (neurons) in the brain is in a true sense a mesh. The data is transmitted over the network via the mesh of synapses seamlessly. Architecturally the presence of the memory and the processing unit as a single abstraction is uniquely advantageous leading to dynamic, self-programmable behaviour in complex environments [9]. The highly stochastic nature of computation in our brain is a very significant divergence from the bit-precise processing of the traditional CPU. The neuromorphic computing hence aspires to move away from the bit-precise computing paradigm towards the probabilistic models of simple, reliable and power and data efficient computing [10] by implementing neuromorphic principles such as spiking, plasticity, dynamic learning and adaptability. This architecture morphs the biological neurons, where the memory and the processing units are present as part of the cell body leading to decentralised presence of memory and processing power over the network. Table 1 lists down some of the fundamental characteristics of the von Neumann and neuromorphic architecture.

With significant commercial interest in sight, research community focused on the commercial scale development of the neuromorphic chips. The most prominent of the neuromorphic chips include the Truenorth [11, 12] from IBM, the Neurogrid [13] developed by the Stanford University and SpiNNaker chip [14] from the University of Manchester, the neuromorphic chip developed in ETH INI, Indiveri et al. [15].

Table 1 A comparison of the key contrasts between von Neumann and neuromorphic computing paradigm [8]

Representation of the data	von Neumann	Neuromorphic
	Sequence of binary numbers	Spike (event) timings
Memory	1. Volatile	1. Long term memory
	2. Non-volatile	2. Short term memory
Plasticity (Learning)	No	Adaptable via:
		1. Long-term potentiation and depression
		2. Short-term potentiation and depression
Processing	1. Deterministic	1. Stochastic
	2. Centralised	2. Decentralised
	3. Sequential	3. Parallel

All of these neuromorphic chips consists of programmable neuron and synapses and uses a multitude of CMOS technologies to achieve the neuromorphic behaviours. The details of the neuromorphic chips are beyond the scope of this article and are well elaborated in [16].

Numerous research [17, 18] has focused on harnessing the theoretical powers of the spiking neural network (SNN). While the majority of the research focus on neurological simulations, its importance in the real world of engineering applications that deal with complex processes (and thus generate the spatio/spectro-temporal data), is yet to be identified. The importance of a computational model to capture and learn spatio/spectro-temporal (SSTD) patterns from data streams is henceforth very significant from an application perspective. Example of problems involving SSTD are: brain cognitive state evaluation based on EEG [19], fMRI data [20], moving object recognition from video data [21], evaluation of the response of a disease on treatment and others. In this article, we discuss a new neuromorphic artificial intelligence paradigm that uses SNN and the first spatio-temporal data machine called NeuCube [22]. This framework has been recently further developed and implemented as an SNN development system for applications on SSTD described in this chapter.

The rest of the article is organised into eight sections. Section 2, briefly describes the existing work on the neuromorphic software implementations including a brief review of the existing neural network simulators. Section 3, concisely presents the architecture of the NeuCube development system as a next-generation pattern recognition system. Section 4 describes the generic SNN prototyping and testing module M1 of NeuCube developed in Matlab. Further down, the implementation of the Neuromorphic hardware and virtual reality environment is presented in Sects. 5 and 6. Section 7 elaborates on the recent development of the NeuCube software in the state of the art Java model view control (MVC) framework. Finally, Sect. 8 summarises and concludes the chapter.

2 Related Work

The number of software implementations that has appeared, as a result of ongoing research in the area of artificial neural networks, is ever growing. Majority of the neural network software is implemented to serve two purposes:

- Data analysis: These software packages are aimed at analysing real-world data derived from practical applications. The data analysis softwares use a relatively simple static architecture, hence are easily configurable and easy to use. Few examples of such software are: multilayer perceptron (MLP) [23], RBF network [24], Probabilistic network (PNN) [25], Self organizing maps (SOM) [26], Evolving connectionist systems, such as DENFIS and EFuNN [27]. These softwares are either available as independent packages, such as NeuCom [28], PyBrain (python) [29], Fast Artificial Neural Network (C++) [30], or as part of a data analytics software like Weka [31], Knime [32], Orange [33] and others.
- Research and development systems: As opposed to the data analysis softwares, they are complex in behaviour, and require background knowledge for usage and configuration. the Majority of the existing SNN softwares, including NeuCube, belong to this class.

We have briefly reviewed some of the key features of the current SNN development systems below.

NEURON [34]: Neuron is aimed at simulating a network of detailed neurological models. Its ability to simulate biophysical properties such as multiple channel types, channel distributions, ionic accumulation and so on renders it well suited for biological modelling [35]. It also supports parallel simulation environment through (1) distributing multiple simulations over multiple processors, and (2) distributing models of individual cells over multiple processors.

PyNEST [36]: The neural simulation tool (NEST) is a software primarily developed in C++ to simulate a heterogeneous network of spiking neurons. NEST is implemented to ideally model neurons in the order of 10^4 and synapses in the order of 10^7 to 10^9 on a range of devices from single core architectures to supercomputers. NEST interfaces with python via implementation of PyNEST. PyNEST allows for greater flexibility in simulation setup, stimuli generation and simulation result analysis. A node and a connection comprise the core elements of the heterogeneous architecture. The flexibility to simulate a neuron, a device or a subnetwork (which can be arranged hierarchically) as a node, provides a major improvement over [37]. Due to the bottom-up approach of network simulation, the software allows for individually configurable neuron states and connection setup.

Circuit Simulator [38]: The circuit simulator is a software developed in C++ for simulation of heterogeneous networks with major emphasis on high-level network modelling and analysis, as opposed to [34]. The C++ core of the software is integrated with Matlab based GUI, for ease of use and analysis. CSIM enables the user to operate both spiking and analog neuron models along with mechanisms of spike and analog signal transmission through its synapse. It also performs dynamic synaptic behaviour by using short and long-term plasticity. In 2009, circuit simulator was further extended to parallel circuit simulator (PCSIM) software with the major extension being implementation on a distributed simulation engine in C++, interfacing with Python based GUI.

Neocortical Simulator [39]: NCS or Neocortical Simulator is an SNN simulation software, mainly intended for simulating mammalian neocortex [35]. During its initial development, NCS was a serial implementation in Matlab but later rewritten in C++ to integrate distributed modelling capability [40]. As reported in [35], NCS could simulate in the order of 106 single compartment neuron and $10^{1}2$ synapses using STP, LTP and STDP dynamics. Due to the considerable setup overhead of the ASCII-based files used for the I/O, a Python-based GUI scripting tool called BRAINLAB [39] was later developed to process I/O specifications for large scale modelling.

Oger Toolbox [41]: Oger toolbox is a Python-based toolbox, which implements modular learning architecture on large datasets. Apart from traditional machine learning methods such as Principal Component Analysis and Independent Component Analysis, it also implements SNN based reservoir computing paradigm for learning from sequential data. This software uses a single neuron as its building block, similar to the implementation in [36]. A Major highlight of this software includes the ability to customise the network with several non-linear functions and weight topologies, and a GPU optimised reservoir using CUDA.

BRIAN [42, 43]: Brian is an SNN simulator application programming interface written in Python. The purpose of developing this API is to provide users with the ability to write quick and easy simulation code [42], including custom neuron models and architecture. The model definition equations are separated from the implementation for better readability and reproducibility. The authors in [43], also emphasises the use of this software in teaching a neuroinformatics course [44]. A major limitation of BRIAN is, however, the requirement of Python knowledge to run the simulation and the lack of GUI for the non-technical user community.

The aforementioned discussion of the existing software highlights the suitability for building highly accurate neurological models but lacks a general framework for modelling temporal or SSTD, such as brain data, ecological and environmental data. Further in the line of the neural network development systems, and more specifically for SNN, where not only an SNN simulator can be developed, but a whole prototype system (also called spatio-temporal data machine) can be generated for solving a complex problem defined by SSTD, the NeuCube framework was proposed [27].

3 The NeuCube Framework and System Architecture

The NeuCube framework for SSTD, illustrated, but not restricted to brain data, is depicted in Fig. 2 [45] and explained below.

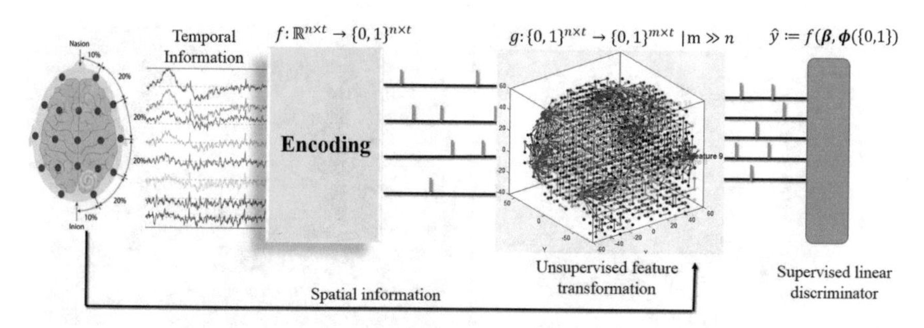

Fig. 2 The NeuCube framework for SSTD. The brain, shown as a source of SSTD is only exemplary, rather than restrictive

- Data encoding: The temporal information generated from the source (e.g. brain, earthquake sites) is passed through a data encoder component using a suitable encoding method, such as BSA [27], Temporal contrast, GAGamma [46]. It transforms the continuous information stream to discrete spike trains ($f : \mathbb{R}^{n \times t} \rightarrow \{0, 1\}^{n \times t}$).
- Mapping spike encoded data and unsupervised learning: The spike trains are then entered into a scalable three dimensional space of hundreds, thousands or millions of spiking neurons, called SNNcube (SNNc), so that the spatial coordinates of the input variables (e.g. EEG channels; seismic sites, and so on) are mapped into spatially allocated neurons in the Cube, and an unsupervised time-dependent learning rule [47, 48] is applied ($g : \{0, 1\}^{n \times t} \rightarrow 0, 1^{m \times t} | m >> n$).
- Supervised learning: After unsupervised learning is applied, the second phase of learning is performed, when the input data is propagated again, now through the trained SNNc, and an SNN output classifier/regressor is trained in a supervised mode ($\hat{y} := h(\beta, \phi(0, 1)))$ [49]. For this purpose, various SNN classifiers, regressors or spike pattern associators can be used, such as deSNN [49] and SPAN [50].

The NeuCube software development system architecture uses the above mentioned core pattern recognition block described in Fig. 2 as the central component and wraps a set of pluggable modules around it. The pluggable modules are mainly developed for: (1) Using fast and scalable hardware components running large scale applications; (2) Immersive model visualisation for in-depth understanding and analysis of the SSTD and its SNN model; (3) Specific applications like personalised modelling, brain computer interfaces and so on (4) Hyperparameter optimisation; and others.

Figure 3a shows the graphical representation of the NeuCube SNN development system for SSTD and Fig. 3b shows the standard configuration in a real life setup in the KEDRI NeuLab. Each module in Fig. 3a is designed to perform an independent task and in some instances, written in a different language and suited to the specific computer platform. All of the modules, however, are integrated via a common communication protocol in module M5. A brief description of the standard modules from Fig. 3a is given below:

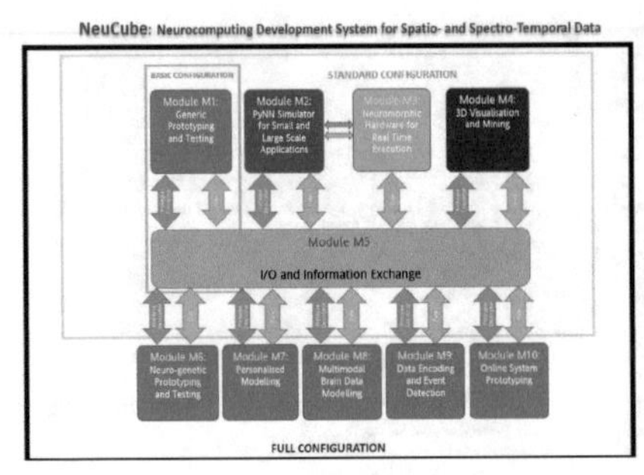

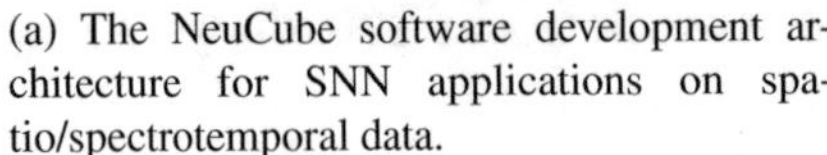

(a) The NeuCube software development architecture for SNN applications on spatio/spectrotemporal data.

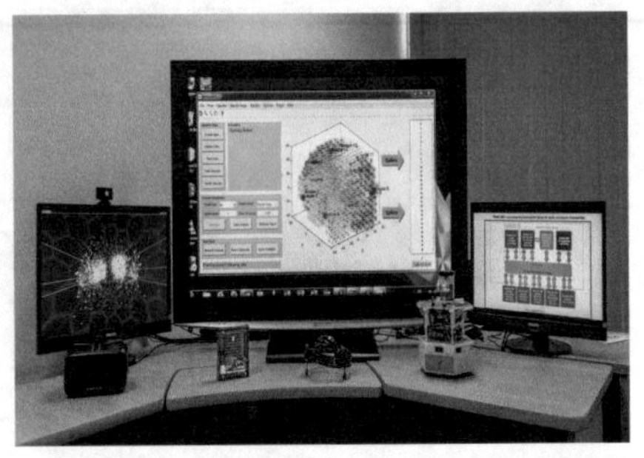

(b) NeuCube Modules M1, M2, M3 and M4 integrated in the KEDRI NeuLab, also showing some application oriented devices, such as Oculus for 3D visualisation, a SpiNNaker small neuromorphic board, Emotiv EEG device, an EEG-controlled mobile robot with Kyushu Institute of Technology

Fig. 3 The NeuCube SNN development system for SSTD

Module M1 It is a generic prototyping and testing module written in Matlab, where an SNN application system can be developed for data mining, pattern recognition and event prediction from temporal or SSTD. Additional functionalities like dynamic visualisation, network analysis toolbox, parameter optimisation are included in this module. Section 4 describes Module M1 in a more elaborate fashion.

Module M2 is a python based simulator of NeuCube for large scale applications or implementation on a neuromorphic hardware (Module M3). This application is developed on top of PyNN package, which is a Python-based simulator-independent language for building SNN. The NeuCube-PyNN [51] module is not only compatible with existing SNN simulators described previously (e.g. Neuron, Brian), but can also be ported to a large neuromorphic hardware such as the SpiNNaker, or on any neuromorphic chip, such as the ETH INI chip, the Zhejiang University chip, and others. The advantage of such hardware lies in the extreme energy efficiency of computation, allowing for the large-scale massively parallel neuromorphic system to run more efficiently.

Module M3 is dedicated for such hardware implementations of NeuCube. Section 5 has a detailed description about Modules M2 and M3.

Module M4 allows for a dynamic visualisation of the 3D structure and connectivity of the NeuCube SNN [52, 53]. Due to the 3-dimensional structure as well as the large number of neurons and connections within NeuCube a simple 2D connectivity/weight matrix or an orthographic 45-degree view of the volume is insufficient. We created a specialised visualisation engine using JOGL (Java Bindings for OpenGL) and GLSL (OpenGL Shading Language). This engine can render the structural connectivity as well as the dynamic spiking activity. Using 3D stereoscopic head-mounted displays such as the Oculus Rift, the perception and understanding of the spatial structure can be improved even further. Section 6 describes module M4 in detail.

Module M5 is the input/output and the information exchange module. This module is responsible for binding all the NeuCube modules together irrespective of the programming language or platform. Experiments that are run on any module produces prototype descriptors containing all the relevant information, which are exported and imported as structured text files, and is compatible with all the modules. We have used language independent JSON (Javascript object notation) format as a structured text, which is lightweight, human readable and can be parsed easily. The present implementation of the I/O module supports the use of three types of data and SNN prototype descriptors. They are: (1) Dataset descriptor, which consist of all the information relevant to the raw and encoded dataset; (2) Parameter descriptor, which is responsible for storing all the user defined and changeable parameters of the software; and, (3) SNN application system descriptor, which stores information related to the NeuCube SNN application system.

Module M6 extends the functionality of module M1, by adding functions for prototyping and testing of neurogenetic data modelling. These functions include models for genetic and proteomic influences in conjunction with brain data. This is provided as an optional feature for specific applications. This module is written in Java and can be further developed as an open source NeuCube SNN development system.

Module M7 facilitates the creation and the testing of a personalised SNN system. It extends module M1 by including additional functionalities for personalised modelling which is based on first clustering of integrated static-dynamic data using new algorithm dWWKNN (dynamic weightedweighted distance K-nearest neighbours) and then learning from the most informative subset of dynamic data for the best possible prediction of output for an individual. This module is for optional use in the context of specific applications [54] (USA patent 2008). This module is for optional use in the context of specific applications [55].

Module M8 is currently being developed as a plugin for multimodal brain data analysis. It aims to integrate different modalities of brain activity information (e.g., EEG, fMRI, MEG) and structural Tractography (DTI) information, in NeuCube, for the purpose of better modelling and learning. This module is also bound to specific applications.

Module M9 is an optional module for data encoding optimisation and event detection. This module includes several data encoding algorithms for mapping analog signal to spike trains based on different data sources [46].

Module M10 provides an additional feature of online learning for real-time data analysis and prediction. In this module, continuous data streams are processed in the form of continuous data blocks.

4 NeuCube Implementations for Prototyping and Testing

The generic prototyping and testing tools or Module M1 are the GUI based implementations for the rapid development of SNN application prototype systems for temporal or spatio/spectro-temporal data. It can also be used for research on SNN

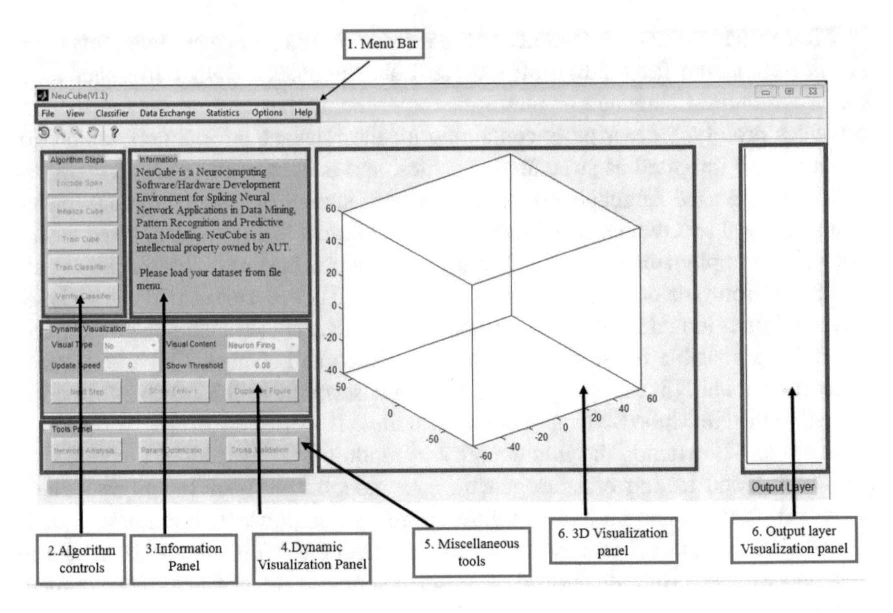

Fig. 4 NeuCube-M1 user interface and panel descriptions

for general purpose pattern recognition from SSTD. Currently, there are two parallel implementations of the NeuCube system in Matlab and Java. This section describes the Matlab system in detail.

The Matlab implementation is designed as a set of continuous signal processing steps as shown in Fig. 2. The user interface of NeuCube M1, as shown in Fig. 4, is built as a logical stepwise process.

4.1 Data Exchange

The data exchange component is used to import or export user defined information to and from the software, which includes temporal or SSTD data, already developed SNN systems, parameters and results. The NeuCube-M1 interacts with the external environment using four data descriptors. They are the following:

- Dataset descriptor: The Dataset descriptor consists of the data (and the metadata), that is to be learned and analysed. In the majority of the cases, a dataset contains a set of time series samples and the output label/value for the sample set. It is also possible to add miscellaneous information like feature name, encoding method and other meta information in the dataset.
- SNNc descriptor: The SNNc descriptor contains all information related to the structure and learning of the SNN. Some of the most important information stored in this

Table 2 Supported file format for descriptors

Descriptor type	Mat	JSON	CSV
Dataset	yes	yes	yes
Cube	yes	yes	no
Parameter	yes	yes	no
Result	yes	no	yes

descriptor are the spatial information of the input and reservoir neurons,structural information of the SNNc and the state of the SNNc during learning.

- Parameter descriptor: The parameter descriptor stores all the user defined parameters including hyperparameters of data encoding algorithms, the unsupervised learning algorithm and the supervised learning algorithm.
- Result descriptor: Result descriptor stores information about the experimental results produced by NeuCube.

NeuCube-M1 supports three different file formats, Mat (binary), JSON (structured text) and CSV (comma separated plain text). Table 2 describes the supported file formats for each of the descriptor type. As a heuristic rule, mat format is recommended for achieving fast I/O. The CSV files are the recommended choice for import/export of dataset and results for later analysis. The JSON format is recommended for intermodular communication. Loading a dataset (temporal information) is the entry point to the software, and can be done by loading a csv or a mat file from the file menu. The import and export of all the descriptors can be performed throughout the lifetime of the experiment.

4.2 Algorithm Interactions

NeuCube-M1, being a general purpose pattern recognition software, allow users to interact with the pattern recognition and signal processing algorithms via the algorithm controls panel, shown in Fig. 4. The algorithm control panel includes a set of buttons for configuring and running the step by step process of data encoding, network initialisation, unsupervised learning (by training the SNNc) and supervised learning. The software uses a guided approach for performing the algorithmic steps by enabling or disabling buttons after every operation. Figure 5a, 5b, 5c and 5d shows the individual user interaction panels for encoding, initialisation, unsupervised and supervised learning respectively. Each panel allows users to choose from a set of algorithms and corresponding hyperparameters. For example, the data encoding panel Fig. (5a), that encode the real-valued signal to spike trains, provide the option of choosing from a set of encoding algorithms and its hyperparameters from the drop down menu. The initialisation panel, as shown in Fig. 5b, initiates the SNNc. NeuCube allows making use of the natural spatial ordering (if any) of the features of the

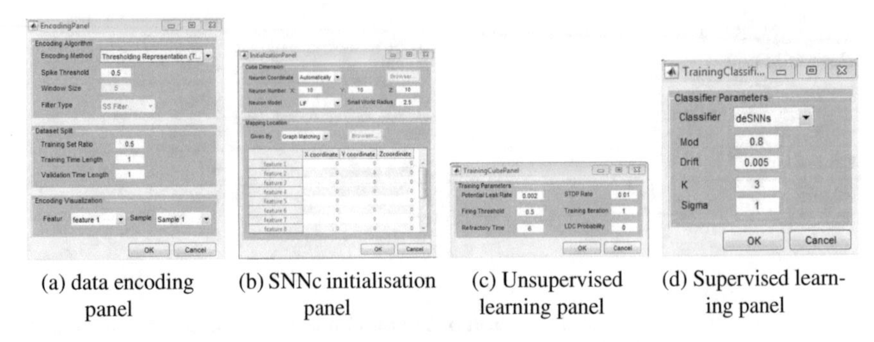

(a) data encoding panel (b) SNNc initialisation panel (c) Unsupervised learning panel (d) Supervised learning panel

Fig. 5 Algorithmic controls through the NeuCube-M1 UI's

data as input to the system. This can be achieved by loading an external csv coordinate file from mapping location option in the encoding panel. If a natural spatial ordering is not present in the data, NeuCube uses the graph matching technique, to enforce a spatial arrangement of the input variables based on their temporal similarities. All the panels for algorithmic control interacts with the visualisation panels, for visualisation and analysis of the data and the models.

4.3 Integrated Visualisation and Network Analysis

Visualisation and model analysis is an integral and unique feature of the NeuCube architecture. As discussed in [45], a NeuCube model acts as a white box, i.e., a learned NeuCube model outputs analysable and interpretable spatio-temporal patterns for knowledge discovery. The visualisations in the M1 module are rendered in the '3D visualisation panel' and 'output layer visualisation panel' and can be manoeuvred by using the controls under the 'dynamic visualisation panel', 'miscellaneous tools' and 'menu bar' in the user interface of Module M1.

Visualisation capabilities of NeuCube M1 module includes: comparative display of real and encoded data; online dynamic visualisation of unsupervised learning; static visualisation of the SNNc model and the output readout layer and visualisation; and SNNc network analysis.

4.3.1 Visualisation of Data Encoding:

The Current version of the M1 module includes an offline encoding and visualisation scheme, but real-time online encoding and visualisation for streamed data is part of another specialised module–Module M10.

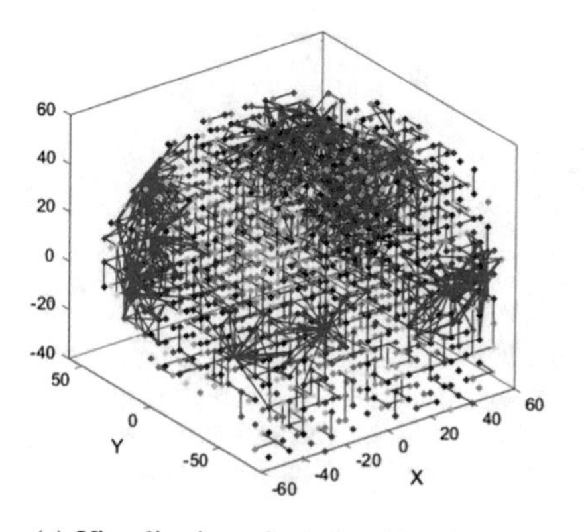

(a) Visualisation of relationships formed in the brain shaped SNNc at the end of unsupervised learning. The blue lines represent positive connections, and the red lines represent negative connections

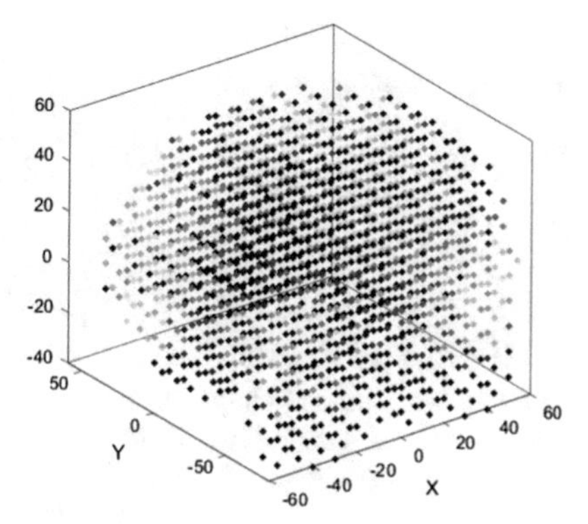

(b) Visualisation of spike emission density. The brighter neurons represent higher spike count

Fig. 6 Static visualisation of SNNc model

4.3.2 Visualisation of 3D SNNc Model:

The SNNc in the NeuCube architecture learns the spatio-temporal patterns over time using the spikes transmitted by the spiking neurons. The SNNc also forms relationships by regulating the connection strengths between the neurons. The visualisation and analysis of this learning process are of utmost importance for knowledge discovery. The unsupervised learning process can be visualised dynamically online, while the system is learning, or can be saved to a movie for later usage and analysis by using dynamic visualisation panel. The plots can be rendered in a continuous, or stepwise fashion. It is also possible to specify the type of activity to be rendered on the go, such as the spiking behaviour, evolution of connection, and others.

The static visualisation of the SNNc represents a snapshot of the final state of the cube at the end of the learning period. The option for static visualisation of the SNNc can be found in the view drop down under the menu bar. Figure 6a shows an example of the static visualisation of the spatial relationships formed via the connections below a defined threshold. The colour of the neurons in Fig. 6b, on the other hand, represent the spike emission count.

4.3.3 SNNc (Network) Analysis:

The network analysis panel can be used for analysis of the SNNc network content. The network analysis consists of two major functionalities: (1) Neuron cluster analysis; (2) Information route analysis. An example of neuron cluster analysis visualisation is shown and described in Fig. 8a, b and c. Information route analysis is used

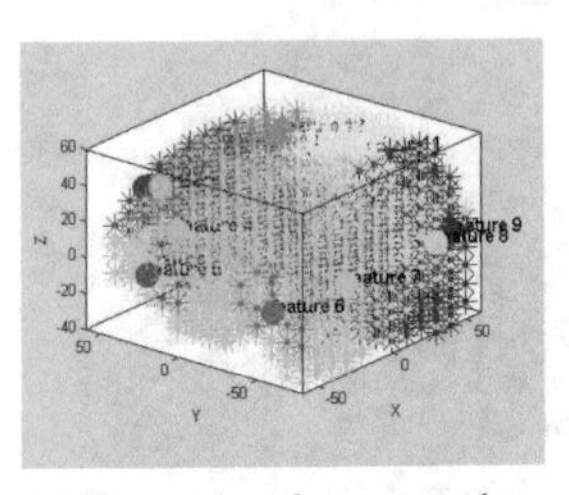

(a) Example of neuron clustering based on connection weights of the network

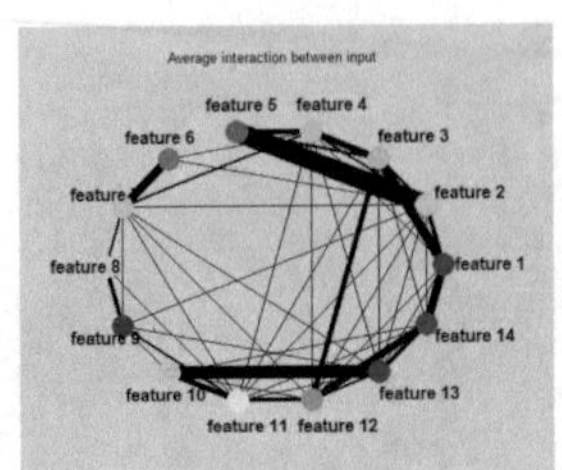

(b) average one to one interaction between the input neuron clusters shown in figure 7a. The thicker lines signify more interaction.

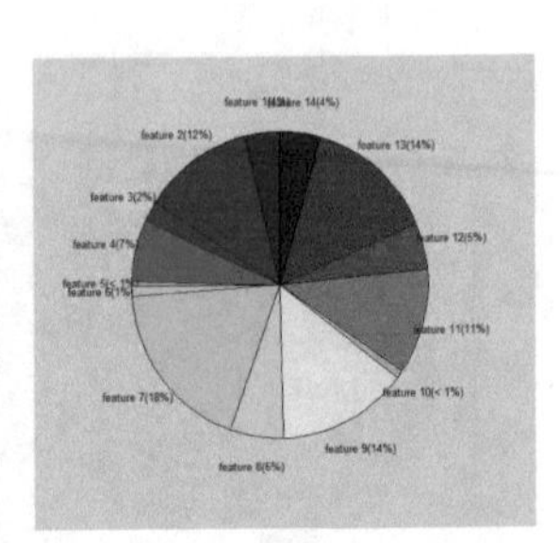

(c) Pie chart depicting the number of neurons belonging to an input cluster group

Fig. 7 Neuron cluster analysis by network analysis toolbox

for analysing information propagation route of the spikes generated by the spiking neurons. This analysis is based on the concept of the rooted tree. The type of information can be chosen by selecting the trace with drop down. Different methods of analysis is described below in brief:

- Max spike gradient: Shows a tree rooted by the input neuron, where a child neuron is chosen to be connected to a parent neuron, if it receives spike from its parents.
- Spreading level: shows a tree from the input neuron to its neighbourhood which reflects the spreading of the spikes. The level number parameter defines the neighbourhood of spread.
- Information amount: Shows a tree rooted by the input neuron, where a child neuron is chosen to be part of the tree, only if it receives a defined percentage of spikes from its parent neuron. The percentage can be specified by the information box, where 0.1 means 10% spikes (Fig. 7).

4.3.4 Output Layer Visualisation

The output layer visualisation is concerned with the K-nearest neighbour based deSNN discriminator used as a supervised learning algorithm in the M1 module. The output layer renders a set of spiking neurons, where each output neuron represents the class label (or the regression value) assigned to one input SSTD sample.

4.4 Parameter Optimisation

Parameter optimisation is developed to allow users to search for the optimal set of hyperparameters that minimises the test accuracy of the NeuCube prototype system (model), either for classification or regression. The computational time for parameter optimisation depends on the number of parameters to be optimised and the size of the NeuCube model. Parameter optimisation in NeuCube Module M1 can be performed using various methods, such as: Grid search; Genetic Algorithm; Differential Evolution; Quantum-Inspired Evolutionary Algorithms, PSO, and so on. The current release of NeuCube M1 includes two methods (Grid search and Genetic algorithm).

5 Implementations on Hardware and Neuromorphic Chip

The NeuCube is also appropriate for implementation on dedicated neuromorphic hardware systems, or distributed computing platforms. Due to the fact that the model is by nature highly scalable, it requires a highly scalable computation platform.

As traditional von-Neumann computational architectures reach their limits [56, 57] in terms of power consumption, transistor size, and communication, new approaches must be sought. Neuromorphic hardware systems, especially designed to solve neuron dynamics and able to be highly accelerated compared to biological time, are a response to these concerns. Systems such as analog VLSI or the SpiNNaker are advantageous by comparison to software based simulations on commodity computing hardware in areas such as biophysical realism; density of neurons per unit of processing power; and significantly lowered power consumption [58, 59]. This is not to say that simulations of the NeuCube cannot occur on traditional computing architectures; merely that dedicated hardware is advantageous in these areas and may be more appropriate for large-scale modelling.

To address this opportunity, a cross-platform version utilising the PyNN API in Python has been written to extend the modular framework established in [51]. This version is targeted primarily towards neuromorphic hardware platforms but is also applicable to commodity distributed hardware systems depending on the simulation backend chosen.

PyNN [60] is a generic SNN simulation markup framework that allows the user to run arbitrary SNN models on a number of different simulation platforms, including software simulators PyNEST and Brian, and some neuromorphic hardware systems such as SpiNNaker and FACETS/BrainScaleS. It provides a write once, run anywhere (where anywhere is the list of simulators it supports) facility for the development of SNN simulations.

One of the possible neuromorphic platforms for the implementation of a NeuCube SNN prototype system developed in module M1 or in any other modules of the NeuCube architecture, is the SpiNNaker device, currently in development at the University of Manchester. SpiNNaker is a general-purpose, scalable, multichip, multicore platform for the real-time massively parallel simulation of large-scale SNN [58]. Each SpiNNaker chip contains 18 ARM968 subsystems responsible for modelling up to one thousand neurons per core, at very low power consumption. These chips communicate through a custom multicast packet link fabric, and an arbitrary number of these chips can be linked together, with the assumption that the networks simulated exhibit some kind of connection locality. The small-world connection structure used in the NeuCube and its scalable nature are appropriate for implementation on this type of hardware.

Alternative implementations of the NeuCube on neuromorphic hardware are currently being pursued on the INI Neuromorphic VLSI systems and the Zhejiang University FPGA system.

6 Dynamic Immersive Visualisation and Interpretation

The complexity of the network within NeuCube, with regards to the sheer number of neurons and connections, and their 3-dimensional structure requires a specialised visualisation system. These connections and their evolution over time is a significant source of information about the data processed therein. A mere orthographic 45-degree view or a 2D connectivity/weight matrix or of the volume would be insufficient for this amount of data. For this reason, we created a specialised visualisation engine for NeuCube datasets that use JOGL (Java Bindings for OpenGL) and GLSL (OpenGL Shading Language) shaders, enabling us to render up to 1.5 million neurons and their connections on relatively recent graphics cards with 60 frames per second or more.

For efficiency, neurons are rendered as stylised spheres that change the size and the colour based on spiking activity. Connections are rendered as lines with different colours depending on their weight (excitatory: green, inhibitory: red). Spiking activity is also visualised as white pulses travelling along these connections. Additional visualisation modes include detection of find hot paths, connection length, and the ability to view the 3D structure in slices. Using a 3D cursor, neurons can be viewed individually, including their configuration parameters and their current activation potential. There are also controls for playing back, pausing, and changing the speed of spiking activity during the training process as well as during online processing. This is a powerful mechanism for further understanding the temporal behaviour or NeuCube.

By using display hardware such as 3D stereoscopic head-mounted displays (HMD) like the Oculus Rift or the HTC Vive, the perception of the spatial structure of the network and the neuron positions is increased. The full potential of the visualisation unfolds in a motion capture space, where the camera perspective and the cursor node position and orientation are controlled by markers that are attached on the actual HMD and a cursor implement (see Fig. 8). This facility enables the user to literally 'walk through' NeuCube, to observe structural and dynamic behaviour, and to select individual neurons with the cursor in a natural manner using a joystick.

Our solution [52, 53] differs from other neural network specific visualisation tools such as BrainGazer [61] and Neuron Navigator (NNG) [62] in that the user can naturally navigate through the 3D space by simply walking and interact with it by gesturing instead of having to use the mouse and memorise keyboard shortcuts. Closer to our visualisation is the work of [63], who are using a Computer Assisted

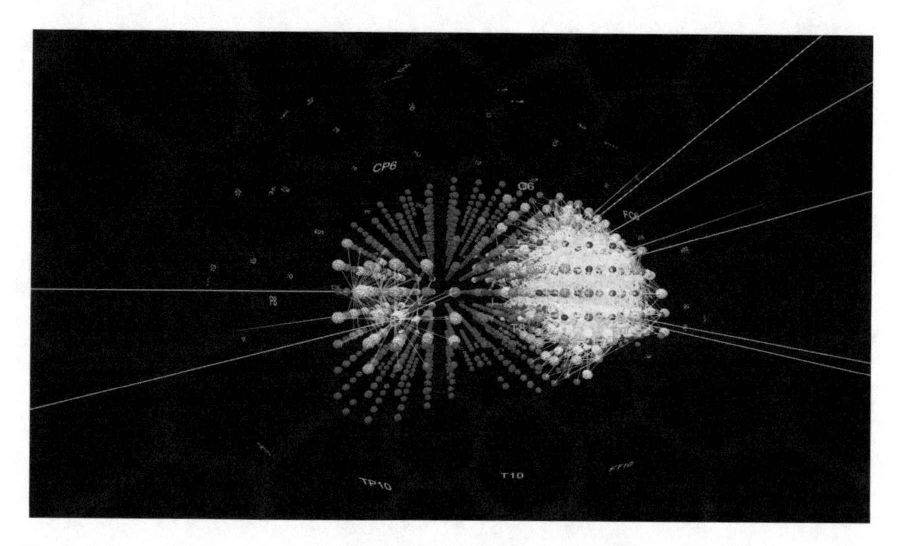

Fig. 8 3D visualisation of the spatial and dynamic structure of a NeuCube application model

Virtual Environment (CAVE) to visualise the spatial structure and activity of an SNN. However, due to the limited space within a CAVE, navigation by simply walking is not possible and requires indirect interaction, e.g., by using a controller. These technologies also do not allow for multiple concurrent users, whereas our system is limited only by the available hardware (HMDs) and can easily be extended due to its modular software architecture.

On the numerous occasions where Module M4 has been used at our facility, we observed that people are quickly starting to walk around and to look at structures. The visualisation and interaction metaphors such as selecting and monitoring individual neurons via the 3D cursor are very intuitive for new and experienced users. Overall, M4 is a vital and very useful tool for facilitating a better understanding of NeuCube through 3D rendering of and intuitive navigation within the structure.

7 Java Implementation of the NeuCube Architecture

The Java implementation integrates the modules M1, for generic prototyping; M4, for a dynamic 3D visualisation of the NeuCube SNN; M5, for data exchange; M6 for neurogenetic data modelling and; M10, for online learning for real time data analysis. Besides the off-line mode, this version incorporates new methods for modelling large and fast on-line multisensory spatiotemporal stream data. Additionally, it integrates a novel approach to map and analyse data that arises from diverse input variables captured by each element of a sensor network.

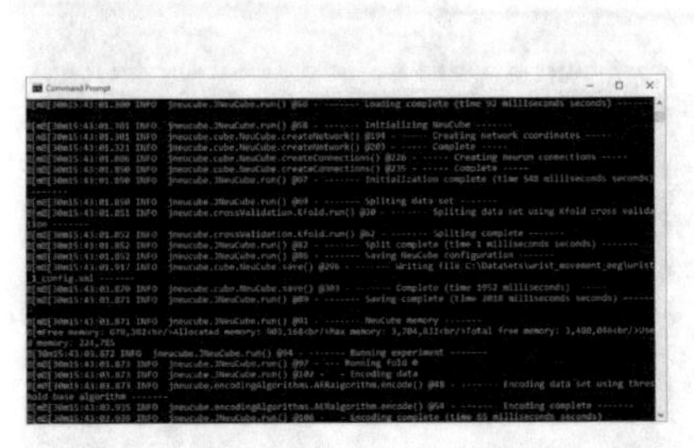

(a) JNeuCube execution through a command line.

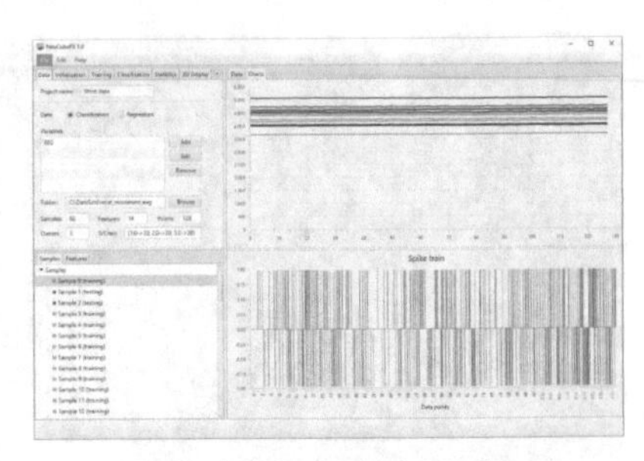

(b) Graphic user interface.

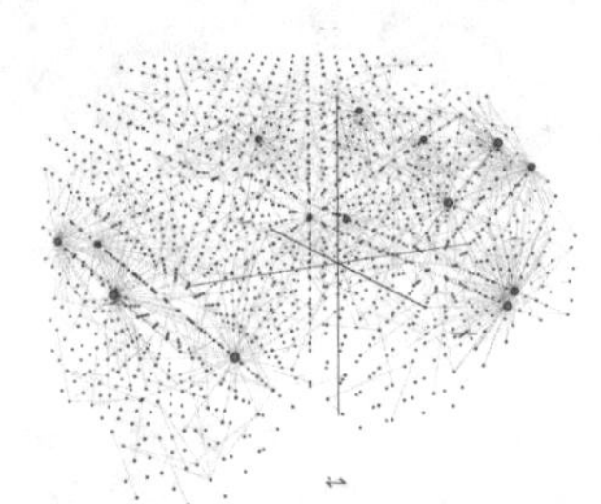

(c) Dynamic 3D visualisation.

Fig. 9 JNeuCube MVC architecture

The Java implementation is based on the model-view-controller (MVC) design pattern. The model, which captures the behaviour of the NeuCube, is developed in Java 8 (JNeuCube) and can be executed in a command line as shown in Fig. 9a; and the view and the controller (user interface), in JavaFX 8 (NeuCubeFX) as shown in Figs. 9b and c. Similar to the Matlab implementation, both Java versions are built as a logical step wise processes that can be analysed individually.

7.1 *Graphic User Interface (GUI)*

The software environment is formed by two main sections: the workflow section, which contains NeuCubes functionality, and the visualisation and network analysis section, which allows analyse and interpret spatiotemporal patterns.

7.1.1 Workflow

This section is formed by four main modules (see Fig. 10) that allow users to interact with the algorithms and methods for solving pattern recognition tasks.

Data Module

In this module, the user can import the SSTD and define the type of pattern recognition problem, classification or regression. Here, the user also defines the number of spatiotemporal variables to work with, e.g. the air speed, atmospheric pressure, temperature, and so on. Every variable can be separately analysed in a different 3D SNN that interact together for solving the problem in hand.

The data exchange (import and export) is a pivotal feature for information analysis. In its architecture, this version has an I/O layer that allows the user to process information in an offline mode (data modelling and prototyping), and in an online mode (online learning and recall). It can also store its state at every step (persistence of the model) and brings the possibility of data transfer to and from the Matlab version. Saving a NeuCube project into a file involves storing all information related to the set of time series, the algorithms and methods utilised for building the structure of the SNN, the validation methods, the training algorithms for both unsupervised and supervised learning, the spiking activity and the connection weights before and after the unsupervised learning, the information about the experimental results, and the parameter values for 3D visualisation.

(a) Data (b) Initialisation (c) Training (d) Classification

Fig. 10 Workflow controls in JNeuCube. Every module involves several steps that can be controlled separately

The current version can support three different file formats for different modules: CSV (comma separated plain text), for loading a data set (temporal information) to work in an offline mode for data modelling and prototyping, and for exporting the SNN structure (spiking activity and connectivity); and XML (structured data and communication protocol), for persistence of the NeuCube at every stage. One of the new features implemented in this Java version is the reading and writing modules for online stream data processing. These modules allow users to implement methods to communicate with any data source and send information to any device.

Initialisation Module

This module allows the user configure the 3D structure of the SNN through three simple steps. In the first step the user can select and configure the properties of the spiking neuron model, so far a simplification of the leaky integrate and fire model, and the Izhikevich model are implemented. However, any other model can easily be implemented. In the second step, the user can define the dimensions of the SNN by choosing the number of neurons per axis (x, y, z) or by choosing a file containing the coordinates of a map (e.g. the Talairach atlas which maps the location of brain structures). A third step is enabled for defining the network connectivity, either using and parameterising a connection algorithm or importing the connections and synaptic weights from files. Finally, in the fourth step, the user can visualise a list with the location and the type (input or reservoir) of all neurons in the SNN. By selecting a neuron in the list, the software shows its details like the number of firings emitted and received, and the pre and post synaptic neurons connected as shown in Fig. 11.

Training Module

After initialization, the user can start to perform experiments. The first step is configuring the cross validation (Monte Carlo or k-fold) to assess the generalisation of the model to an independent data set. Then, selecting the encoding algorithm transforms the spatio-temporal data into spike trains that NeuCube can process. Hidden patterns can be extracted through an offline or online unsupervised training. In an offline mode, it is possible to record every state of the NeuCube during learning and analyse its behaviour at any point of interest. In an online mode, only the states before and after learning can be analysed. Currently, the spike time dependent plasticity (STDP) rule is implemented, but other learning rules can also be implemented. The last step of this module corresponds to the supervised training. This learning method recognises the temporal pattern produced after the unsupervised learning for classification or regression tasks. Similar to previous modules, different training methods can be implemented; at present, only the deSNN is available.

Classification

This is the last module that directly involves the algorithms and methods utilised by the NeuCube for pattern recognition. Here, an offline and an online classification

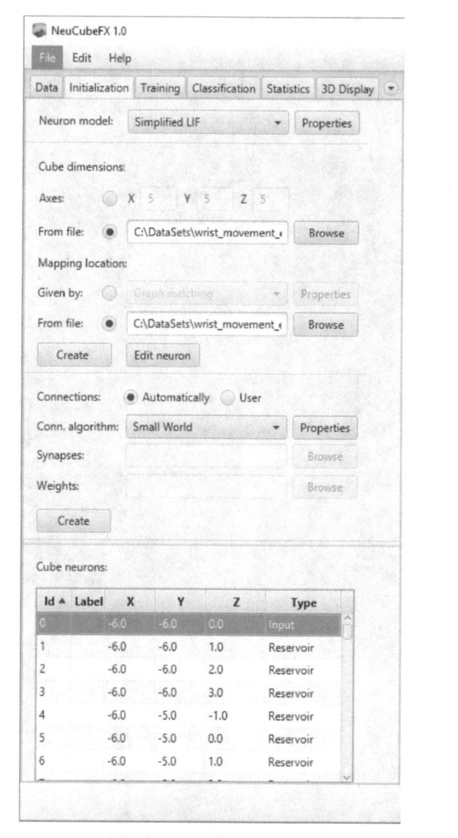

(a) Initialisation module

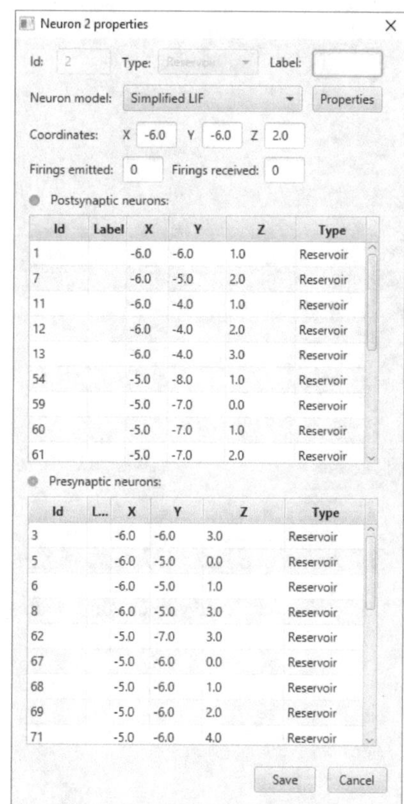

(b) Neuron's details

Fig. 11 Selection and visualisation of the detail of a neuron in the reservoir of JNeuCube

of new data can be performed. It is also possible to visualise the spiking activity produced while the data is introduced to the SNN. Finally, the user can visualise statistical results such as the confusion matrix and its derivations for generalisation analysis.

7.1.2 Visualisation and Network Analysis

The aim of this section is to communicate information clearly and efficiently via statistical graphics, plots, and a 3D dynamic visualisation of the NeuCube SNN. This is an effective module that helps users analyse and reason about data, spatiotemporal patterns and learning processes.

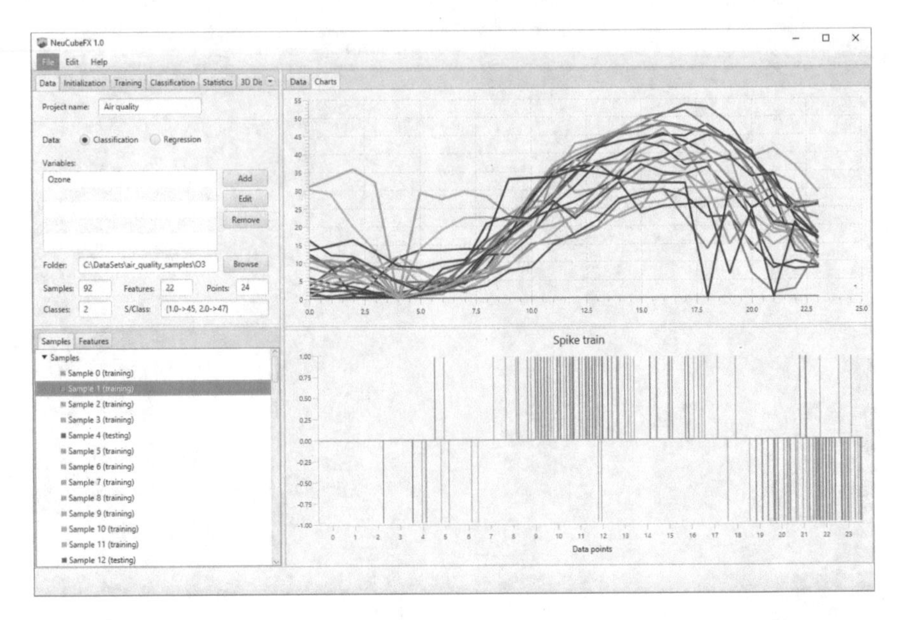

Fig. 12 JNeuCube Data visualisation before (top chart) and after (bottom chart) encoding

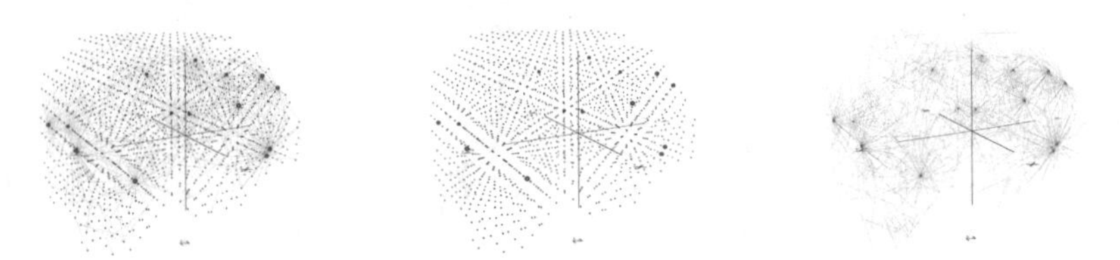

(a) Shows the complete spiking neural network.

(b) Shows the reservoir (blue spheres) and input neurons (dark green spheres).

(c) Shows the connectivity among neurons. The blue and red lines respectively represent positive and negative connections.

Fig. 13 Snapshots of the JNeuCube dynamic 3D visualisation of the NeuCube model using the Talairach atlas for EEG data processing

Dataset Visualisation

With this module, users can analyse every sample and feature of a data set. The NeuCubeFX makes data more accessible, understandable and usable by including tables and charts to visualise raw and encoded (spike trains) data as shown in Fig. 12.

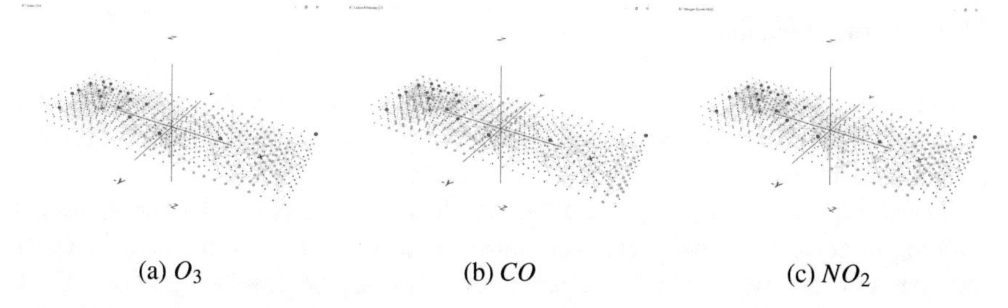

(a) O_3 (b) CO (c) NO_2

Fig. 14 Example of modelling a greenhouse gases sensor network. Different spiking activity and connection evolution are produced by every input variable (ozone, carbon monoxide and nitrogen dioxide)

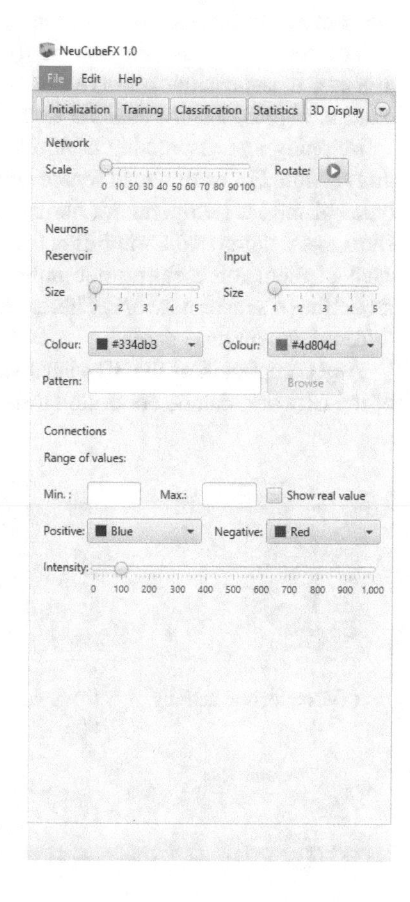

Fig. 15 3D display control in JNeuCube

3D Dynamic Visualisation

A characteristic feature of the NeuCube is the 3D dynamic visualisation module. Here, research scientists can explore how the data is being processed and interpreted by the NeuCube model.

The unsupervised learning can be dynamically visualised while the system is learning in an online mode (streamed data in real time). In an offline mode, every moment and state of the NeuCube can be saved during the learning process. Then, the user can go to any moment of such process and explore the spiking activity and/or the evolution of connections. At this point, the user can also rotate the NeuCube and analyse different areas of interest, study the properties of a particular neuron, or how the connection between two neurons changed from one time to another.

The recall process can also be dynamically visualised, during the stimulation process, it is possible observe the spiking activity produced by the data sample pattern. Figure 13 shows the 3D visualisation of a brain shaped SNNc.

Mapping and visualising multiple input variables is a novel feature included in this version. For example, in sensor network applications, a node can provide various types of output (temperature, humidity, air pressure, and so on.) at the same time. Here, we decompose a multidimensional SNNc model into several SNNc models, each of them representing one input variable. In Fig. 14 we show an example of modelling a sensor network. There, every node measures the concentration of three different greenhouse gases.

Any component of the 3D visualisation such as neurons, connection or the scale of the network, can be controlled through the 3D display tab (see Fig. 15).

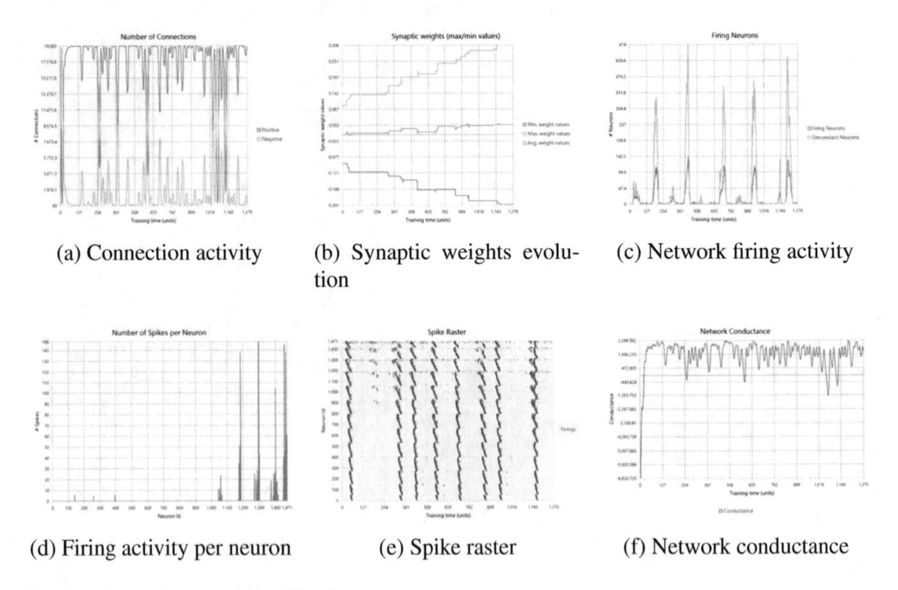

(a) Connection activity (b) Synaptic weights evolution (c) Network firing activity

(d) Firing activity per neuron (e) Spike raster (f) Network conductance

Fig. 16 Network analysis in JNeuCube

7.2 *Network Analysis*

Analysing and interpreting spatiotemporal patterns produced during learning is a unique feature that makes NeuCube acts as a white box. The statistics module allows the user to discover new knowledge related to: the number of positive and negative connections (Fig. 16a), the minimum and maximum values of the synaptic weights (Fig. 16b), the number of fired neurons and their descendants (Fig. 16c), the number of spikes produced by each neuron (Fig. 16d), the spike raster (Fig. 16e), and the network conductance described (Fig. 16f).

8 Conclusion

The chapter describes the main principles and applications of the first neuromorphic spatio-temporal data machine NeuCube. NeuCube is also a development system for a wide scope of applications. Its current implementation consists of 10 modules, written in 3 languages that are compatible through a common interface and shared data and structure formats. A free copy of the main NeuCube module as a limited and trial version is available from: http://www.kedri.aut.ac.nz/neucube/.

Acknowledgements The NeuCube development is funded by the Auckland University of Technology SRIF grant. Nikola Kasabov and Giacomo Indiveri from ETH and University of Zurich were granted an EU Marie Curie grant in 2011–2012 to start a preliminary research on SNN for spatio-temporal data. The research groups lead by Zeng-Guang Hou and Jie Yang from China contributed to the earlier software implementation of the NeuCube development system.

References

1. David, B.: The Advent of the Algorithm: The 300-Year Journey From an Idea to the Computer. Houghton Mifflin Harcourt (2001)
2. Warren, S., McCulloch, Walter, P.: A logical calculus of the ideas immanent in nervous activity. Bull. Math. Biophy. **5**(4), 115–133 (1943)
3. LeCun, Y., Bengio, Y., Hinton, G.: Deep learning. Nature **521**(7553), 436–444 (2015)
4. Robert, R.S.: Moore's law: past, present and future. IEEE Spectr. **34**(6), 52–59 (1997)
5. Brian R.: The Origins of Digital Computers: Selected Papers. Springer (2013)
6. Toumey, C.: Less is moore. Nature Nanotechnol. **11**, 2–3 (2016)
7. Mead, C.: Neuromorphic electronic systems. Proc. IEEE **78**(10), 1629–1636 (1990)
8. Nikola, K., Neelava, S., Nathan, S.: From von neumann, John atanasoff and abc to neuromorphic computation and the neucube spatio-temporal data machine. In: 2016 IEEE 8th International Conference on Intelligent Systems (IS), pp. 15–21. IEEE (2016)
9. Ivan, S.: Neuromorphic Computing: From Materials To Systems Architecture. Accessed 16 July 2016
10. Calimera, A., Macii, E., Poncino, M.: The human brain project and neuromorphic computing. Function. Neurol. **28**(3), 191–196 (2013)
11. Hsu, J.: Ibm's new brain [news]. IEEE Spectr. **51**(10), 17–19 (2014)

12. Merolla, P.A., Arthur, J.V., Rodrigo, A.-I., Cassidy, A.S., Sawada, J., Akopyan, F., Jackson, B.L., Imam, N., Guo, C., Nakamura, Y. et al.: A million spiking-neuron integrated circuit with a scalable communication network and interface. Science **345**(6197), 668–673 (2014)
13. Ben, V.B., Peiran, G., McQuinn, E., Swadesh, C., Anand, R.C., Bussat, J.-M., Rodrigo, A.-I., John, V.A., Paul, A.M., Kwabena, B.N.: A mixed-analog-digital multichip system for large-scale neural simulations. Proc. IEEE **102**(5), 699–716 (2014)
14. Steve, B., Furber, D., Lester, R., Luis, Plana, A., Jim, D., Garside, E.P., Steve, T., Andrew, D.B.: Overview of the spinnaker system architecture. IEEE Trans. Comput. **62**(12), 2454–2467 (2013)
15. Indiveri, G., Linares-Barranco, Bernabé, H., Tara, J., Van Schaik, A., Etienne-Cummings, R., Delbruck, T., Liu, Shih-Chii, Dudek, Piotr, Häfliger, Philipp, Renaud, S., et al.: Neuromorphic silicon neuron circuits. Front. Neurosci. **5**, 73 (2011)
16. Indiveri, Giacomo: Liu, S.-C. Memory and information processing in neuromorphic systems. Proc. IEEE **103**(8), 1379–1397 (2015)
17. Maass, W.: Networks of spiking neurons: the third generation of neural network models. Neural Netw. **10**(9), 1659–1671 (1997)
18. Maass, W., Bishop, C.M.: Pulsed Neural Networks. MIT press, New york (2001)
19. Elisa, C., Nikola, K., Grace, Y., Wang et al.: Analysis of connectivity in neucube spiking neural network models trained on EEG data for the understanding of functional changes in the brain: a case study on opiate dependence treatment. Neural Netw. **68**:62–77 (2015)
20. Maryam Gholami, D., Elisa, C., Nikola, K.: Classification and segmentation of fmri spatio-temporal brain data with a neucube evolving spiking neural network model. In: 2014 IEEE Symposium on Evolving and Autonomous Learning Systems (EALS), pp. 73–80. IEEE (2014)
21. Delbruck, T., Patrick, L.: Fast sensory motor control based on event-based hybrid neuromorphic-procedural system. In: Circuits and Systems, 2007. ISCAS 2007. IEEE International Symposium on, pp. 845–848. IEEE (2007)
22. Nikola, K., Nathan, M.S., Enmei, T., Stefan, M., Neelava, S., Elisa, C., Muhaini, O., Maryam, G., Doborjeh, Norhanifah M., Reggio, H., et al.: Evolving spatio-temporal data machines based on the neucube neuromorphic framework design methodology and selected applications. Neural Netw. **78**, 1–14 (2016)
23. Eric, B.B.: On the capabilities of multilayer perceptrons. J. Complex. **4**(3), 193–215 (1988)
24. Sandberg, W.: Universal approximation using radial-basis-function networks. Neural Computat **3**(2), 246–257 (1991)
25. Donald, F.: Specht. Probabilistic neural networks. Neural networks **3**(1), 109–118 (1990)
26. Kohonen, T.: The self-organizing map. Neurocomputing **21**(1), 1–6 (1998)
27. Nikola, K.: Evolving connectionist systems: the knowledge engineering approach. Springer Science & Business Media (2007)
28. Neucom: http://www.theneucom.com. Accessed 15 Aug 2015
29. Schaul, Tom: Bayer, Justin, Wierstra, Daan, Sun, Yi, Felder, Martin, Sehnke, Frank, Rückstieß, Thomas, Schmidhuber, Jürgen: Pybrain. J. Mach. Learn. Res. **11**, 743–746 (2010)
30. Steffen, N., Evan, N.: Fast artificial neural network library. leenissen.dk/fann/html/files/fann-h.html. (2000)
31. Mark, H., Eibe, F., Geoffrey, H., Bernhard, P., Peter, R., Ian, H.W.: The Weka data mining software: an update. ACM SIGKDD Explorat. Newslett. **11**(1):10–18 (2009)
32. Michael, R.B., Nicolas, C., Fabian, D., Thomas, R., Gabriel, Tobias, K., Thorsten, M., Peter, O., Christoph, S., Kilian, T., Bernd, W.K.: The konstanz information miner. In: Data Analysis, Machine Learning and Applications, pp. 319–326. Springer (2008)
33. Demšar, J., Zupan, B., Leban, G., Tomaz, C.: From experimental machine learning to interactive data mining. Springer, Orange (2004)
34. Michael, L., Hines, N., Carnevale, T.: The neuron simulation environment. Neural Comput. **9**(6), 1179–1209 (1997)
35. Romain, B., Michelle, R., Ted, C., Hines, M., Beeman, D., Bower, J.M., Diesmann, M., Morrison, A., Goodman, P.H., Harris, Jr., Frederick, C., et al.: Simulation of networks of spiking neurons: a review of tools and strategies. J. Comput. Neurosci. **23**(3), 349–398 (2007)

36. Jochen Martin, E., Moritz, H., Eilif, M., Markus, D., Marc-Oliver, G.: Pynest: a convenient interface to the nest simulator. Front. Neuroinformat. **2** (2008)
37. Dejan, P., Thomas, N.,Klaus, S.: Pcsim: a parallel simulation environment for neural circuits fully integrated with python. Front. Neuroinformat. **3** (2009)
38. Thomas, N., Henry, M., Wolfgang, M.: Computer models and analysis tools for neural microcircuits. In: Neuroscience Databases, pp. 123–138. Springer (2003)
39. Rich, D.: Brainlab: a toolkit to aid in the design, simulation, and analysis of spiking neural networks with the NCS environment. PhD thesis, University of Nevada Reno (2005)
40. E Courtenay, W.: *Parallel implementation of a large scale biologically realistic neocortical neural network simulator*. PhD thesis, University of Nevada Reno (2001)
41. Dejan, P.: Oger: Modular learning architectures for large-scale sequential processing
42. Goodman, Dan, F.M.: Code generation: a strategy for neural network simulators. Neuroinformatics **8**(3), 183–196 (2010)
43. Goodman, Dan, F.M., Brette, R.: The brian simulator. Front. Neurosci. **3**(2), 192 (2009)
44. Diesmann, M.: Gewaltig, Marc-Oliver, Aertsen, Ad: Stable propagation of synchronous spiking in cortical neural networks. Nature **402**(6761), 529–533 (1999)
45. Nikola, K.: Kasabov. Neucube: A spiking neural network architecture for mapping, learning and understanding of spatio-temporal brain data. Neural Netw. **52**, 62–76 (2014)
46. Neelava Sengupta, Nathan Scott, Nikola, K.: Framework for knowledge driven optimisation based data encoding for brain data modelling using spiking neural network architecture. In: Proceedings of the Fifth International Conference on Fuzzy and Neuro Computing (FANCCO-2015), pp. 109–118. Springer (2015)
47. Fusi, S.: Spike-driven synaptic plasticity for learning correlated patterns of mean firing rates. Rev. Neurosci. **14**(1–2), 73–84 (2003)
48. Song, Sen: Kenneth D Miller, and Larry F Abbott. Competitive hebbian learning through spike-timing-dependent synaptic plasticity. Nature Neurosci. **3**(9), 919–926 (2000)
49. Kasabov, N.: Dhoble, K., Nuntalid, N., Indiveri, G.: Dynamic evolving spiking neural networks for on-line spatio-and spectro-temporal pattern recognition. Neural NetW. **41**, 188–201 (2013)
50. Mohemmed, Ammar: Schliebs, Stefan, Matsuda, Satoshi, Kasabov, Nikola: Span: Spike pattern association neuron for learning spatio-temporal spike patterns. Int. J. Neural Syst. **22**(04), 1250012 (2012)
51. Nathan S., Nikola K., Giacomo Indiveri.: Neucube neuromorphic framework for spatio-temporal brain data and its python implementation. In: Neural Information Processing, pp. 78–84. Springer (2013)
52. Marks, S., Javier, E., Nathan, S.: Immersive Visualisation Of 3-dimensional Neural Network Structures. (2015)
53. Stefan Marks. Immersive Visualisation Of 3-dimensional Spiking Neural Networks. Evolving Syst. pp. 1–9 (2016)
54. Kasabov, N., Yingjie, H.: Integrated optimisation method for personalised modelling and case studies for medical decision support. Int. J. Function. Informat. Personal. Med. **3**(3), 236–256 (2010)
55. Maryam Gholami, D., Nikola, K.: Personalised modelling on integrated clinical and EEG spatio-temporal brain data in the neucube spiking neural network system. In: 2016 International Joint Conference on Neural Networks (IJCNN), pp. 1373–1378. IEEE (2016)
56. Hadi, E., Emily, B., Renee, S.T., Amant, K.S, Doug, B.: Dark silicon and the end of multicore scaling. ACM SIGARCH Comput. Architect. News **39**(3):365 (2011)
57. Perrin, D.: Complexity and high-end computing in biology and medicine. Advanc. Experiment. Med. Biol. **696**, 377–84 (2011)
58. Furber, S.: To build a brain. IEEE Spect. **49**(8), 44–49 (2012)
59. Indiveri, G., Linares-Barranco, Bernabé, Tara Julia, H., André van Schaik, Ralph Etienne-Cummings, Tobi Delbruck, Shih-Chii Liu, Piotr Dudek, Philipp, Häfliger, Sylvie, R., Johannes, S., Gert, C., John, A., Kai, H., Fopefolu, F., Sylvain, S., Teresa, S.-G., Jayawan, W., Yingxue, W., Kwabena, B.: Neuromorphic silicon neuron circuits. Front. Neurosci. **5**, 73 (2011)

60. Andrew, P., Davison, D., Brüderle, Jochen, E., Jens, K., Eilif, M., Dejan, P., Laurent, P., Pierre, Y.: PyNN: A common interface for neuronal network simulators. Front. Neuroinformat. **211** (2008)
61. Bruckner, S., Solteszova, V., Groller, E., Hladuvka, J., Buhler, K., Yu, J., Dickson, B.: BrainGazer–visual queries for neurobiology research. IEEE Trans. Visualizat. Comput. Graph. **15**(6), 1497–1504 (2009)
62. Lin, C.-Y. Tsai, ,K.-L., Wang, S.-C., Hsieh, C.-H., Chang, H.-M., Chiang, A.-S.: The neuron navigator: exploring the information pathway through the neural maze. In: IEEE Pacific Visualization Symposium (PacificVis) 20, pp. 35–42 (2011)
63. von Kapri, A., Rick, T., Potjans, T.C., Diesmann, M., Kuhlen, T.: Towards the visualization of spiking neurons in virtual reality. Stud. Health Technol. Informat. **163**, 685–87 (2011)

Applications of Computational Intelligence in Industrial and Environmental Scenarios

Ruggero Donida Labati, Angelo Genovese, Enrique Muñoz, Vincenzo Piuri and Fabio Scotti

1 Introduction

In industrial and environmental monitoring and control applications, experts in the field usually design physical models and perform a statistical analysis to infer knowledge about the observed phenomena. With such knowledge, it is possible to design and implement the corresponding monitoring and control methods. However, taking into account all the variables of the observed phenomenon can make the modeling and design process complex, with the result of obtaining inaccurate systems. Variations in the observed processes may also occur (e.g., wearing of the industrial machinery during the monitoring of industrial processes and meteorological anomalies while performing environmental monitoring), introducing the need of manual and complex adjustments to the models, or even making the system inadequate. Monitoring and control systems for industrial and environmental applications are frequently implemented in non-ideal operational conditions, which introduce noise in the signals acquired by the sensors and increase the complexity of the designed model [29].

R. Donida Labati (✉) · A. Genovese · E. Muñoz · V. Piuri · F. Scotti
Department of Computer Science, Università Degli Studi di Milano,
via Bramante 65, 26013 Crema, Italy
e-mail: ruggero.donida@unimi.it

A. Genovese
e-mail: angelo.genovese@unimi.it

E. Muñoz
e-mail: enrique.munoz@unimi.it

V. Piuri
e-mail: vincenzo.piuri@unimi.it

F. Scotti
e-mail: fabio.scotti@unimi.it

V. Sgurev et al. (eds.), *Learning Systems: From Theory to Practice*, Studies in Computational Intelligence 756, https://doi.org/10.1007/978-3-319-75181-8_2

Computational Intelligence (CI) techniques permit to design and implement intelligent systems for industrial and environmental monitoring and control applications that (i) do not require any modeling of the process based on expert knowledge and (ii) can infer a model by learning from examples. The CI methods most commonly used are Artificial Neural Networks (ANN) (such as, feed-forward networks, radial basis function networks, self-organizing maps, and convolutional neural networks), Support Vector Machines (SVM), Fuzzy Logic, and Evolutionary Computation (EC) [54].

These techniques can dynamically update the parameters of the model as the system is running, for instance by providing more recent examples from which the model can be updated. Intelligent monitoring systems have also the advantage of increasing the robustness to noisy data with respect to traditional approaches. The constant increase of computational power of processing architectures, along with the reduced size and cost of the devices, permits to use a higher number of sensors [48, 70] and therefore a more widespread diffusion of CI-based intelligent monitoring systems for a greater number of applications [22, 113]. Examples of industrial applications include production monitoring [10], control of robots [77], detection of faults in the machinery [52, 84], and control of the quality of industrial processes [9]. In

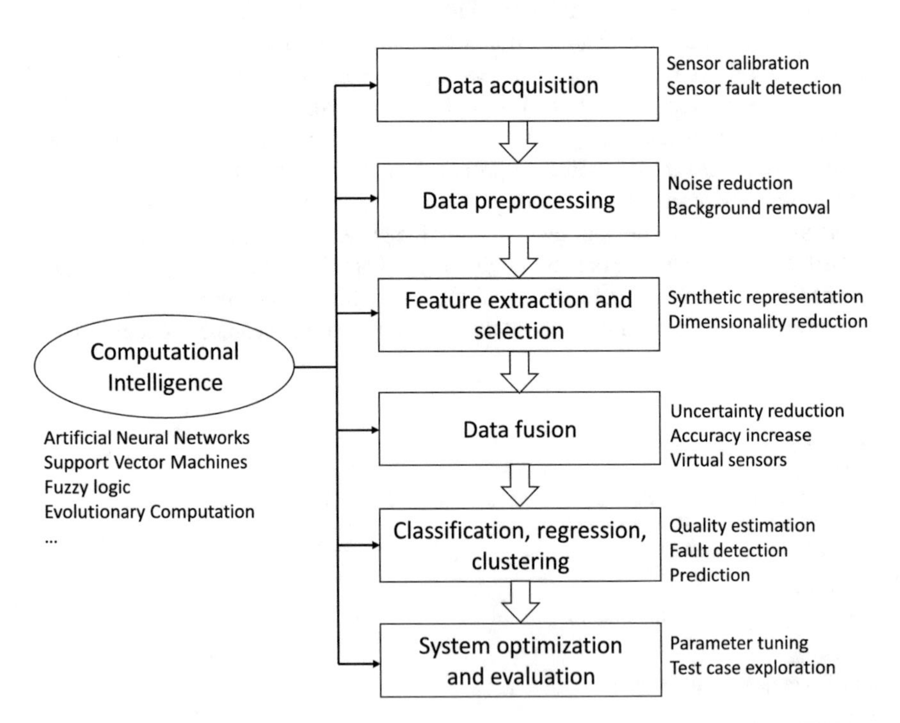

Fig. 1 Outline of an intelligent monitoring system for industrial and environmental applications. The figure enumerates the main applications of CI techniques for each step of monitoring systems

environmental scenarios, applications include forecasting [51, 73] and detection of hazardous situations (e.g., oil spill [92]).

Intelligent monitoring and control systems based on CI techniques share a similar architecture that comprises the following steps: (i) data acquisition, (ii) data preprocessing, (iii) feature extraction and selection, (iv) data fusion, (v) classification, regression, or clustering, (vi) system optimization and evaluation. Studies in the literature proposed applications of CI techniques for each of these steps [9, 13, 72]. Figure 1 illustrates the steps of the intelligent monitoring and control systems, together with the main applications of CI techniques.

This paper presents an overview on CI techniques for the design of intelligent monitoring systems in industrial and environmental scenarios. The paper is organized in as follows. Section 2 overviews the main CI techniques proposed in the literature. Section 3 presents applications of industrial and environmental monitoring. Section 4 describes the architecture of intelligent monitoring and control systems, by overviewing the methods in the literature using CI techniques for each step of the systems. Section 5 concludes the paper.

2 Overview of Computational Intelligence Techniques

CI techniques are methods that enable intelligent behavior in monitoring and control systems, with the features of being flexible and adaptive [35]. Among the advantages of CI techniques, there are the capabilities of working with noisy or incomplete data, providing approximate solutions, adapting to changes in operational and environmental conditions, and featuring a low computational complexity. These advantages have altogether contributed to the increasing use of CI techniques for industrial and environmental applications [9, 61].

This section presents the CI techniques mostly used in the field of industrial and environmental monitoring, such as Artificial Neural Networks (ANN), Support Vector Machines (SVM), Fuzzy Logic, and Evolutionary Computation (EC) [54].

ANNs exploit several layers of interconnected neurons to reproduce and mimic the reasoning ability of the human brain [8]. A learning procedure permits to adjust the weights associated to each neuron and the corresponding transfer functions. In this manner, the ANN can adapt itself to the training examples and process data to achieve approximate solutions with fault tolerance capabilities and low computational complexity. Several topologies of ANNs appear in the literature, presenting differences in the structure of the network and in the interconnections between the neurons. The topologies mostly used are feed-forward neural networks, radial basis function networks, and self-organizing maps [36, 49]. Recently, convolutional neural networks and other deep learning approaches have also been attracting the attention of the academic and industrial communities [64].

SVMs are a particular type of kernel-based methods that have the purpose of projecting the input data onto a space with a higher dimensionality to facilitate the learning process. In particular, the learning process in SVMs optimizes a convex surface without ending in local optima. One of the main advantages of SVMs consists in having a limited number of parameters, resulting in a faster tuning process [19].

Methods based on Fuzzy Logic have the peculiarity of dealing with information expressed with various degrees of uncertainty. For this reason, they are especially well suited to deal with incomplete or imprecise data. Advantages of Fuzzy systems include the use of linguistic concepts and the capability to deal with conflicting objectives by designing knowledge bases [97].

Methods using EC perform a global optimization by mimicking the biological evolution. In particular, the algorithms start with an initial population of solutions that iteratively interact among themselves by following principles derived from natural evolution, to improve their overall quality. The process of creating the initial population involves the assignment of random values to the parameters, or the use of heuristics tailored for the application scenario. Methods for updating the population include random permutations, crossover, or selection. Several algorithms based on EC have been proposed in the literature, including genetic algorithms and swarm intelligence (e.g., ant colony optimization and particle swarm optimization) [91].

3 Applications of Industrial and Environmental Monitoring

Nowadays, hardware processing architectures feature increased computing power, reduced size, and limited cost. For example, current smartphones and mini-PCs can easily exceed the computational capacity of desktop workstations made few years ago [82]. This increased availability of powerful and low-cost architectures allowed a greater number of monitoring infrastructures to be deployed in several novel scenarios [22, 113]. Furthemore, advances in the manufacturing of CMOS sensors have enabled the use of long-range vision systems using off-the-shelf equipment [20, 111]. Current devices have high computational power, extended battery life, and advanced networking capabilities [42], which are key factors to enable the deployment of sensors for intelligent monitoring in novel scenarios such as industrial and environmental applications, where resilience and robustness are of paramount importance. Therefore, the research is moving towards a greater use of CI techniques, due to their capability to aggregate inputs from different sensors and learn the relationship between the sensed data and the parameters of the observed phenomenon. At the same time, CI methods can adapt to variations in operational conditions. Thanks to these characteristics, CI techniques permit an accurate, robust, and fast detection of anomalies in industrial and environmental monitoring applications [9, 37, 61].

3.1 Industrial Monitoring

Intelligent monitoring systems can be classified according to their architecture in two main categories: (i) centralized systems and (ii) distributed systems.

Centralized monitoring systems typically employ vision-based sensors (e.g., CMOS cameras) to sense data at long distances. Vision-based systems are used since they can perform a touchless and non-destructive monitoring of the industrial process [90]. A single pan-and-tilt camera could cover a large area [46], thus reducing the need of having multiple sensing nodes. Moreover, it is possible to extract multiple features from the images captured, further reducing the need and cost of deploying multiple sensors. For instance, vision-based systems can monitor volume [30], surface defects [44, 83], or analyze the particle size distribution [27, 31, 38] of raw materials used in manufacturing processes.

Distributed monitoring systems often use Wireless Sensor Networks (WSNs) to build low-cost, robust, and pervasive monitoring networks composed by several inexpensive sensor nodes. The use of multiple and redundant sensor nodes allows the WSN to reconfigure itself in cases of changes of the physical placement of the nodes or sensor failures [48, 52].

CI techniques have widely shown their suitability to perform monitoring tasks in industrial environments. Examples of their applications include the detection of anomalies in tools and machinery [14, 52, 53, 84, 103], analysis of quantities of interest of raw materials or final products (e.g., shape, volume, or weight) [27, 30, 38], analysis of the quality of the manufactured product [4, 7, 28, 40, 44, 83, 85, 86].

3.2 Environmental Monitoring

Environmental monitoring systems can be classified according to their architecture into three types [10, 22, 23]: (i) centralized systems, (ii) distributed systems, and (iii) remote sensing systems.

Centralized systems typically employ a single vision-based sensing node, to provide long-range detection capabilities in situations that require to monitor wide environments. Example of centralized environmental monitoring systems are wildfire detection technologies based on a low-cost camera [32, 47].

Distributed systems often use WSNs since large-scale sensor networks composed by small and inexpensive nodes can be deployed easily in areas difficult to reach (e.g., volcanoes and oceans) or on wide environments (e.g., mountain range). The advanced networking capabilities of the nodes and a battery life sufficient to carry out the monitoring task for long time periods allow the sensor nodes to self-organize even if they are deployed automatically from a distance, for example, when launched from an airplane [61, 70]. Environmental monitoring systems using a distributed architecture can monitor climate change [68], structural health [106], water quality [78, 79, 105], or meteorology [51].

Remote sensing systems acquire data using satellite imagery to monitor environmental phenomena on a planetary scale. Examples of remote sensing systems for environmental monitoring include methods for detecting seismic activity [107], air pollution [110], and water pollution [92]. Satellite imagery can also detect meteorological phenomena and perform weather forecasts [73].

CI techniques are often used for environmental monitoring. They are widely used to classify features extracted from sensed data [32, 73, 92, 107, 110], perform forecasts from time series [51, 79], and aggregate data obtained from heterogeneous sensors (e.g., humidity, vibrations, temperature, chemical concentrations) in WSNs [61, 68, 78, 106].

4 Architecture of Intelligent Monitoring and Control Systems

The structure of industrial and environmental monitoring systems typically includes six main steps [9], as illustrated in Fig. 1. CI techniques can be used to perform or enhance each step of the system. In the following, we overview the applications of CI for each single step:

1. *Data acquisition*: the monitoring system usually collects one-dimensional signals, two-dimensional signals, three-dimensional models, or frame sequences. In this step, CI techniques can perform data linearization, sensor calibration, and system diagnosis.
2. *Data preprocessing*: the noise in the collected data is reduced, and the pattern of interest is extracted from the input multi-dimensional signals (e.g., image segmentation).
3. *Feature extraction and selection*: the features of interest for the monitoring application are extracted from the enhanced data. In this step, CI techniques can be used to obtain a synthetic representation of the phenomenon and reduce the dimensionality of the extracted features.
4. *Data fusion*: the system aggregates data or features coming from different sources, obtaining a compact description of the phenomenon of interest. In this step, CI techniques can increase the accuracy of traditional methods and reduce the uncertainty of virtual measurements.
5. *Classification, regression or clustering*: the monitoring system analyzes the features to derive patterns of interest for the monitored phenomenon. In this step, CI techniques can be used to perform fault detection, quality estimation, or prediction.
6. *System optimization and evaluation*: the parameters of the monitoring system are adjusted to improve its performance. In this step, CI techniques can be used to tune the parameters of the system, modify the used model according to new operational conditions, or evaluate the performance of the system by exploring different test cases.

Table 1 Summary of the different CI techniques used for each step of monitoring systems and the corresponding industrial and environmental applications

Step	Application	CI technique						
		ANN				SVM	Fuzzy	EC
		FF	RBF	SOM	CNN			
Data acquisition	Sensor calibration	[24, 57]	–	–	–	[74]	–	–
	Sensor fault detection	–	–	–	–	–	[75]	[89]
Data preprocessing	Noise reduction	[5, 6, 17, 98]	–	–	–	[102]	[96]	[81]
	Background removal	[88]	–	–	–	–	[2, 27]	–
Feature extraction and selection	Synthetic representation	[43, 65, 116]	–	[67, 87, 104]	[53, 115]	–	–	–
	Dimensionality reduction	–	–	–	–	–	–	[58, 108, 109]
Data fusion	Uncertainty reduction	[69]	–	[63]	–	[69]	[50, 59, 93]	–
	Accuracy increase	[95]	–	–	–	[14]	[103]	–
	Virtual sensors	[71]	–	–	–	–	[55]	–
Classification, regression, clustering	Quality estimation	[7, 28]	–	–	–	[85]	[60]	[60]
	Fault detection	[103]	–	–	[66]	[14]	[103]	–
	Prediction	[12, 16, 32, 79, 92, 107]	[3]	–	[11, 25]	[45, 62, 105]	[33, 55, 76, 114]	[105]
System optimization and testing	Parameter tuning	–	–	–	–	–	–	[1, 18, 112]
	Test case exploration	–	–	–	–	–	–	[94, 99, 100]

Note ANN: Artificial Neural Network; FF: Feed-Forward Neural Network; RBF: Radial Basis Function Neural Network; SOM: Self-Organizing Maps; CNN: Convolutional Neural Network; SVM: Support Vector Machine; EC: Evolutionary Computation

Table 1 summarizes the different CI techniques that have been proposed for each step performed by the monitoring systems and the corresponding industrial and environmental applications. In the remainder of this section, we elaborate more on the use of CI techniques in the different steps.

4.1 Data Acquisition

The data acquisition step collects information in the form of one-dimensional (e.g., time series) or multi-dimensional signals (e.g., images, three-dimensional models, frame sequences). To collect data, industrial and environmental monitoring systems can use sensors mimicking the human senses and sensors collecting physical quantities, including pressure, humidity, temperature, acceleration, or presence of chemicals. CI techniques can be used to enhance the data acquisition step by performing sensor calibration [24, 57, 74, 89] and sensor fault detection [37, 75, 80]. CI techniques widely used in this contex are ANNs [24, 57], SVMs [74], Fuzzy Logic [75], and EC [89].

To achieve a more pervasive monitoring, the extracted information can be exchanged using cloud-based infrastructure and internet of things. We note, however, that they can introduce security and privacy concerns, especially regarding industrial trade secrets [21].

4.2 Data Preprocessing

Data acquired in industrial and environmental scenarios can be affected by noise introduced by different non-ideal conditions, such as adverse operational conditions, variations in the environmental parameters, and interferences in the communication channels. The preprocessing step has then the purpose of enhancing the quality of the acquired data, remove the noise in the signals, and extract the pattern of interest.

Typically, ad-hoc filters process one-dimensional or multi-dimensional signals to enhance the quality of the data according to the characteristics of the monitoring application, by removing frequencies that do not represent the information to be analyzed or by restoring the original signal from the acquired noisy data. CI techniques are also adopted for data filtering, using methods based on ANNs [5, 6, 17, 98], SVMs [102], Fuzzy Logic [96], and EC [81].

In systems based on multi-dimensional signals (e.g., images or frame sequences) it is particularly important to segment the regions of interest from the background, so that the subsequent processing steps can avoid to consider unnecessary information [54]. To this purpose, CI techniques can be applied to train a model able to recognize the foreground from the background by learning from examples. In particular, ANNs have been used to extract regions of interest of bee foraging using images acquired from satellites [88], and methods based on Fuzzy Logic have been applied to select wood particles for panel production [27] or periods of interest from time-series of chemical processes [2].

4.3 Feature Extraction and Selection

Data captured by the sensors have usually a high dimensionality (e.g., acquisitions of long time series or large images from a high number of sensors) and cannot be directly processed to accurately infer the model of the observed phenomena. The feature extraction and selection steps have then the purpose of reducing the dimensionality of the preprocessed data to obtain a synthetic representation, usually called a feature vector [54]. Feature extraction and selection methods based on CI techniques permit to significantly reduce the amount of data to be processed by the subsequent steps of the systems, thus reducing the probability of overfitting problems and the need of computational resources [26]. Feature extraction and feature selection are typically performed in sequence. First, feature extraction methods compute a feature set with lower size with respect to the input multidimensional signal (e.g., the entire size of the captured image). Second, feature selection methods further reduce the dimensionality of the feature space by selecting a subset of the most discriminative features.

Traditional techniques for feature extraction include the linear discriminant analysis, principal component analysis, independent component analysis, polynomial approximation, and multidimensional scaling [54]. In industrial and environmental monitoring, examples of feature vectors include abstract representations of the gray-level variations that describe irregularities of a surface [34], the shape of a moving region [32], and anomalies in frequency ranges of the input signal [28, 72].

CI techniques based on ANNs have been widely used to perform feature extraction in industrial and environmental monitoring applications, with methods using feed-forward neural networks [43, 65, 116], self-organizing maps [67, 87, 104], radial basis function networks [15], and convolutional neural networks [53, 115].

Feature selection techniques search for the subset of features that best describes the observed phenomenon, by using strategies more efficient with respect to the exhaustive search, which can be unfeasible in terms of computational complexity in the case of feature vectors with high dimensionality. Traditional techniques are frequently based on Sequential Floating Search or Sequential Forward (Backward) Selection [54]. CI techniques can also perform feature selection: in particular, methods based on EC are especially well suited for this purpose [58, 108, 109].

4.4 Data Fusion

In monitoring applications, the data fusion step can reduce uncertainty in the collected data, for instance, by aggregating information obtained from several sensors of the same type [56], and reducing possible problems due to conflicting data and outliers. Data fusion techniques can increase the accuracy of the monitoring process by fusing information of heterogeneous sensors to achieve a more comprehensive description of the phenomenon [39]. Data fusion methods can also create virtual (or soft) sensors to indirectly measure physical quantities by inferring knowledge from the fusion of other information [41].

Fusion methods for uncertainty reduction are especially used in WSNs to aggregate the data collected by the heterogeneous sensors. CI techniques are especially suited to combine numerous sources of information to derive knowledge and perform an intelligent monitoring [61]. In industrial monitoring applications, Fuzzy Logic has been used to detect faults [59]. In environmental monitoring scenarios, self-organizing maps have been used to derive information about flood levels [63] and Fuzzy Logic has been used to monitor the temperature [93].

In industrial and environmental applications, several CI techniques have been used to increase the accuracy of the monitoring process. In particular, methods based on ANNs are widely used in industrial scenarios to combine the information obtained by different sensors. For example, ANNs have been used with the purpose of detecting the wearing of the tools used in the production machinery [95]. Similarly, methods based on SVM have been used to determine faults in the motors [14]. A combination of ANNs and Fuzzy Logic has also been used to model and monitor anomalies in hybrid energy systems [103]. In environmental monitoring, methods for aggregating data to reduce uncertainty in detecting floods have applied both ANNs and SVMs [69]. Techniques based on Fuzzy Logic have been used to detect landslides by aggregating geophysical data [50].

CI techniques have also been applied for the creation of virtual sensors for the indirect monitoring of physical quantities. When measuring the desired physical quantity is too complex or expensive, CI techniques can be applied to infer knowledge about the desired physical quantity, for example, by aggregating information acquired by different sources. In industrial applications, virtual sensors based on a Neuro-Fuzzy system have been proposed to monitor a hydrogen electrolyzer [55], while in environmental scenarios ANNs have been used to perform the localization of odor sources [71].

4.5 Classification, Regression and Clustering

The classification, regression, and clustering steps have the purpose of creating a model of the industrial or environmental phenomenon under analysis. These steps process the extracted features by associating to each observation a discrete label (classification) or a continuous value (regression), or by grouping them according to certain distance measures (clustering). CI techniques have extensively demonstrated their suitability for these tasks, since they can obtain flexible and adaptive models of the phenomenon through training and are robust to noisy or incomplete data. CI methods can approximate a model by using a finite number of examples, provide solutions by using a limited amount of resources, and mimic the human generalization ability [8, 49].

In classification, a discrete label is assigned to sets of features sharing some characteristics that differentiate them from the rest of the samples [101] (e.g., low-quality products, machinery faults, and meteorological anomalies). In industrial monitoring applications, CI techniques can detect low-quality outcomes of the production processes. The techniques mostly used are ANNs [7, 28], SVMs [85], and Evolutionary-Fuzzy systems [60]. CI methods can also perform fault detection in industrial scenarios, using convolutional neural networks [66], Neuro-Fuzzy systems [103] or SVMs [14]. As for environmental monitoring applications, ANNs can predict natural disasters [32, 107] and pollution [92]. Recent applications based on deep learning techniques have also addressed the problem of disaster prediction [11, 25].

In regression tasks, the goal is to estimate a continuous value from the set of features associated to each sample. In industrial monitoring applications, CI techniques often use regression to predict energy levels or quality parameters. In particular, ANNs can predict energy consumption [12] and Fuzzy Logic can estimate quality parameters in industrial processes [114]. Fuzzy Logic [55] and radial basis function networks [3] using regression-based learning procedures have also been proposed to create virtual sensors for industrial monitoring. As for environmental monitoring applications, regression-based techniques can predict meteorological phenomena or pollution levels, with ANNs used to forecast the water quality in rivers [79] and a combination of SVMs and EC used to predict air quality in urban areas [45, 105].

In clustering, the objective is to group samples according to certain distance measures. Differently from classification techniques, clustering algorithms typically use unsupervised learning procedures that do not require any a-priori knowledge on the semantics of the groups of similar data present in the considered set of samples. In industrial monitoring applications, CI techniques using clustering procedures based on Fuzzy Logic have been used to group the energy demands in different periods of time for prediction purposes [76]. Techniques based on SVMs have also been used to predict the optimal maintenance scheduling [62]. In environmental monitoring applications, clustering techniques based on ANNs have been proposed to predict seismic activities [16] and Fuzzy clustering methods have been used to estimate pollution [33].

4.6 System Optimization and Evaluation

Monitoring and control systems for industrial and environmental applications may include many models and a huge amount of parameters. Tuning these parameters can became a very difficult process, because the variables can interact in unpredictable ways. Therefore, tuning the systems based on the exhaustive exploration of all the possible combinations of parameters is almost impossible. Other simple techniques, such as trial and error, can only find sub-optimal configurations [9]. Differently, CI techniques based on EC, such as tabu search or ant colony optimization, represent effective strategies to find near-optimal solutions [18]. For instance, advanced EC techniques have been used to optimize the parameters of milling machines [112] and multi-generation energy systems [1].

Another important task to optimize the behavior of industrial and environmental monitoring and control systems consists in testing their modules. This task permits to avoid expensive errors and improve the overall performance of the system. In most cases, this task cannot be performed only by human operators because of the large domain of potential test cases, frequently too complex to be explored exhaustively. Moreover, testing procedures performed only by human operators can increase the cost of the systems, because they could miss important errors.

EC-based techniques have emerged as a viable solution to automate the testing process. This kind of techniques can generate effective test cases [100], and have been successfully applied to complex control systems, such as railway automation [94] and car production [99].

5 Conclusions

The design of industrial and environmental monitoring systems is a complex task because it has to face intricate processes, complex phenomena, noisy and missing data. Methods based on mathematical models or statistical analysis require careful engineering, and in many cases obtain incomplete or inaccurate models. In addition, such methods are difficult to apply when the data are noisy or incomplete. In contrast, Computational Intelligence (CI) methods such as Artificial Neural Networks, Support Vector Machines, Fuzzy Logic, or Evolutionary Computation can obtain flexible models that can cope with noise, data incompleteness, and varying operational conditions, while having a limited computational complexity.

This work has reviewed recent advances in the application of CI techniques to industrial and environmental monitoring scenarios. We have provided a taxonomy of the methods by grouping them according to the step of the design process that they cover. In particular, we considered the following steps: data acquisition, data preprocessing, feature extraction, data fusion, classification, regression, clustering and system optimization. In this way, we provided a complete design methodology. We have also illustrated the suitability of CI methods to face industrial and environmental monitoring and control problems. In particular, we have shown that CI approaches can obtain better performance compared with traditional methods and can provide adaptable and robust solutions. Recent advances in CI, such as deep learning and hybrid systems, make us believe that the applications of CI to industrial and environmental problems will continue to grow, with further improvements in terms of adaptability and performance.

Acknowledgements This work was supported in part by the EC within the H2020 program under grant agreement 644597 (ESCUDO-CLOUD).

References

1. Ahmadi, P., Dincer, I., Rosen, M.A.: Thermodynamic modeling and multi-objective evolutionary-based optimization of a new multigeneration energy system. Energy Convers. Manag. **76**, 282–300 (2013)
2. Alaei, H.K., Salahshoor, K., Alaei, H.K.: A new integrated on-line fuzzy clustering and segmentation methodology with adaptive PCA approach for process monitoring and fault detection and diagnosis. Soft Comput. **17**(3), 345–362 (2013)
3. Alexandridis, A.: Evolving RBF neural networks for adaptive soft-sensor design. Int. J. Neural Syst. **23**(6), 1350029 (2013)
4. Alippi, C., Braione, P.: Classification methods and inductive learning rules: what we may learn from theory. IEEE Trans. Syst. Man Cybern. Part C (Applications and Reviews) **36**(5), 649–655 (2006)
5. Alippi, C., Casagrande, E., Fumagalli, M., Scotti, F., Piuri, V., Valsecchi, L.: An embedded system methodology for real-time analysis of railways track profile. In: Proceedings of the 19th IEEE Instrumentation and Measurement Technology Conference (IMTC), pp. 747–751 (2002)
6. Alippi, C., Casagrande, E., Scotti, F., Piuri, V.: Composite real-time image processing for railways track profile measurement. IEEE Trans. Instrum. Meas. **49**(3), 559–564 (2000)
7. Alippi, C., D'Angelo, G., Matteucci, M., Pasquettaz, G., Piuri, V., Scotti, F.: Composite techniques for quality analysis in automotive laser welding. In: Proceedings of the 2003 IEEE International Symposium on Computational Intelligence for Measurement Systems and Applications (CIMSA), pp. 72–77 (2003)
8. Alippi, C., Ferrero, A., Piuri, V.: Artificial intelligence for instruments and measurement applications. IEEE Instrum. Meas. Mag. **1**(2), 9–17 (1998)
9. Alippi, C., Roveri, M., Piuri, V., Scotti, F.: Computational intelligence in industrial quality control. In: Proceedings of the 2005 IEEE International Workshop on Intelligent Signal Processing (WISP), pp. 4–9. Faro, Portugal (2005)
10. Amigoni, F., Brandolini, A., Caglioti, V., Di Lecce, V., Guerriero, A., Lazzaroni, M., Lombardi, F., Ottoboni, R., Pasero, E., Piuri, V., Scotti, O., Somenzi, D.: Agencies for perception in environmental monitoring. IEEE Trans. Instrum. Meas. **55**(4), 1038–1050 (2006)
11. Amit, S.N.K.B., Shiraishi, S., Inoshita, T., Aoki, Y.: Analysis of satellite images for disaster detection. In: Proceedings of the 2016 IEEE International Geoscience and Remote Sensing Symposium (IGARSS), pp. 5189–5192 (2016)
12. Azadeh, A., Ghaderi, S.F., Sohrabkhani, S.: Annual electricity consumption forecasting by neural network in high energy consuming industrial sectors. Energy Convers. Manag. **49**(8), 2272–2278 (2008)
13. Azar, T.A., Vaidyanathan, S.: Computational intelligence applications in modeling and control. Springer International Publishing (2015)
14. Banerjee, T.P., Das, S.: Multi-sensor data fusion using support vector machine for motor fault detection. Inf. Sci. **217**, 96–107 (2012)
15. Bellocchio, F., Borghese, N.A., Ferrari, S., Piuri, V.: 3D Surface Reconstruction: multi-scale hierarchical approaches. Springer (2013)

16. Braeuer, B., Bauer, K.: A new interpretation of seismic tomography in the southern Dead Sea basin using neural network clustering techniques: interpretation of tomography in the SDSB. Geophys. Res. Lett. **42**(22), 9772–9780 (2015)
17. Burger, H.C., Schuler, C.J., Harmeling, S.: Image denoising: Can plain neural networks compete with BM3D? In: Proceedings of the 2012 IEEE Conference on Computer Vision and Pattern Recognition (CVPR), pp. 2392–2399 (2012)
18. Cadenas, J.M., Garrido, M.C., Muñoz, E.: Facing dynamic optimization using a cooperative metaheuristic configured via fuzzy logic and SVMs. Appl. Soft Comput. **11**(8), 5639–5651 (2011)
19. Campbell, C.: An introduction to kernel methods. In: Howlett, R.J., Jain, L.C. (eds.) Radial basis function networks: design and applications. Springer, Berlin (2000)
20. Charfi, Y., Wakamiya, N., Murata, M.: Challenging issues in visual sensor networks. IEEE Wirel. Commun. **16**(2), 44–49 (2009)
21. De Capitani di Vimercati, S., Foresti, S., Livraga, G., Samarati, P.: Data privacy: definitions and techniques. Int. J. Uncertain. Fuzziness Knowl. Based Syst. **20**(6), 793–817 (2012)
22. De Capitani di Vimercati, S., Genovese, A., Livraga, G., Piuri, V., Scotti, F.: Privacy and security in environmental monitoring systems: issues and solutions. In: Vacca J.R (ed.) Computer and Information Security Handbook, 2nd edn, pp. 835–853. Morgan Kaufmann, Boston (2013)
23. De Capitani di Vimercati, S., Livraga, G., Piuri, V., Scotti, F.: Privacy and security in environmental monitoring systems. In: Proceedings of the 2012 IEEE 1st AESS European Conference on Satellite Telecommunications (ESTEL), pp. 1–6 (2012)
24. Di Natale, C., Davide, F.A.M., D'Amico, A., Göpel, W., Weimar, U.: Sensor arrays calibration with enhanced neural networks. Sens. Actuators B: Chem. **19**(1), 654–657 (1994)
25. Ding, A., Zhang, Q., Zhou, X., Dai, B.: Automatic recognition of landslide based on CNN and texture change detection. In: Proceedings of the 2016 31st Youth Academic Annual Conference of Chinese Association of Automation (YAC), pp. 444–448 (2016)
26. Domingos, P.: A few useful things to know about machine learning. Commun. ACM **55**(10), 78–87 (2012)
27. Donida Labati, R., Genovese, A., Muñoz, E., Piuri, V., Scotti, F., Sforza, G.: Improving OSB wood panel production by vision-based systems for granulometric estimation. In: Proceedings of the 2015 IEEE 1st International Forum on Research and Technologies for Society and Industry (RTSI), pp. 557–562 (2015)
28. Donida Labati, R., Genovese, A., Muñoz, E., Piuri, V., Scotti, F., Sforza, G.: Analyzing images in frequency domain to estimate the quality of wood particles in OSB production. In: Proceedings of the 2016 IEEE Internationa Conference on Computational Intelligence and Virtual Environments for Measurement Systems and Applications (CIVEMSA). Budapest, Hungary (2016)
29. Donida Labati, R., Genovese, A., Muñoz, E., Piuri, V., Scotti, F., Sforza, G.: Computational intelligence for industrial and environmental applications. In: Proceedings of the 2016 IEEE 8th International Conference on Intelligent Systems (IS), pp. 8–14 (2016)
30. Donida Labati, R., Genovese, A., Piuri, V., Scotti, F.: Low-cost volume estimation by two-view acquisitions: a computational intelligence approach. In: Proceedings of the 2012 International Joint Conference on Neural Networks (IJCNN), pp. 1–8 (2012)
31. Donida Labati, R., Genovese, A., Piuri, V., Scotti, F.: A virtual environment for the simulation of 3D wood strands in multiple view systems for the particle size measurements. In: Proceedings of the 2013 IEEE International Conference on Computational Intelligence and Virtual Environments for Measurement Systems and Applications (CIVEMSA), pp. 162–167 (2013)
32. Donida Labati, R., Genovese, A., Piuri, V., Scotti, F.: Wildfire smoke detection using computational intelligence techniques enhanced with synthetic smoke plume generation. IEEE Trans. Syst. Man Cybern. Syst. **43**(4), 1003–1012 (2013)
33. D'Urso, P., Di Lallo, D., Maharaj, E.A.: Autoregressive model-based fuzzy clustering and its application for detecting information redundancy in air pollution monitoring networks. Soft Comput. **17**(1), 83–131 (2013)

34. Dutta, S., Datta, A., Chakladar, N.D., Pal, S.K., Mukhopadhyay, S., Sen, R.: Detection of tool condition from the turned surface images using an accurate grey level co-occurrence technique. Precis. Eng. **36**(3), 458–466 (2012)
35. Engelbrecht, A.: Computational Intelligence: an introduction. Wiley (2007)
36. Ferrari, S., Frosio, I., Piuri, V., Borghese, N.A.: Automatic multiscale meshing through HRBF networks. IEEE Trans. Instrum. Meas. **54**(4), 1463–1470 (2005)
37. Ferrari, S., Piuri, V.: Neural networks in intelligent sensors and measurement systems for industrial applications. In: Ablameyko, S., Goras, L., Gori, M., Piuri, V. (eds.) Neural networks for instrumentation, measurement, and related industrial applications, pp. 19–42. IOS Press (2003)
38. Ferrari, S., Piuri, V., Scotti, F.: Image processing for granulometry analysis via neural networks. In: Proceedings of the 2008 IEEE International Conference on Computational Intelligence for Measurement Systems and Applications (CIMSA), pp. 28–32 (2008)
39. Foo, P.H., Ng, G.W.: High-level information fusion: an overview. J. Adv. Inf. Fus. **8**(1), 33–72 (2013)
40. Fortuna, L., Giannone, P., Graziani, S., Xibilia, M.G.: Virtual instruments based on stacked neural networks to improve product quality monitoring in a refinery. IEEE Trans. Instrum. Meas. **56**(1), 95–101 (2007)
41. Fortuna, L., Graziani, S., Xibilia, M.: Soft sensors for product quality monitoring in debutanizer distillation columns. Control Eng. Pract. **13**(4), 499–508 (2005)
42. Fowler, K.: Sensor survey Part 2: Sensors and sensor networks in five years. IEEE Instrum. Meas. Mag. **12**(2), 40–44 (2009)
43. Fuente, M.J., Garcia-Alvarez, D., Sainz-Palmero, G.I., Vega, P.: Fault detection in a wastewater treatment plant based on neural networks and PCA. In: Proceedings of the 2012 20th Mediterranean Conference on Control Automation (MED), pp. 758–763 (2012)
44. Gamassi, M., Piuri, V., Scotti, F., Roveri, M.: Genetic techniques for pattern extraction in particle boards images. In: Proceedings of the 2006 IEEE International Conference on Computational Intelligence for Measurement Systems and Applications (CIMSA), pp. 129–134 (2006)
45. García Nieto, P.J., Combarro, E.F., del Coz Díaz, J.J., Montañés, E.: A SVM-based regression model to study the air quality at local scale in Oviedo urban area (Northern Spain): a case study. Appl. Math. Comput. **219**(17), 8923–8937 (2013)
46. GE Oil and Gas: Ca-Zoom® digital PTZ industrial inspection cameras. https://www.gemeasurement.com/inspection-ndt/remote-visual-inspection/ca-zoom-industrial-ptz-cameras
47. Genovese, A., Donida Labati, R., Piuri, V., Scotti, F.: Wildfire smoke detection using computational intelligence techniques. In: Proceedings of the 2011 IEEE International Conference on Computational Intelligence for Measurement Systems and Applications (CIMSA), pp. 1–6 (2011)
48. Gungor, V.C., Hancke, G.P.: Industrial wireless sensor networks: challenges, design principles, and technical approaches. IEEE Trans. Ind. Electron. **56**(10), 4258–4265 (2009)
49. Haykin, S.: Neural Networks and Learning Machines. v.10. Prentice Hall (2009)
50. Hibert, C., Grandjean, G., Bitri, A., Travelletti, J., Malet, J.P.: Characterizing landslides through geophysical data fusion: example of the La Valette landslide (France). Eng. Geol. **128**, 23–29 (2012)
51. Hossain, M., Rekabdar, B., Louis, S.J., Dascalu, S.: Forecasting the weather of nevada: a deep learning approach. In: Proceedings of the 2015 International Joint Conference on Neural Networks (IJCNN), pp. 1–6 (2015)
52. Hou, L., Bergmann, N.W.: Novel industrial wireless sensor networks for machine condition monitoring and fault diagnosis. IEEE Trans. Instrum. Meas. **61**(10), 2787–2798 (2012)
53. Ince, T., Kiranyaz, S., Eren, L., Askar, M., Gabbouj, M.: Real-time motor fault detection by 1-D convolutional neural networks. IEEE Trans. Ind. Electron. **63**(11), 7067–7075 (2016)
54. Jain, A.K., Duin, R.P.W., Mao, J.: Statistical pattern recognition: a review. IEEE Trans. Pattern Anal. Mach. Intell. **22**(1), 4–37 (2000)

55. Karri, V., Ho, T., Madsen, O.: Artificial neural networks and neuro-fuzzy inference systems as virtual sensors for hydrogen safety prediction. Int. J. Hydrog. Energy **33**(11), 2857–2867 (2008)
56. Khaleghi, B., Khamis, A., Karray, F.O., Razavi, S.N.: Multisensor data fusion: a review of the state-of-the-art. Inf. Fusion **14**(1), 28–44 (2013)
57. Khan, S.A., Shahani, D.T., Agarwala, A.K.: Sensor calibration and compensation using artificial neural network. ISA Trans. **42**(3), 337–352 (2003)
58. Kothari, V., Anuradha, J., Shah, S., Mittal, P.: A Survey on Particle Swarm Optimization in Feature Selection, pp. 192–201. Springer, Berlin, Heidelberg (2012)
59. Kreibich, O., Neuzil, J., Smid, R.: Quality-based multiple-sensor fusion in an industrial wireless sensor network for MCM. IEEE Trans. Ind. Electron. **61**(9), 4903–4911 (2014)
60. Krömer, P., Platoš, J., Snášel, V.: Mining multi-class industrial data with evolutionary fuzzy rules. In: Proceedings of the 2013 IEEE International Conference on Cybernetics (CYBCONF), pp. 191–196 (2013)
61. Kulkarni, R.V., Forster, A., Venayagamoorthy, G.K.: Computational intelligence in wireless sensor networks: a survey. IEEE Commun. Surv. Tutor. **13**(1), 68–96 (2011)
62. Langone, R., Alzate, C., De Ketelaere, B., Vlasselaer, J., Meert, W., Suykens, J.A.K.: LS-SVM based spectral clustering and regression for predicting maintenance of industrial machines. Eng. Appl. Artif. Intell. **37**, 268–278 (2015)
63. Larios, D.F., Barbancho, J., Rodríguez, G., Sevillano, J.L., Molina, F.J., León, C.: Energy efficient wireless sensor network communications based on computational intelligent data fusion for environmental monitoring. IET Commun. **6**(14), 2189 (2012)
64. Lecun, Y., Bengio, Y., Hinton, G.: Deep learning. Nature **521**(7553), 436–444 (2015)
65. Lerner, B., Guterman, H., Aladjem, M., Dinstein, I.: A comparative study of neural network based feature extraction paradigms. Pattern Recognit. Lett. **20**(1), 7–14 (1999)
66. Li, G., Rong, M., Wang, X., Li, X., Li, Y.: Partial discharge patterns recognition with deep Convolutional Neural Networks. In: Proceedings of the 2016 International Conference on Condition Monitoring and Diagnosis (CMD), pp. 324–327 (2016)
67. Li, P., Li, N., Cao, M.: Meteorology features extraction for transmission line icing process based on Kohonen Self-Organizing Maps. In: Proceedings of the 2010 International Conference on Computer Design and Applications (ICCDA), pp. 430–433 (2010)
68. Liu, Q., Jin, D., Shen, J., Fu, Z., Linge, N.: A WSN-based prediction model of microclimate in a greenhouse using extreme learning approaches. In: Proceedings of the 2016 18th International Conference on Advanced Communication Technology (ICACT), pp. 1–2 (2016)
69. Longbotham, N., Pacifici, F., Glenn, T., Zare, A., Volpi, M., Tuia, D., Christophe, E., Michel, J., Inglada, J., Chanussot, J., Du, Q.: Multi-modal change detection, application to the detection of flooded areas: outcome of the 2009–2010 data fusion contest. IEEE J. Sel. Top. Appl. Earth Obs. Remote Sens. **5**(1), 331–342 (2012)
70. Makhtar, A.K., Yussof, H., Al-Assadi, H., Yee, L.C., Othman, M.F., Shazali, K.: Wireless sensor network applications: A study in environment monitoring system. In: Proceedings of the 2012 International Symposium on Robotics and Intelligent Sensors (IRIS), vol. 41, pp. 1204–1210 (2012)
71. Marco, S., Gutiérrez-Gálvez, A., Lansner, A., Martinez, D., Rospars, J.P., Beccherelli, R., Perera, A., Pearce, T.C., Verschure, P.F.M.J., Persaud, K.: A biomimetic approach to machine olfaction, featuring a very large-scale chemical sensor array and embedded neuro-bio-inspired computation. Microsyst. Technol. **20**(4–5), 729–742 (2014)
72. Marwala, T.: Condition Monitoring using Computational Intelligence Methods: applications in mechanical and electrical systems. Springer Science and Business Media (2012)
73. Marzano, F.S., Rivolta, G., Coppola, E., Tomassetti, B., Verdecchia, M.: Rainfall nowcasting from multisatellite passive-sensor images using a recurrent neural network. IEEE Trans. Geosci. Remote Sens. **45**(11), 3800–3812 (2007)
74. Mohamed, R., Ahmed, A., Eid, A., Farag, A.: Support vector machines for camera calibration problem. In: Proceedings of the 2006 International Conference on Image Processing (ICIP), pp. 1029–1032 (2006)

75. Mourot, G., Bousghiri, S., Kratz, F.: Sensor fault detection using fuzzy logic and neural networks. In: Proceedings of the International Conference on Systems, Man and Cybernetics (SMC), pp. 369–374 (1993)
76. Muñoz, E., Ruspini, E.H.: Simulation of fuzzy queueing systems with a variable number of servers, arrival rate, and service rate. IEEE Trans. Fuzzy Syst. **22**(4), 892–903 (2014)
77. Nakama, T., Muñoz, E., LeBlanc, K., Ruspini, E.: Generalizing and formalizing precisiation language to facilitate human-robot interaction. In: Computational Intelligence, pp. 381–397. Springer (2016)
78. Nor, A.S.M., Faramarzi, M., Yunus, M.A.M., Ibrahim, S.: Nitrate and sulfate estimations in water sources using a planar electromagnetic sensor array and artificial neural network method. IEEE Sens. J. **15**(1), 497–504 (2015)
79. O'Connor, E., Smeaton, A.F., O'Connor, N.E., Regan, F.: A neural network approach to smarter sensor networks for water quality monitoring. Sensors **12**(4), 4605 (2012)
80. Palade, V., Bocaniala, C.D., Jain, L.: Computational Intelligence in Fault Diagnosis. Springer (2006)
81. Paulinas, M., Ušinskas, A.: A survey of genetic algorithms applications for image enhancement and segmentation. Inf. Technol. control **36**(3) (2015)
82. phoneArena: A modern smartphone or a vintage supercomputer: which is more powerful? http://www.phonearena.com/news/A-modern-smartphone-or-a-vintage-supercomputer-which-is-more-powerful_id57149 (2014)
83. Piuri, V., Scotti, F., Roveri, M.: Visual inspection of particle boards for quality assessment. In: Proceedings of the IEEE International Conference on Image Processing (ICIP), pp. 521–524 (2005)
84. Prieto, M.D., Cirrincione, G., Espinosa, A.G., Ortega, J.A., Henao, H.: Bearing fault detection by a novel condition-monitoring scheme based on statistical-time features and neural networks. IEEE Trans. Ind. Electron. **60**(8), 3398–3407 (2013)
85. Qiao, T., Ren, J., Craigie, C., Zabalza, J., Maltin, C., Marshall, S.: Quantitative prediction of beef quality using visible and NIR spectroscopy with large data samples under industry conditions. J. Appl. Spectrosc. **82**(1), 137–144 (2015)
86. Ribeiro, B.: Support vector machines for quality monitoring in a plastic injection molding process. IEEE Trans. Syst. Man Cybern. Part C (Applications and Reviews) **35**(3), 401–410 (2005)
87. Sagheer, A.: Piecewise one dimensional Self Organizing Map for fast feature extraction. In: Proceedings of the 2010 10th International Conference on Intelligent Systems Design and Applications (ISDA), pp. 633–638 (2010)
88. Sammouda, R., Adgaba, N., Touir, A., Al-Ghamdi, A.: Agriculture satellite image segmentation using a modified artificial Hopfield neural network. Comput. Hum. Behav. **30**, 436–441 (2014)
89. Sarcevic, P., Pletl, S., Kincses, Z.: Evolutionary algorithm based 9DOF sensor board calibration. In: Proceedings of the 12th IEEE International Symposiumon Intelligent Systems and Informatics (SISY), pp. 187–192 (2014)
90. Shirvaikar, M.: Trends in automated visual inspection. J. Real Time Image Process. **1**(1), 41–43 (2006)
91. Simon, D.: Evolutionary Optimization Algorithms. Wiley (2013)
92. Singha, S., Bellerby, T.J., Trieschmann, O.: Satellite oil spill detection using artificial neural networks. IEEE J. Sel. Top. Appl. Earth Obs. Remote Sens. **6**(6), 2355–2363 (2013)
93. Su, I.J., Tsai, C.C., Sung, W.T.: Area temperature system monitoring and computing based on adaptive fuzzy logic in wireless sensor networks. Appl. Soft Comput. **12**(5), 1532–1541 (2012)
94. Szenkovits, A., Gaskó, N., Jahier, E.: Environment-model based testing with differential evolution in an industrial setting, pp. 819–830 (2016)
95. Teti, R., Segreto, T., Simeone, A., Teti, R.: Multiple sensor monitoring in nickel alloy turning for tool wear assessment via sensor fusion. In: Proceedings of the 8th CIRP Conference on Intelligent Computation in Manufacturing Engineering, pp. 85–90 (2013)

96. Toh, K.K.V., Isa, N.A.M., Ashidi, N.: Noise adaptive fuzzy switching median filter for salt-and-pepper noise reduction. IEEE Signal Process. Lett. **17**(3), 281–284 (2010)
97. Trillas, E., Eciolaza, L.: Fuzzy Logic: an introductory course for engineering students. Springer Publishing Company, Incorporated (2015)
98. Vaseghi, S.V.: Advanced Digital Signal Processing and Noise Reduction, 4th edn. Wiley (2008)
99. Vos, T.E.J., Baars, A.I., Lindlar, F.F., Windisch, A., Wilmes, B., Gross, H., Kruse, P.M., Wegener, J.: Industrial case studies for evaluating search based structural testing. Int. J. Softw. Eng. Knowl. Eng. **22**(08), 1123–1149 (2012)
100. Vos, T.E.J., Lindlar, F.F., Wilmes, B., Windisch, A., Baars, A.I., Kruse, P.M., Gross, H., Wegener, J.: Evolutionary functional black-box testing in an industrial setting. Softw. Qual. J. **21**(2), 259–288 (2013)
101. Wang, L., Fu, X.: Data mining with computational intelligence. In: Advanced Information and Knowledge Processing. Springer, Berlin, New York (2005)
102. Wang, X.Y., Yang, H.Y., Zhang, Y., Fu, Z.K.: Image denoising using SVM classification in nonsubsampled contourlet transform domain. Inf. Sci. **246**, 155–176 (2013)
103. Wijayasekara, D., Linda, O., Manic, M., Rieger, C.: FN-DFE: Fuzzy-neural data fusion engine for enhanced resilient state-awareness of hybrid energy systems. IEEE Trans. Cybern. **44**(11), 2065–2075 (2014)
104. Wu, J.L., Li, I.J.: A SOM-based dimensionality reduction method for KNN classifiers. In: Proceedings of the 2010 International Conference on System Science and Engineering (ICSSE), pp. 173–178 (2010)
105. Xiang, Y., Jiang, L.: Water quality prediction using LS-SVM and Particle Swarm Optimization. In: Proceedings of the 2nd International Workshop on Knowledge Discovery and Data Mining (WKDD), pp. 900–904 (2009)
106. Xie, X., Guo, J., Zhang, H., Jiang, T., Bie, R., Sun, Y.: Neural-network based structural health monitoring with wireless sensor networks. In: Proceedings of the 2013 9th International Conference on Natural Computation (ICNC), pp. 163–167 (2013)
107. Xu, F., Song, X., Wang, X., Su, J.: Neural network model for earthquake prediction using DMETER data and seismic belt information. In: Proceedings of the 2010 2nd WRI Global Congress on Intelligent Systems (GCIS), vol. 3, pp. 180–183 (2010)
108. Xue, B., Zhang, M., Browne, W.N.: Particle swarm optimization for feature selection in classification: a multi-objective approach. IEEE Trans. Cybern. **43**(6), 1656–1671 (2013)
109. Xue, B., Zhang, M., Browne, W.N., Yao, X.: A survey on evolutionary computation approaches to feature selection. IEEE Trans. Evol. Comput. **20**(4), 606–626 (2016)
110. Yao, L., Lu, N., Jiang, S.: Artificial neural network (ANN) for multi-source PM2.5 estimation using surface, MODIS, and meteorological data. In: Proceedings of the 2012 International Conference on Biomedical Engineering and Biotechnology (iCBEB), pp. 1228–1231 (2012)
111. Ye, Y., Ci, S., Katsaggelos, A.K., Liu, Y., Qian, Y.: Wireless video surveillance: a survey. IEEE Access **1**, 646–660 (2013)
112. Yildiz, A.R.: A new hybrid differential evolution algorithm for the selection of optimal machining parameters in milling operations. Appl. Soft Comput. **13**(3), 1561–1566 (2013)
113. Yin, S., Ding, S.X., Xie, X., Luo, H.: A review on basic data-driven approaches for industrial process monitoring. IEEE Trans. Ind. Electron. **61**(11), 6418–6428 (2014)
114. Zhang, M., Liu, X.: A soft sensor based on adaptive fuzzy neural network and support vector regression for industrial melt index prediction. Chemom. Intell. Lab. Syst. **126**, 83–90 (2013)
115. Zhao, W., Du, S.: Spectral-spatial feature extraction for hyperspectral image classification: A dimension reduction and deep learning approach. IEEE Trans. Geosci. Remote Sens. **54**(8), 4544–4554 (2016)
116. Zhao, Z., Liu, F.: Industrial monitoring based on moving average PCA and neural network. In: Proceedings of the 30th Annual Conference of IEEE Industrial Electronics Society (IECON), vol. 3, pp. 2168–2171 (2004)

Real, Imaginary and Complex Sets

Vassil Sgurev

1 Introduction

Problems exist in the abstract mathematical and information technologies structures that cannot be solved in the framework of these scientific directions. This imposes expansion of those structures.

In the real numbers a problem have emerged about solving an equation of the type

$$x^2 = -1;\ x = \sqrt{-1}; \tag{1}$$

which cannot be solved in the framework of real numbers as no real number exists which when squared to give result -1. This have enforced extension of the class real numbers to complex ones through the imaginary unit $i = \sqrt{-1}$. The general form of the class of complex numbers z is:

$$z = a + ib, \tag{2}$$

where a and b are real numbers.

In an analogical way in [1] the imaginary logical value i is introduced taking values from the set $\{i, \neg i\}$. This makes possible the logical equation $F \wedge i = T$ through the relation

$$F \wedge i = T$$

where $\{F, T\}$ are the two states of the classical propositional logic.

V. Sgurev (✉)
Institute of Information and Communication Technologies – Bulgarian Academy of Sciences, Acad. G. Bonchev str., bl. 2, 1113 Sofia, Bulgaria
e-mail: vsgurev@gmail.com

V. Sgurev et al. (eds.), *Learning Systems: From Theory to Practice*, Studies in Computational Intelligence 756, https://doi.org/10.1007/978-3-319-75181-8_3

2 Real, Imaginary and Complex Sets

In the present work a solution of the set equation

$$\mathrm{A} \cup \mathrm{B} = \mathrm{C}; \tag{3}$$

where A, B and C are sets and $\varnothing$ is the empty set.

The symbols $\cup, \cap, \backslash, \Delta$ denote the operations union, intersection, difference, and symmetric difference respectively in the Cantor's set theory. If we assume that non-empty sets A and C are given in (3) such that $\mathrm{A}\backslash\mathrm{C} \neq \varnothing$ and $\mathrm{C}\backslash\mathrm{A} \neq \varnothing$; then as it is seen from Fig. 1 no such set B exists in the framework of the classical set theory [2–4] such that it satisfies the requirements of (3).

If in Eq. (3) $\mathrm{A} \neq \varnothing$ and $\mathrm{C} \neq \varnothing$ then no set A will be found which united with any other set B would produce an empty set $\mathrm{C} = \varnothing$ [7, 8].

An exit from this may be found if analogically to the complex numbers imaginary elements are introduced, sets and subsets

$$ia \in i\mathrm{A};\ i\mathrm{A} \subseteq i\mathrm{B}; \tag{4}$$

such that at interaction between them and the real sets of the classical set theory

$$a \in \mathrm{A}; \mathrm{A} \subseteq \mathrm{B}; \tag{5}$$

the following requirement to be observed: for each $ia \in i\mathrm{A}$ and $a \in \mathrm{A}$

$$a \cup ia = \varnothing \text{ and } \mathrm{A} \cup i\mathrm{A} = \varnothing. \tag{6}$$

It follows from the above said that any element $a \in \mathrm{A}$ may be in one of the two opposite states: real—*1a* or imaginary—*ia*. The set $\{1, i\}$ contains the symbols denoting the two states—real *1* and imaginary—*i* in one only of which the element $a \in \mathrm{A}$ may be found. Further on the symbol *1* will not be put before the element $a \in \mathrm{A}$ when it is in real state. The symbol *i* will be always present before any imaginary element or imaginary set. The symbol *i* may be considered as a peculiar analog to $i = \sqrt{-1}$ in the complex numbers but as a distinction it has no concrete value and it is just a symbol to denote the imaginary state of the separate elements [5, 6].

Definition Each element of a given set may be in only one of the two possible states $\{1, i\}$ but never in both simultaneously.

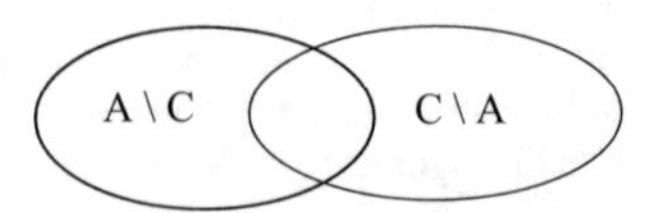

Fig. 1 Infeasible intersection of two relative complements

Definition The non-empty set A will be called *normed* if in it each element $a \in \mathrm{A}$ is found in one only of the two states $\{1, i\}$, i.e. no subset D of A exists such that $\mathrm{D} = \{a, ia\} \subseteq \mathrm{A}$.

Sets that do not keep the latter definition will be called *non-normed*. Norming of an arbitrary set $\mathrm{A} \subset \mathrm{U}$ where U is the universal set is carried out by keeping the requirement (6).

Definition If $a \in \mathrm{A}$ and $ib \in i\mathrm{B}$ then

$$d = a \cup ib \text{ and } \mathrm{C} = \mathrm{A} \cup i\mathrm{B} \tag{7}$$

will be called a complex element d and a complex set C respectively.

On the base of the imaginary and complex sets introduced a more general solution of Eq. (3) may be pointed out: if $\mathrm{A} \neq \varnothing$; $\mathrm{C} \neq \varnothing$ and $\mathrm{C} \subset \mathrm{A}$ then it follows from (3)

$$\mathrm{B} = \mathrm{C} \cup i(\mathrm{A} \backslash \mathrm{C}).$$

If in the equality above it is assumed that $\mathrm{C} \neq \varnothing$ then in this case $\mathrm{B} = i\mathrm{A}$ and the relation from (6) is observed namely $\mathrm{A} \cup i\mathrm{A} \neq \varnothing$.

Definition Two complex sets

$$\mathrm{C}_1 = \mathrm{A}_1 \cup i\mathrm{B}_1 \text{ and } \mathrm{C}_2 = \mathrm{A}_2 \cup i\mathrm{B}_2 \tag{8}$$

are equal between each other, i.e. $\mathrm{C}_1 = \mathrm{C}_2$ if $a \in \mathrm{A}_1$ is an element of A_2 also and vice versa, and $ib_1 \in \mathrm{B}_1$ is an element of $i\mathrm{B}_2$ and vice versa.

Definition The set C_1 from (8) is a subset of C_2 i.e $\mathrm{C}_1 \subseteq \mathrm{C}_2$ if each element $a \in \mathrm{A}_1$ is an element of A_2 also and $ib_1 \in \mathrm{B}_1$ is an element of $i\mathrm{B}_2$. This may be put down

$$\mathrm{A}_1 \subseteq \mathrm{A}_2;\ i\mathrm{B}_1 \subseteq i\mathrm{B}_2;\ (\mathrm{A}_1 \cup i\mathrm{B}_1) \subseteq (\mathrm{A}_2 \cup i\mathrm{B}_2).$$

The method of mutual inclusions in the complex sets is of the following kind:

$$\mathrm{C}_1 = \mathrm{C}_2 \text{ if } \mathrm{C}_1 \subseteq \mathrm{C}_2 \text{ and } \mathrm{C}_2 \subseteq \mathrm{C}_1. \tag{9}$$

Operations union, intersection, relative complement, symmetric difference, and splitting for the complex sets from (8) will be defined observing the following requirements:

$$\mathrm{A} \cap \mathrm{B} = \varnothing \text{ and } \mathrm{A} \cap i\mathrm{B} = \varnothing. \tag{10}$$

The expediency of these assumptions will be proved further through relations from (18) to (22).

1. The union of two complex sets $C_1 \cup C_2$ from (8) is equal to

$$\begin{aligned}(A_1 \cup iB_1) \cup (A_2 \cup iB_2) &= (A_1 \cup A_2) \cup i(B_1 \cup B_2)\\ &= \{a/a \in A_1 \vee a \in A_2\} \cup \{ib/ib \in B_1 \vee ib \in iB_2\}. \quad (11)\end{aligned}$$

2. The intersection of the same sets $C_1 \cap C_2$ leads to the following chain of equalities:

$$\begin{aligned}(A_1 \cup iB_1) \cap (A_2 \cup iB_2) &= (A_1 \cap A_2) \cup (A_1 \cap iB_2) \cup (A_2 \cap iB_1) \cup i\,(B_1 \cap B_2)\\ &= (A_1 \cap A_2) \cup i\,(B_1 \cap B_2)\\ &= \{a/a \in A_1 \wedge a \in A_2\} \cup \{ib/ib \in iB_1 \wedge ib \in iB_2\} \quad (12)\end{aligned}$$

at that it follows from assumption (10) that:

$$A_1 \cap iB_2 = \varnothing \text{ and } A_2 \cap iB_1 = \varnothing.$$

3. The difference $C_1 \backslash C_2$ is equal to:

$$\begin{aligned}(A_1 \cup iB_1) \backslash (A_2 \cup iB_2) &= (A_1 \backslash (A_2 \cup iB_2)) \cup (iB_1 \backslash (A_2 \cup iB_2))\\ &= ((A_1 \backslash A_2) \cap (A_1 \cup iB_2)) \cup ((iB_1 \backslash A_2) \cap (iB_1 \backslash iB_2))\\ &= ((A_1 \backslash A_2) \cap A_1) \cup (iB_1 \cap i\,(B_1 \backslash B_2)) = (A_1 \backslash A_2) \cup i\,(B_1 \backslash B_2)\\ &= \{a/a \in A_1 \wedge a \notin A_2\} \cup \{ib/ib \in iB_1 \wedge ib \notin iB_2\}; \quad (13)\end{aligned}$$

In analogical way it may be proved:

$$(A_2 \cup iB_2) \backslash (A_1 \cup iB_1) = (A_2 \backslash A_1) \cup i\,(B_2 \backslash B_1).$$

4. The symmetric difference $C_1 \Delta C_2$ may be defined by using the differences from (13), namely $C_1 \backslash C_2$ and $C_2 \backslash C_1$:

$$\begin{aligned}(A_1 \cup iB_1) \Delta (A_2 \cup iB_2) &= ((A_1 \cup iB_1) \backslash (A_2 \cup iB_2)) \cup ((A_2 \cup iB_2) \backslash (A_1 \cup iB_1))\\ &= (A_1 \backslash A_2) \cup i\,(B_1 \backslash B_2) \cup (A_2 \backslash A_1) \cup i\,(B_2 \backslash B_1)\\ &= ((A_1 \backslash (A_2) \cup (A_2 \backslash (A_1)) \cup i\,(B_1 \backslash B_2) \cup i\,(B_2 \backslash B_1). \quad (14)\end{aligned}$$

5. Splitting is an operation which in complex sets may be accomplished in the following way:
   Let for the complex sets from (8) be true:

$$C = C_1 \cup C_2 = A \cup iB \quad (15)$$

$$A_1 \cup A_2 = A;\ iB_1 \cup iB_2 = iB; \quad (16)$$

$$A_1 \cap A_2 = \varnothing;\ iB_1 \cap iB_2 = \varnothing \quad (17)$$

   Then it may be stated for the complex set C that it is split into two subsets C_1 and C_2.
   In Figs. 2, 3, 4, 5 and 6 the respective dashed parts are shown for operations $\cup$, $\cap$, $\backslash$, Δ, and splitting.

The term “normalization” will be extended on the complex sets in the following way.

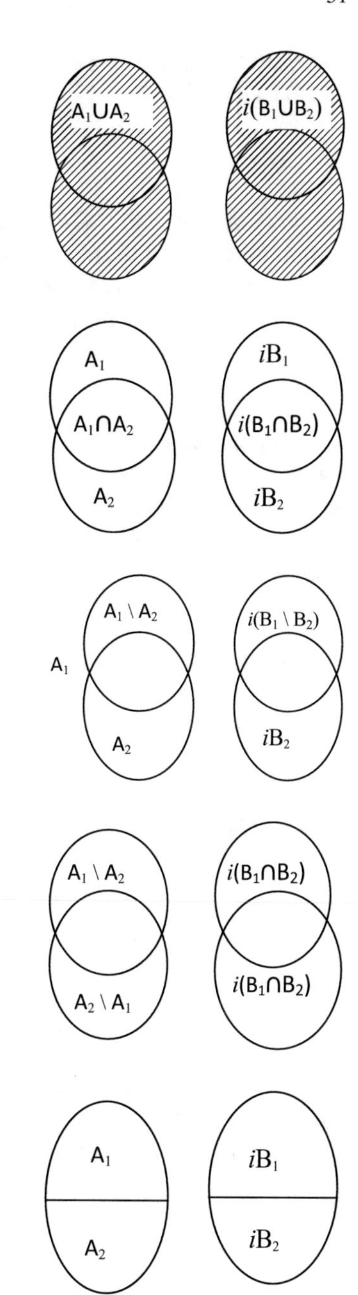

Fig. 2 $\cup$

Fig. 3 $\cap$

Fig. 4 $\setminus$

Fig. 5 Δ

Fig. 6 Splitting

Definition The complex set $C = A \cup iB$ will be called *normalized* if it satisfies the following condition:

$$A \cap B = \varnothing. \tag{18}$$

Proposition *Condition (18) is necessary and sufficient for normalization of the complex set $C = A \cup iB$, i.e. keeping the requirements (6) for the same set. Non-keeping the conditions (18) means that at least one element d exists in the set D, for which*

$$A \cap B = D \neq \varnothing; d \in D. \tag{19}$$

It follows from this that the element d is in both in real and imaginary state which leads to a contradiction and proves the impossibility of (19) and the truth of (18). And vice versa—if assumed that in (19) D = Ø then this immediately results in (18). So the complex set C = A ∪ iB is normed if in it no element is contained that is simultaneously in two different states—real and imaginary, i.e. requirement (18) is observed.

Corollary *The following relations are equivalent to (18)*

$$A \backslash B = A; B \backslash A = B; \tag{20}$$

$$A \cup B = A \,\Delta\, B. \tag{21}$$

It will be assumed that (18) is observed. Then we may put down:

$$A = (A \backslash B) \cup (A \cap B) = A \backslash B;$$
$$B = (B \backslash A) \cup (A \cap B) = B \backslash A;$$

$$A \cup B = ((A \backslash B) \cup (A \cap B)) \cup ((B \backslash A) \cap (A \cap B)) = (A \backslash B) \cup (B \backslash A) = A \,\Delta\, B.$$

The converse is also true—from the truth of (20) and (21), (18) follows. This means that if at least one of the equalities (20) and (21) is observed then (18) is also observed and the respective complex set (A ∪ iB) is normed.

Proposition *The requirement*

$$A \cap iB = \varnothing; \tag{22}$$

is necessary but it is not sufficient condition for normalization of the complex set $C = A \cup iB$.

A case will be considered when (22) is observed but requirement (18) is not observed. Let

$$A = (A' \cup a); B = (B' \cup a); \tag{23}$$

for the complex set $C = A \cup iB$. As by definition the intersection of the two classes of sets—real {A} and imaginary {iB} is an empty set then (22) is true. On the other hand the intersection of sets A and B from (23) results in

$$A \cap B = (A' \cup a) \cap (B' \cup a) \neq \varnothing$$

and not keeping (18) which means that (22) is not sufficient condition for normalization of C.

Condition (18) will be supposed to be observed. Then condition (22) evidently is always observed, which proves its necessity for (18).

Corollary *On the base of the previous proposition it may be proved that the following relations are necessary but not sufficient conditions for normalization of the complex set C.*

$$A \backslash iB = A; iB \backslash A = iB; \tag{24}$$

$$A \cup iB = A \,\Delta\, iB. \tag{25}$$

The proof for this may be carried out in the same manner like in the previous proposition.

It follows from results received that operations on complex sets may be reduced to either only real sets {A}—if in the different operations the imaginary parts are reduced to empty set, or only imaginary sets—if in the operations the real parts {A} are converted into empty set, or to complex sets.

Analogically, like in the Cantor's sets universal sets may be defined and the complements of the complex set.

The following complex universal set U will be defined consisting of a real universal set U' and universal imaginary set iU'' for which

$$U = U' \cup U'' \tag{26}$$

and either

$$U' \subseteq U'' \text{ or } U'' \subseteq U', \text{ or } U' = U''. \tag{27}$$

At that the complex universal set U is not subject to norming through relations of the type (6) i.e. for it in general requirements (6) and (18) are not in effect. It is expedient to define the complements in the following way:

$$\bar{A} = U \backslash A; \overline{iB} = U \backslash iB; \tag{28}$$

$$\bar{C} = U \backslash C = U \backslash (A \cup iB) = \bar{A} \cap i\bar{B}. \tag{29}$$

The following relations exist for the complements of complex sets.

Proposition *For the complements $\bar{C}$, $\bar{A}$, $\bar{B}$ it may be put down:*

$$\overline{\bar{A} \cap \overline{iB}} = \bar{\bar{C}} = A \cup iB; \tag{30}$$

which follows from

$$\begin{aligned}\overline{\bar{A} \cap \overline{iB}} &= U \backslash (\bar{A} \cap i\bar{B}) = (U \backslash \bar{A}) \cup (U \backslash \overline{iB}) \\ &= (U \backslash (U \backslash A)) \cup (U \backslash (U \backslash i\bar{B}) \\ &= (U \cap A) \cup (U \cup iB) = A \cup iB. \end{aligned} \tag{31}$$

Relations (29) to (31) show that the well-known law of double negation from propositional logic is observed for the complex sets, namely:

$$\overline{\overline{A \cup iB}} = \overline{\bar{A} \cap \overline{iB}} = A \cup iB. \tag{32}$$

It is observed if C is only real or imaginary set, i.e.:

(a) if $iB = \varnothing$, then $C = A$; $\bar{\bar{A}} = A$;
(b) if $A = \varnothing$, then $C = iB$; $\overline{\overline{iB}} = iB$.

It will be shown that for the universal set U and any complex sets C_1 and C_2 both De Morgan laws are observed, i.e.,

$$\overline{C_1 \cup C_2} = \overline{C_1} \cap \overline{C_2}; \tag{33}$$
$$\overline{C_1 \cap C_2} = \overline{C_1} \cup \overline{C_2}. \tag{34}$$

Relations exit:

$$\overline{C_1 \cup C_2} = U \backslash (C_1 \cup C_2) = (U \backslash C_1) \cap (U \backslash C_2) = \overline{C_1} \cap \overline{C_2};$$

$$\overline{C_1 \cap C_2} = U \backslash (C_1 \cap C_2) = (U \backslash C_1) \cup (U \backslash C_2) = \overline{C_1} \cup \overline{C_2}.$$

Both De Morgan laws are observed as particular cases both for real and imaginary sets separately:

(a) If $A_1 = A_2 = \varnothing$, then

$$\overline{iB_1 \cup iB_2} = \overline{iB_1} \cap \overline{iB_2} \text{ and } \overline{iB_1 \cap iB_2} = \overline{iB_1} \cup \overline{iB_2};$$

(b) If $iB_1 = iB_2 = \varnothing$, then

$$\overline{A_1 \cup A_2} = \overline{A_1} \cap \overline{A_2} \text{ and } \overline{A_1 \cap A_2} = \overline{A_1} \cup \overline{A_2};$$

(c) If $A_2 = \varnothing$ and $iB_1 = \varnothing$ then

$$\overline{A_1 \cup iB_2} = \overline{A_1} \cap \overline{iB_2} \text{ and } \overline{A_1 \cap iB_2} = \overline{A_1} \cup \overline{iB_2};$$

(d) If $A_1 = \varnothing$ and $iB_2 = \varnothing$ then

$$\overline{A_2 \cup iB_1} = \overline{A_2} \cap \overline{iB_1} \text{ and } \overline{A_2 \cap iB_1} = \overline{A_2} \cup \overline{iB_1}.$$

Let the complements of the real and imaginary sets, respectively, from (26) to the universal set be denoted by $\overline{A'} = (U' \backslash A)$ and $\overline{iB'} = (U'' \backslash iB)$. Then the following relations are true:

$$\begin{aligned}
\bar{C} &= \left((U' \cup U'') \backslash (A \cup iB)\right) = \left((U' \backslash A) \cap (iU' \backslash iB)\right) \cup \left((iU'' \backslash A) \cap (iU'' \backslash iB)\right) \\
&= \left((U' \backslash A) \cap U'\right) \cup \left((iU'' \backslash iB) \cap iU''\right) \\
&= (U' \backslash A) \cup (iU'' \backslash iB) = \overline{A'} \cup \overline{iB'}; \qquad (35) \\
\bar{A} &= (U' \cup U'') \backslash A = (U' \backslash A) \cup (iU'' \backslash A) = (U' \backslash A) \cup iU'' = \overline{A'} \cup iU''; \qquad (36) \\
\overline{iB} &= (U' \cup U'') \backslash iB = (U' \backslash iB) \cup (iU'' \backslash iB) = (iU'' \backslash iB) \cup U' = \overline{iB}' \cup U'; \qquad (37)
\end{aligned}$$

Comparison of relations (28) and (29) to those from (35) to (37) will give the following results:

$$\bar{C} = \bar{A} \cap \bar{B} = \overline{A'} \cup \overline{iB'}; \qquad (38)$$

$$\bar{A} = \bar{A'} \cup \overline{iU}'' \cup \overline{iB} = \overline{iB}' \cup U' \qquad (39)$$

The classic Cantor's sets have algebraic properties characteristic to the Boolean algebra and the lattice [2–4]. It is evident that such properties have the sets which are either in real or imaginary states only and in this case they are identical to the Cantor's sets. It will be shown that that under some assumptions the complex sets thus defined have also similar properties. These algebraic properties that correspond to the complex numbers are shown in Table 1. The requirements from #23 and #24 that satisfy (6) are characteristic for the complex sets only.

Relations from (11) to (14) show that for complex sets the requirements for commutativity #1 and distributivity of these sets are observed, and from the results from (24) to (34) follow the two De Morgan's laws for double negation and the requirements from #7 to #17 from Table 1. The validity of the remaining rules from the same table may be proved in analogical way with the exception of simultaneously keeping the rules #3 and #4 for associativity and #23 and #24 for norming. Validity of the other rules of the same table may be proved in an analogical way except rules #3 and #4 for associativity and #23 and #24 for norming. In this case ambiguity may appear in the results which follow from the following:

Proposition *Let* $C_3 = A_3 \cup iB_3$. *If*

Table 1 Validity of set operations for imaginary sets

1	$C_1 \cup C_2 = C_2 \cup C_1$	13	$C_1 \cup \overline{C_1} = U$
2	$C_1 \cap C_2 = C_2 \cap C_1$	14	$C_1 \cap \overline{C_1} = \varnothing$
3	$C_1 \cup (C_2 \cup C_3) = (C_1 \cup C_2) \cup C_3$	15	$C_1 \cup C_1 = C_1$
4	$C_1 \cap (C_2 \cap C_3) = (C_1 \cap C_2) \cap C_3$	16	$C_1 \cap C_1 = C_1$
5	$C_1 \cap (C_2 \cup C_3) = (C_1 \cap C_2) \cup (C1 \cap C3)$	17	$\overline{\overline{C_1}} = C_1$
6	$C_1 \cup (C_2 \cup C_3) = (C_1 \cup C_2) \cap (C1 \cup C3)$	18	$C_1 \backslash C_2 = C_1 \cap \overline{\overline{C_2}}$
7	$\overline{C_1 \cup C_2} = \overline{C_1} \cap \overline{C_2}$	19	$C_1 \Delta C_2 = (C_1 \cup C_2) \backslash (C_1 \cap C_2)$
8	$\overline{C_1 \cap C_2} = \overline{C_1} \cup \overline{C_2}$	20	$(C_1 \Delta C_2) \Delta C_3 = C_1 \Delta (C_2 \Delta C_3)$
9	$C_1 \cup \varnothing = C_1$	21	$C_1 \Delta C_2 = C_2 \Delta C_1$
10	$C_1 \cap \varnothing = \varnothing$	22	$C_1 \cap (C_2 \Delta C_3) = (C_1 \cap C_2) \Delta (C_2 \cap C_3)$
11	$C_1 \cap U = C_1$	23	$A \cup iA = \varnothing$
12	$C_1 \cup U = U$	24	$C_1 \Delta C_2 = C_1 \cup C_2$

$$C_1 = A_1 \cup iB_1; C_2 = B_1 \cup iA_1 \; and \; C_3 = C_2; \tag{40}$$

then

$$(C_1 \cup C_2) \cup C_3 \neq (C_2 \cup C_3) \cup C_1. \tag{41}$$

According to (40) the left-hand side of (41) is equal to C_3 as it follows from (6) and (40) $C_1 \cup C_2 = \varnothing$. But the right-hand side is an empty set as according to (40) and the requirement for idempotence $C_2 \cup C_3 = C_2$ and united with C_1 results in Ø. Hence the two sides of (41) are not equal between each other and that means that for different order of fulfillment of the associative operations at norming (6) different results are achieved.

To avoid this in the complex sets it is necessary to keep the following.

Rule for Normalization: Normalization of any complex set is to be carried out only after fulfillment of all possible associative operations, i.e. norming (6) should be done last in order. A similar requirement exists in the propositional logic also where the operation *negation* in the beginning of any expression is executed last in order and for the whole expression simultaneously when all other possible logical operations are realized.

In many cases receiving results in complex sets, analogically to the classical Cantor's set theory, is a corollary of the fact that operations union, intersection, difference, symmetric difference, and negation from (11) to (14) for the complex sets are carried out as a rule separately for the real and the imaginary sets and in the framework of the same classical theory. The real and imaginary sets interact mostly through the requirements for normalization (6) and after their execution they do not intersect between each other.

For each complex set C the set of all subsets of C may be formed. It is called **boolean** of the set C and it is denoted by 2^{C}.

$$2^{C} = \{X/X \subseteq C\} \tag{42}$$

where X is a subset of the set C.

Let the elements of C, A, and iB be denoted by |C|, |A|, and |iB| respectively. Then reckoning with the specificity of the complex sets the possible number of elements of the boolean 2^{C} may be defined in the following way

$$2^{|C|} = 2^{|A|} \times 2^{|iB|}. \tag{43}$$

The following example will be considered:
Let the complex set C be defined in the following way:

$$A = \{a_1, a_2\}\,;\ |A| = 2;\ iB = \{ib_1, ib_2, ib_3\}\ |iB| = 3. \tag{44}$$

$$C = \{a_1, a_2, ib_1, ib_2, ib_3\}\,;\ |C| = 5. \tag{45}$$

Then according to (43)

$$2^5 = 2^2 \times 2^3 = 4 \times 8 = 32. \tag{46}$$

Hence from the considered as an example complex set from (44) to (46) 32 subsets may be formed in general including the empty set ∅.

For defining the operation "Cartesian product of two complex sets" will be used the following concepts:

The ordered pair (a, b) consists of two elements—a, and b, located in a given order. Both pairs (a, b) and (c, d) are equal if

$$a = c \text{ and } b = d. \tag{47}$$

The binary relation φ will be called the set of ordered pairs. If $(a, b) \in \varphi$, then $a\,\varphi\,b$. Direct (Cartesian) product $C_1 \times C_2$ of two complex sets C_1 and C_2 will be called the aggregate of all ordered pairs (a, b) such that $a \in C_1$ and $b \in C_2$.

Like in the classic set theory for the complex sets C_1 and C_2 it is true

$$C_1 \times C_2 \neq C_2 \times C_1. \tag{48}$$

Table 2 Multiplication of complex numbers

	b	ib
a	(a, b)	$i(a, b)$
ia	$i(a, b)$	(a, b)

Table 3 Multiplication of real numbers

	b	$-b$
a	(a, b)	$-(a, b)$
$-a$	$-(a, b)$	(a, b)

Table 4 Logical equivalence

	T	F
T	T	F
F	F	T

Under multiplication of two complex sets a set of new elements is formed $\{(a, b)/(a, b) \in \varphi\}$ and due to that it is necessary to determine in what state—real or imaginary each of the ordered sets $(a, b) \in \varphi$ is found. It is suggested under multiplication of the two complex sets C_1 and C_2 the following Table 2 to be used for defining the states of each ordered pair $\{(a, b)/(a, b) \in \varphi\}$.

In analogical way the *positive* and *negative* states of real numbers are defined as well as the states *true* or *false* in the logical operation *equivalence*. This similarity follows from comparing Tables 2 to 3 (multiplication of real numbers) and Table 4 for the logical operation *equivalence*.

As it follows from (2) in the complex numbers the imaginary component i has a concrete numerical value $\sqrt{-1}$ due to which the product $(ia \times ib)$ is equal to $-(a, b)$. This does not correspond to Tables 2, 3 and 4.

The concept of **function** may be defined on the complex sets in a way, analogical to the classical sets.

Definition The binary relation f is called function if from $(a, b) \in f$ and $(a, c) \in f$ follows that $b = c$. According to (6) the imaginary set iA plays the role of a negative set as at its union with the set A will be the empty set $\varnothing$. So iA may be considered as a negative set in regard to A. As in general more imaginary states are possible, as in the quaternions of the complex numbers, it is preferable the set $\{i\text{A}\}$ to be considered as imaginary and to be called in this way.

3 Instead of Conclusion

1. A class of sets is proposed in the present work, called imaginary which when united with the classical Cantor's sets, called in the present work real, lead to a new class of sets called complex sets.

2. It is shown that the complex sets possess some new unique properties and may be considered as a kind of extension of the classical Cantor's sets.
3. It is shown how through the complex sets a number of set equations such as (3) and (6) may be solved which have no solution in the framework of the classical real sets.
4. Models of physical systems are proposed on the base of the complex sets in which the union of a series of their elements (particles) of different opposite charge results in their neutralization (annihilation).
5. A series of relations related to the defined operations union, intersection, negation, difference and symmetric difference are obtained for the complex sets.
6. It is shown that the imaginary and complex sets obey the requirements of the algebraic structure lattice and Boolean algebra and some features of the associativity are pointed out related to the introduced concept of *normalization of complex sets*. It is proved that both De Morgan laws are in effect for the complex sets, the double negation law, for idempotence, and distributivity, commutativity, as well as for a number of results evolving from the defined complements to the complex sets. Necessary and sufficient conditions as well as a rule for normalization of a given complex set are received.
7. The concepts of function, binary relation and Cartesian product of complex sets are defined. At that multiplication the pair of elements received belongs to one of the sets—real or imaginary. This is defined by the table proposed for multiplication analogical to the multiplication of real numbers or to the concept of *equivalence* in the propositional logic.

References

1. Sgurev, V.: An essay on complex valued propositional logic. Artif. Intel. Decision Making **2**, 85–101. RAS, ISSN 2071-8594 (2014)
2. Fraenkel, A., Bar-Hillel, Y., Levi, A.: Foundations of Set Theory, Amsterdam, North Holland (1973)
3. Kuratowski, K.: Introduction to Set Theory and Topology. Warszawa, PSP (1979)
4. Kuratowski, K., Mostowski, A.: Set Theory. North Holland, Amsterdam, PSP, Warszawa (1967)
5. Sgurev, V.: An approach for building of imaginary and complex sets. Artif. Intel. Decision Making, **4** (2016) (in print)
6. Sgurev, V.: Features of the imaginary and complex sets. Comptes Rendus de l'Aademy Bulgare des Sciences, **1**. ISSN 1310-1331 (2017) (in print)
7. Petrovsky, A.: An axiomatic approach to metrization of multiset space. In: Tzeng G., Wang U., Yu P. (eds.) Multiple Criteria Decision Making, pp. 129–140. Springer, N.Y. (1994)
8. Vopenka, P.: Mathematics in the Alternative Set Theory, Teubner-Texth, Leipzig (1979)

Intercriteria Analysis over Patterns

Krassimir Atanassov

AMS Classification: 03E72

1 Introduction

The intercriteria analysis was introduced in [1, 2]. It is based on the apparatus of Intuitionistic Fuzzy Sets (IFSs, see, e.g., [3]) and of the Index Matrices (IMs, see [1]). Here, for the first time we discuss the possibility to apply the intercriteria analysis over patterns.

First, we give short remarks on the concept of an Intuitionistic Fuzzy Pair (IFP, see, e.g., [4]). It is an object of the form $\langle a, b\rangle$, where $a, b \in [0, 1]$ and $a + b \leq 1$, that is used as an evaluation of some object or process and which components (a and b) are interpreted as degrees of membership and non-membership, or degrees of validity and non-validity, or degree of correctness and non-correctness, etc. One of the geometrical interpretations of the IFPs is shown on Fig. 1.

Let us have two IFPs $x = \langle a, b\rangle$ and $y = \langle c, d\rangle$. We define the relations

$$\begin{array}{lll} x < y & \text{iff} & a < c \text{ and } b > d \\ x > y & \text{iff} & a > c \text{ and } b < d \\ x \geq y & \text{iff} & a \geq c \text{ and } b \leq d \\ x \leq y & \text{iff} & a \leq c \text{ and } b \geq d \\ x = y & \text{iff} & a = c \text{ and } b = d \end{array}$$

K. Atanassov (✉)
Department of Bioinformatics and Mathematical Modelling,
Institute of Biophysics and Biomedical Engineering, Bulgarian Academy of Sciences,
105 Acad. G. Bonchev Str., 1113 Sofia, Bulgaria
e-mail: krat@bas.bg

K. Atanassov
Laboratory of Intelligent Systems, Prof. Dr. Asen Zlatarov University Burgas,
1 "Prof. Yakimov" Blvd, 8010 Burgas, Bulgaria

V. Sgurev et al. (eds.), *Learning Systems: From Theory to Practice*, Studies in Computational Intelligence 756, https://doi.org/10.1007/978-3-319-75181-8_4

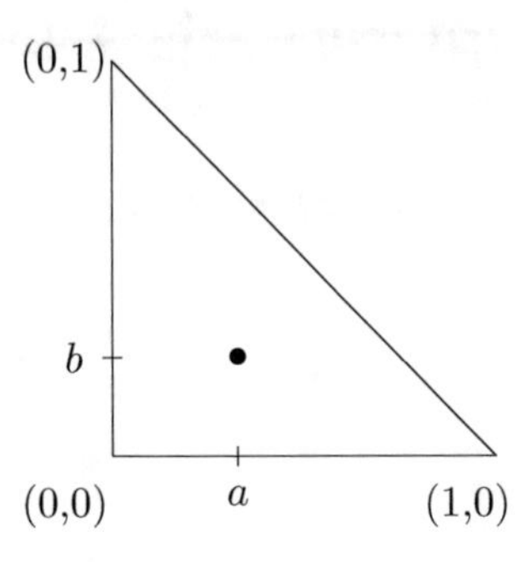

Fig. 1 A geometrical interpretation of an IFP

Second, we discuss the concept of an Index Matrix (IM, see [1, 5]).

Let I be a fixed set of indices and $\mathcal{R}$ be the set of the real numbers. By IM with index sets K and L ($K, L \subset I$), we denote the object:

$$[K, L, \{a_{k_i,l_j}\}] \equiv \begin{array}{c|cccc} & l_1 & l_2 & \dots & l_n \\ \hline k_1 & a_{k_1,l_1} & a_{k_1,l_2} & \dots & a_{k_1,l_n} \\ k_2 & a_{k_2,l_1} & a_{k_2,l_2} & \dots & a_{k_2,l_n} \\ \vdots & \vdots & \vdots & \ddots & \vdots \\ k_m & a_{k_m,l_1} & a_{k_m,l_2} & \dots & a_{k_m,l_n} \end{array},$$

where $K = \{k_1, k_2, ..., k_m\}$, $L = \{l_1, l_2, ..., l_n\}$, for $1 \le i \le m$, and $1 \le j \le n$: $a_{k_i,l_j} \in \mathcal{R}$.

In [1, 5], different operations, relations and operators are defined over IMs.

When elements a_{k_i,l_j} are IFPs, the IM is called Intuitionistic Fuzzy IM (IFIM). It has the form:

$$= \begin{array}{c|ccccc} & l_1 & \dots & l_j & \dots & l_n \\ \hline k_1 & \langle \mu_{k_1,l_1}, \nu_{k_1,l_1} \rangle & \dots & \langle \mu_{k_1,l_j}, \nu_{k_1,l_j} \rangle & \dots & \langle \mu_{k_1,l_n}, \nu_{k_1,l_n} \rangle \\ \vdots & \vdots & \ddots & \vdots & \ddots & \vdots \\ k_i & \langle \mu_{k_i,l_1}, \nu_{k_i,l_1} \rangle & \dots & \langle \mu_{k_i,l_j}, \nu_{k_i,l_j} \rangle & \dots & \langle \mu_{k_i,l_n}, \nu_{k_i,l_n} \rangle \\ \vdots & \vdots & \ddots & \vdots & \ddots & \vdots \\ k_m & \langle \mu_{k_m,l_1}, \nu_{k_m,l_1} \rangle & \dots & \langle \mu_{k_m,l_j}, \nu_{k_m,l_j} \rangle & \dots & \langle \mu_{k_m,l_n}, \nu_{k_m,l_n} \rangle \end{array},$$

where for every $1 \le i \le m, 1 \le j \le n$: $V(a_{k_i,l_j}) = \langle \mu_{k_i,l_j}, \nu_{k_i,l_j} \rangle$ and $0 \le \mu_{k_i,l_j}, \nu_{k_i,l_j}, \mu_{k_i,l_j} + \nu_{k_i,l_j} \le 1$.

2 Intercriteria Analysis Applied over Patterns

Let us have some original pattern (e.g., the one on Fig. 2) that must be compared with some alternative patterns (alternatives). The original must be situated in a rectangle, so that it touches each one of the sides.

Each one of these alternatives (e.g., alternative *A* from Fig. 3) must be modified in a way that it is situated in a rectangle in the described above sense (see Fig. 4).

Now, we modify the alternative in a way, that its rectangle coincides with the original pattern's rectangle, as it is shown on Fig. 5.

Third, we unite both rectangles and the figures in them, as it is shown on Fig. 6.

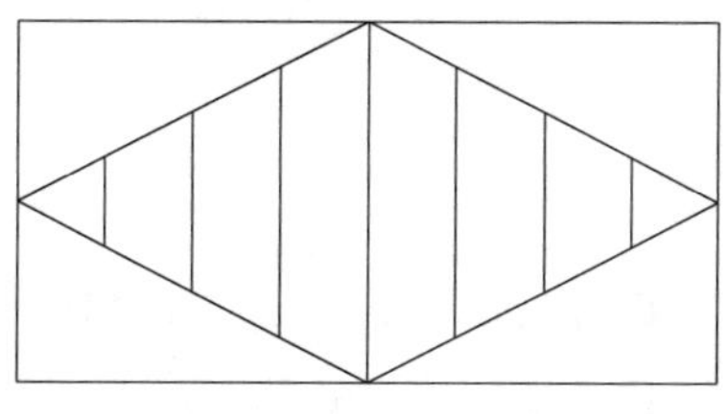

Fig. 2 The original pattern

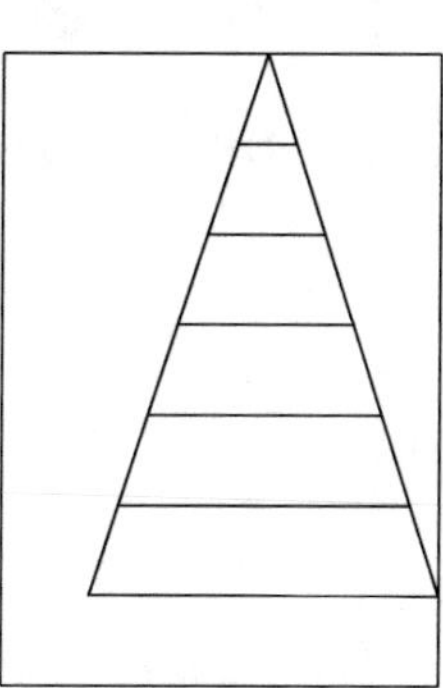

Fig. 3 An alternative pattern

Fig. 4 First modification of the alternative pattern

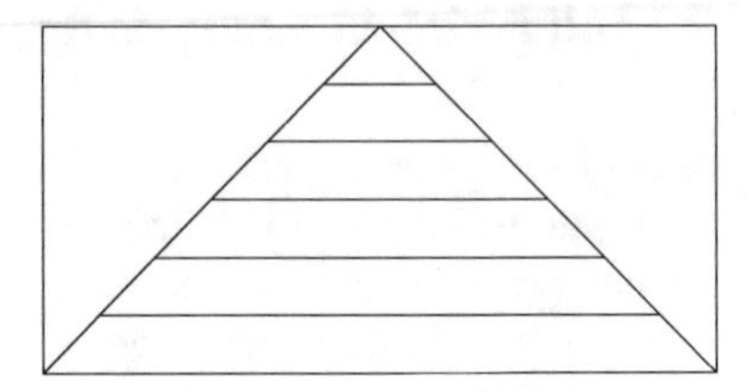

Fig. 5 Second modification of the alternative pattern

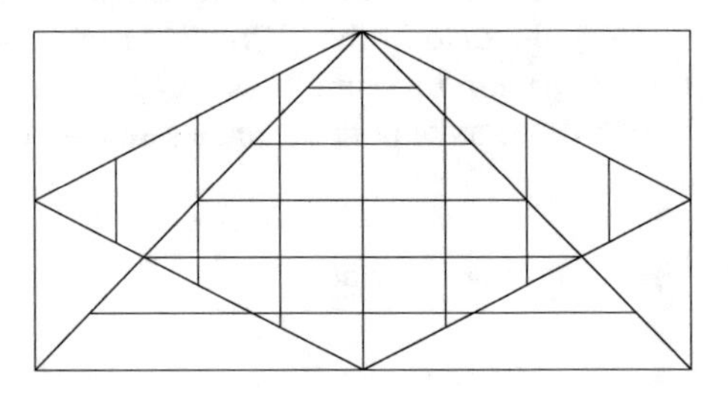

Fig. 6 Composition of the original pattern and the i-th alternative pattern

Fourth, let S be the area of the joint figure F, M be the area of F, marked by , N be the area of F, marked by and P be the area of Fe, marked by . Then to F we can juxtapose the pair $\langle \mu(F), \nu(F) \rangle$ for which

$$\mu(F) = \frac{M}{S},$$

and

$$\nu(F) = \frac{N}{S}.$$

Therefore, this pair is an IFP. Obviously,

$$\pi(F) = 1 - \mu(F) - \nu(F) = 1 - \frac{M}{S} - \frac{N}{S} = \frac{P}{S}.$$

Let us call the discussed above patterns "elementary patterns". Now, we will call a "pattern" an object composed by elementary patterns. Among the set of all patterns, we can determine one of them (by previously given criterion or determined in an arbitrary way) that we call "original pattern". Its elementary patterns will be also original elementary patterns. The remaining patterns can be called "alternative pattern". The elementary patterns of the original pattern will be used to modify all other elementary patterns of the alternative patterns by the above described procedure. Therefore, as a result, we obtain set of patterns, having equal rectangles, in which the figures are modified by the above way.

Let us order the elementary patterns of the original pattern in a row and after this, by the same way, let us order the elementary patterns of all other patterns (in rows).

So, we obtain an IM with elements patterns. This is the first case, when the elements of an IM are patterns, but this is not forbidden in IM theory—see [1], where the object an Extended IM is defined. The elements of this IM can be of arbitrary nature.

Therefore, if we construct the IM with elements the patterns and if we number the rows by the names of the patterns, e.g., $C_1, C_2, ..., C_m$, and the columns that contain i-th elements of the linearly ordered elementary patterns of the different patterns—with, e.g., symbols $O_1, O_2, ..., O_n$, then we obtain an extended IM that has the form

$$A = \begin{array}{c|ccccccc} & O_1 & \cdots & O_i & \cdots & O_j & \cdots & O_n \\ \hline C_1 & a_{C_1,O_1} & \cdots & a_{C_1,O_i} & \cdots & a_{C_1,O_j} & \cdots & a_{C_1,O_n} \\ \vdots & \vdots & \ddots & \vdots & \ddots & \vdots & \ddots & \vdots \\ C_k & a_{C_k,O_1} & \cdots & a_{C_k,O_i} & \cdots & a_{C_k,O_j} & \cdots & a_{C_k,O_n} \\ \vdots & \vdots & \ddots & \vdots & \ddots & \vdots & \ddots & \vdots \\ C_l & a_{C_l,O_1} & \cdots & a_{C_l,O_i} & \cdots & a_{C_l,O_j} & \cdots & a_{C_l,O_n} \\ \vdots & \vdots & \ddots & \vdots & \ddots & \vdots & \ddots & \vdots \\ C_m & a_{C_m,O_1} & \cdots & a_{C_m,O_i} & \cdots & a_{C_m,O_j} & \cdots & a_{C_m,O_n} \end{array},$$

where for every p, q $(1 \leq p \leq m\,,\ 1 \leq q \leq n)$:

(1) C_p is a criterion, taking part in the evaluation,
(2) O_q is an object, being evaluated.
(3) a_{C_p,O_q} is the q-th elementary patterns of the p-th pattern.

Now, we bring sequentially the elementary patterns from k-th and from l-th rows, compare them in the above described manner and as a results, for a_{C_k,O_i}, a_{C_k,O_j}, a_{C_l,O_i} and a_{C_l,O_j} obtain the IFPs $\langle \alpha_{C_k,O_i}, \beta_{C_k,O_i} \rangle$, $\langle \alpha_{C_k,O_j}, \beta_{C_k,O_j} \rangle$, $\langle \alpha_{C_l,O_i}, \beta_{C_l,O_i} \rangle$ and $\langle \alpha_{C_l,O_j}, \beta_{C_l,O_j} \rangle$, for each i $(1 \leq i \leq n)$, j $(1 \leq j \leq m)$.

Let $S^{\mu}_{k,l}$ be the number of cases in which

$$\langle \alpha_{C_k,O_i}, \beta_{C_k,O_i} \rangle < \langle \alpha_{C_k,O_j}, \beta_{C_k,O_j} \rangle$$

and

$$\langle \alpha_{C_l,O_i}, \beta_{C_l,O_i} \rangle < \langle \alpha_{C_l,O_j}, \beta_{C_l,O_j} \rangle,$$

or

$$\langle \alpha_{C_k,O_i}, \beta_{C_k,O_i} \rangle > \langle \alpha_{C_k,O_j}, \beta_{C_k,O_j} \rangle$$

and

$$\langle \alpha_{C_l,O_i}, \beta_{C_l,O_i} \rangle > \langle \alpha_{C_l,O_j}, \beta_{C_l,O_j} \rangle$$

are simultaneously satisfied.

Let $S^{\nu}_{k,l}$ be the number of cases in which

$$\langle \alpha_{C_k,O_i}, \beta_{C_k,O_i} \rangle > \langle \alpha_{C_k,O_j}, \beta_{C_k,O_j} \rangle$$

and

$$\langle \alpha_{C_l,O_i}, \beta_{C_l,O_i} \rangle < \langle \alpha_{C_l,O_j}, \beta_{C_l,O_j} \rangle,$$

or

$$\langle \alpha_{C_k,O_i}, \beta_{C_k,O_i} \rangle < \langle \alpha_{C_k,O_j}, \beta_{C_k,O_j} \rangle$$

and

$$\langle \alpha_{C_l,O_i}, \beta_{C_l,O_i} \rangle > \langle \alpha_{C_l,O_j}, \beta_{C_l,O_j} \rangle$$

are simultaneously satisfied.

Obviously,

$$S^{\mu}_{k,l} + S^{\nu}_{k,l} \leq \frac{n(n-1)}{2}.$$

For every k, l, such that $1 \leq k < l \leq m$ and for $n \geq 2$, we define

$$\mu_{C_k,C_l} = 2\frac{S^{\mu}_{k,l}}{n(n-1)}, \quad \nu_{C_k,C_l} = 2\frac{S^{\nu}_{k,l}}{n(n-1)}.$$

Hence,

$$\mu_{C_k,C_l} + \nu_{C_k,C_l} = 2\frac{S^{\mu}_{k,l}}{n(n-1)} + 2\frac{S^{\nu}_{k,l}}{n(n-1)} = \frac{2}{n(n-1)}(S^{\mu}_{k,l} + S^{\nu}_{k,l}) \leq 1.$$

Therefore, $\langle \mu_{C_k,C_l}, \nu_{C_k,C_l} \rangle$ is an IFP.

Now, we can construct the IM

$$\begin{array}{c|ccc} & C_1 & \cdots & C_m \\ \hline C_1 & \langle \mu_{C_1,C_1}, \nu_{C_1,C_1} \rangle & \cdots & \langle \mu_{C_1,C_m}, \nu_{C_1,C_m} \rangle \\ \vdots & \vdots & \ddots & \vdots \\ C_m & \langle \mu_{C_m,C_1}, \nu_{C_m,C_1} \rangle & \cdots & \langle \mu_{C_m,C_m}, \nu_{C_m,C_m} \rangle \end{array},$$

that determines the degrees of correspondence between patterns $C_1, \ldots, C_m$.

3 Intercriteria Analysis and Intuitionistic Fuzzy Histograms

The concept of an intuitionistic fuzzy histogram was introduced in [6] and discussed in details in [3].

Let us give an example: let us have n series of m events which elements can be symbols A, B or an empty symbol $*$. On Fig. 7, $m = 5, n = 6$.

Now, we can change symbol A with , symbol $*$ with and symbol B with an empty cell and we obtain Fig. 8.

Fig. 7 An example

*	A	B	B	A	*
B	B	*	*	B	B
*	B	*	A	*	*
B	*	B	B	*	*
A	A	A	A	A	A

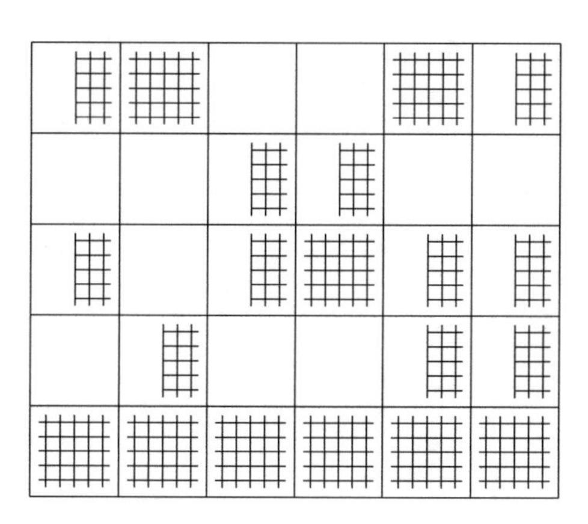

Fig. 8 First modification of the example from Fig. 7

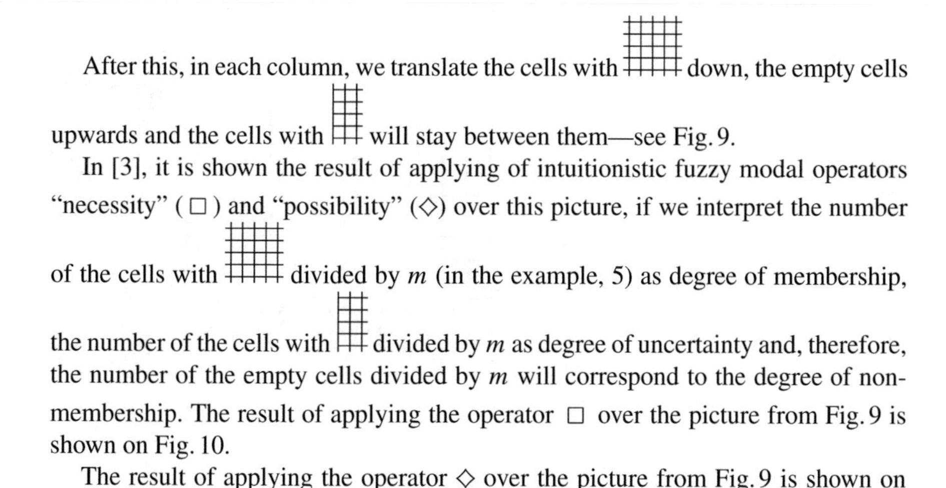

After this, in each column, we translate the cells with ▦ down, the empty cells upwards and the cells with ▥ will stay between them—see Fig. 9.

In [3], it is shown the result of applying of intuitionistic fuzzy modal operators "necessity" (□) and "possibility" (◇) over this picture, if we interpret the number of the cells with ▦ divided by m (in the example, 5) as degree of membership, the number of the cells with ▥ divided by m as degree of uncertainty and, therefore, the number of the empty cells divided by m will correspond to the degree of non-membership. The result of applying the operator □ over the picture from Fig. 9 is shown on Fig. 10.

The result of applying the operator ◇ over the picture from Fig. 9 is shown on Fig. 11.

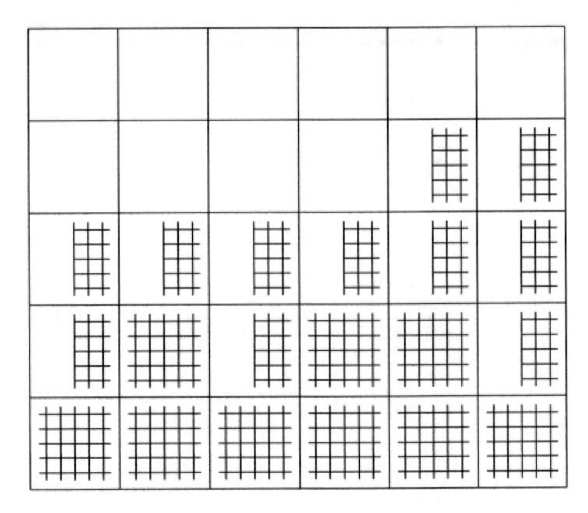

Fig. 9 Second modification of the example from Fig. 7

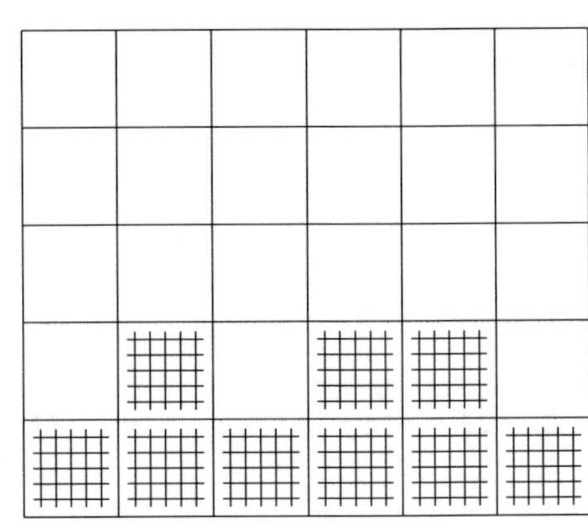

Fig. 10 The result of applying the operator $\Box$ over the picture from Fig. 9

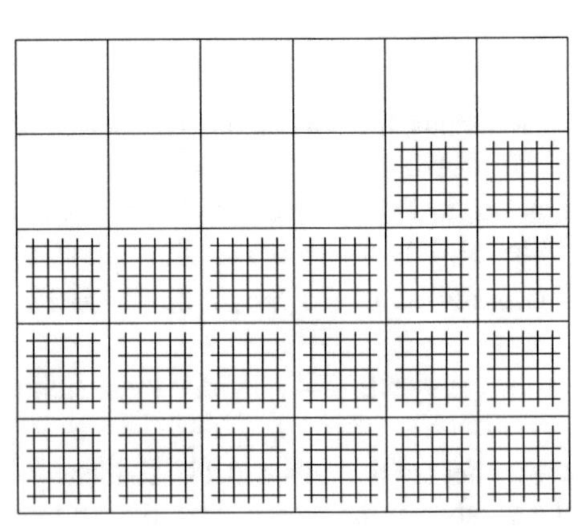

Fig. 11 The result of applying the operator $\Diamond$ over the picture from Fig. 9

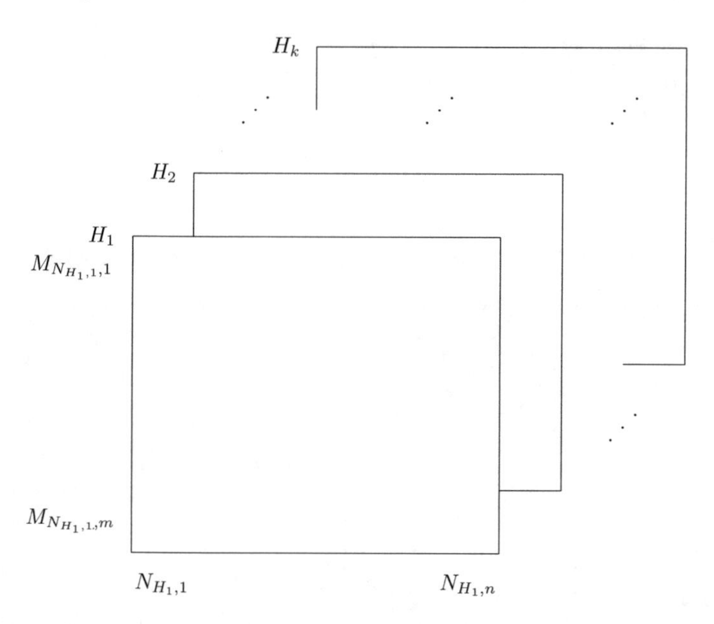

Fig. 12 Three-dimensional cube with intuitionistic fuzzy histograms

In [3] a lot of operators, extending the above mentioned modal operators, are defined. Each one of them will modify Fig. 9 in different way. So, we can obtain set of patterns, over which we can apply the intercriteria analysis in the described in Sect. 2 sense, searching similarities and differences between the separate patterns.

The more interesting case is, when we have different intuitionistic fuzzy histograms, over which we apply the intercriteria analysis. Let they be H_1, H_2, ..., H_k and for simplicity, let each one of them has n columns $N_{H_i,1}$, $N_{H_i,2}$, ..., $N_{H_i,n}$ for i-th histogram, and with m cells $M_{N_{H_i,j},1}$, $M_{N_{H_i,j},2}$, ..., $M_{N_{H_i,j},m}$ for j-th column of the i-th histogram, in them. Therefore, we obtain a three-dimensional cube, e.g., this one from Fig. 12.

By the moment, the intercriteria analysis works with two-dimensional IMs, while, each three-dimensional cube is represented by a three-dimentional IM. Following [1], we can represent the three-dimensional IM by a two-dimensional IM, e.g., with the form from Fig. 13 and after this, to use the procedure from Sect. 2 to compare the histograms.

$$
\begin{array}{c|cccccc}
 & M_{N_{H_1,1},1} & M_{N_{H_1,1},2} & \ldots & M_{N_{H_1,1},m} & M_{N_{H_1,2},1} \ \ldots & M_{N_{H_k,n},m} \\
\hline
H_1 & & & & & & \\
H_2 & & & & & & \\
\vdots & & & & & & \\
H_k & & & & & &
\end{array}
$$

Fig. 13 Two-dimensional IM corresponding to the three-dimensional cube from Fig. 12

Fig. 14 Map of Bulgaria

4 Conclusion

In the presented research, an Intercriteria Analysis over very simple in shape patterns is discussed. In future, we will extend the described procedure for complex patterns. For example, the original pattern can be the map of a country, e.g., of Bulgaria (see Fig. 14). Then the above idea can be used, e.g., in linguistics for searches proximity of the individual word in use, or in meteorology for searches of regions with similar climatic parameters, etc. In these cases, the elementary patterns are the Bulgaria regions. A next step of extension of the introduced procedure is related to replacing of the black-white patterns with coloured ones. As it is shown in [3], the colours from the RGB-diagram can be represented by one of the IFS extensions, called temporal IFS. So, by using it, we will replace IFS-evaluations with Temporal IFS-evaluations.

Another idea for future is related to compare simultaneously 3 instead of 2 criteria, as it is described in [7–9].

Also, in a next research, the IMs, used here, will be replaced by 3-dimensional IMs. In this case, we will have the capability to describe more complex problems, related, e.g., to analysis of scenes.

Acknowledgements The author is thankful for the support provided by the Bulgarian National Science Fund under Grant Ref. No. DFNI-I-02-5.

References

1. Atanassov, K.: Index Matrices: Towards an Augmented Matrix Calculus. Springer, Cham (2014)
2. Atanassov, K., Mavrov, D., Atanassova, V.: Intercriteria decision making: a new approach for multicriteria decision making, based on index matrices and intuitionistic fuzzy sets. Issues Intuit. Fuzzy Sets General. Nets **11**, 1–8 (2014)
3. Atanassov, K.: On Intuitionistic Fuzzy Sets Theory. Springer, Berlin (2012)
4. Atanassov, K., Szmidt, E., Kacprzyk, J.: On intuitionistic fuzzy pairs. Notes Intuit. Fuzzy Sets **19**(3), 1–13 (2013)
5. Atanassov, K.: Generalized index matrices. Comptes rendus de l'Academie Bulgare des Sciences **40**(11), 15–18 (1987)
6. Atanassova, L., Gluhchev, G., Atanassov, K.: On intuitionistic fuzzy histograms. Notes Intuit. Fuzzy Sets **16**(4), 31–36 (2010)
7. Atanassova, V., Doukovska, L., Michalikova, A., Radeva, I.: Intercriteria analysis: from pairs to triples. Notes Intuit. Fuzzy Sets **22**(5), 98–110 (2016)
8. Roeva, O., Pencheva, T., Angelova, M., Vassilev, P.: InterCriteria analysis by pairs and triples of genetic algorithms application for models identification. Stud. Comput. Intel. **655**, 193–218 (2016)
9. Vassilev, P., Todorova, L., Andonov, V.: An auxiliary technique for InterCriteria Analysis via a three dimensional index matrix. Notes Intuit. Fuzzy Sets **21**(2), 71–76 (2015)

From Primary to Conventional Science

Boris Stilman

1 The Primary Language

In 1957, J. von Neumann hypothesized existence of the Primary Language of the human brain [60]. He suggested that the external language (including multiplicity of natural languages as well as mathematics and computer science) that we use in communicating with each other might be quite different from the internal language used for computation by the human brain. He argued that we are still unaware of the nature of the Primary Language for mental calculation. He writes, “It is only proper to realize that [human] language is largely an historical accident. The basic human languages are traditionally transmitted to us in various forms, but their very multiplicity proves that there is nothing absolute and necessary about them. Just as languages like Greek or Sanskrit are historical facts and not absolute logical necessities, it is only reasonable to assume that logics and mathematics are similarly historical, accidental forms of expression. They may have essential variants, i.e., they may exist in other forms than the ones to which we are accustomed. … The language here involved may well correspond to a short code in the sense described earlier, rather than to a complete code [in modern terms, he means high-level vs. low-level programming languages]: when we talk mathematics, we may be discussing a secondary language, built on the primary language truly used by the central nervous system.” Over the last 60 years since J. von Neumann hypothesized existence of the Primary Language, the nature of this language has been unknown.

The existence of a singular primary language for internal computation by the human brain should be considered yet another application of the principles of

B. Stilman (✉)
University of Colorado Denver, Denver, CO, USA
e-mail: boris@stilman-strategies.com

B. Stilman
STILMAN Advanced Strategies, Denver, CO, USA

V. Sgurev et al. (eds.), *Learning Systems: From Theory to Practice*, Studies
in Computational Intelligence 756, https://doi.org/10.1007/978-3-319-75181-8_5

simplicity and reuse by the nature. This means that a successful solution found by the nature should spread over space and time. The major example of such solution is the spread of the universal DNA-based life over the history of the whole planet. Going further once developed as a result of evolution of human intelligence the Primary Language should have been used by all the human species, including our direct predecessors Homo Sapience as well as our past relatives such as Neanderthals and Denisovans. In addition, as was suggested by von Neumann, the Primary Language should have been reused as a singular universal foundation for the secondary (symbolic) languages and sciences. Moreover, the same components of the Primary Language that served humanity successfully over millennia before appearance of the secondary languages should have been reused as "semantic" components of those languages.

Extensive investigation of the mental visual streams (mental movies) [39–50, 55, 56], as a foundation of the algorithms essential for evolution of human intelligence and development of humanity led us to observe their powerful universal nature. We suggest that the Primary Language is the Language of Visual Streams (LVS). Following von Neumann, we call the LVS a language although it is not a language in mathematical sense, i.e., it is not a set of strings of symbols. The Primary Language is the engine with the ability to generate visual streams, focus them in desired direction, and reason about them. We will demonstrate that the LVS is the underlying foundation of all the human languages and sciences as it should have been according to [60].

2 Where Those Streams Came from

Thought experiments allow us, by pure reflection, to draw conclusions about the laws of nature [9, 10]. For example, Galileo before even starting dropping stones from the Tower in Pisa used pure imaginative reasoning to conclude that two bodies of different masses fall at the same speed. The Albert Einstein's thought experiments that inspired his ideas of the special and general relativity are known even better [12, 13, 22]. The efficiency and the very possibility of thought experiments show that our mind incorporates animated models of the reality, e.g., laws of physics, mathematics, human activities, etc. Scientists managed to decode some of the human mental images by visualizing their traces on the cortex [9]. It was shown that when we imagine a shape "in the mind's eye," the activity in the visual areas of the brain sketches the contours of the imagined object; thus, mental images have the analogical nature. It appears that we simulate the laws of nature by physically reflecting the reality in our brain. The human species and even animals would have had difficulty to survive without even minimal "understanding" of the laws of environment. Over the course of evolution and during development of every organism, our nervous system learns to comprehend its environment, i.e., to "literally take it into ourselves" in the form of mental images, which is a small-scale reproduction of the laws of nature. Neuropsychologists discovered that "we carry within ourselves a universe of mental objects whose laws imitate those of physics and geometry" [9]. In [40], we suggested that we also carry the laws of the major human relations including the laws of optimal warfighting. The laws of nature and human relations manifest themselves in many different ways. However, the clearest manifestation is in perception and in action. For example, we can say that the sensorimotor system of the human brain "understands kinematics" when it anticipates the trajectories of objects. It is fascinating that these same "laws continue to be applicable in the absence of any action or perception when we merely imagine a moving object or a trajectory on a map" [9]. This observation, of course, covers actions of all kinds of objects, natural and artificial. Scientists have shown that the time needed to rotate or explore these mental images follows a linear function of the angle or distance traveled as if we really traveled with a constant speed. They concluded that mental trajectory imitates that of a physical object [9, 10].

The idea of the major role of the visual streams came from our investigation of the Algorithm of Discovery (AD), Sect. 9. We utilized experiences of great scientists including [1–3, 6–8, 12–14, 15–17, 23, 25, 57, 58, 60–63]. In addition, we considered research results in cognitive science and neuroscience [4, 9, 10, 20, 22, 24] and our own research experience [26–56]. Based on all those results we suggested that the essence of the discoveries is in the visual streams. We further suggested that these are movie-like visual (for the mind's "eye") mental processes. Sometimes, they reflect past reality. More frequently, they reflect artificial mentally constructed reality. This reality could be an artificial world with artificial laws of nature including space and time. For example, this could be laws of real or completely different physics or mathematics. This world is populated with realistic and/or artificial animated entities. Those entities are mentally constructed, and their prototypes may not exist in real life. When we run a visual stream, we simply observe life events, realistic or artificial, in this world. Of course, those animated events happen according to the laws of this world and we are "lucky" to be present and see what is happening. Usually, we have the power to alter this world by reconstructing the laws, the entities, etc. according to our wishes. Then, the visual stream becomes a movie (or play) showing in the end a solution to the task staged in this artificial world. Usually, this solution comes without a proof because it was not proved but played in this world. When the solution is known, the proof itself could be broken into small subtasks staged in the altered world and eventually discovered.

Within the brain, the visual streams run consciously and subconsciously and may switch places from time to time (in relation to conscious/subconscious use). We may run several visual streams concurrently, morph them, and even use logic for such morphing, although this use is auxiliary.

We suggested that these mental artificial worlds could be investigated and modeled employing algorithms, and, consequently, implemented in software. For several years, this was our road to the AD. Recently, we expanded this hypothesis and suggested that the Primary Language is the LVS. Following this expansion, in this paper, for the first time, we consider various types of thought experiments and visual streams as the foundation of the Primary Language.

3 Our Approach

Our approach to discovering the nature of the Primary Language is analogous to an attempt to understand the algorithm of a program while watching its execution. In addition, we are trying to discover the instruction set of the "computer" running this program, i.e., the means of the human brain to run it. In our investigation, we assume that this program simulates sequences of events while its interface includes color movies demonstrating visually those events. With multiple published introspections of great writers, artists, and scientists, we can recreate clips from various movies, i.e., their imaginary thought experiments. What really helps is the assumption that all those movies were demonstrated by the programs running essentially the same algorithm. With our own developments in Computer Science, we have additional power of asking what-if questions via morphing our own movies and getting answers by watching those morphed movies until the very end. Unfortunately, we do not have this power with the creations of others. However, we could utilize their published and oral introspections.

We suggested that there should be several major algorithms crucial for survival and development of humanity that are based directly on the Primary Language because they were developed at the time when human symbolic languages and sciences did not exist. Due to their origin, it should be easier to reveal the base of these algorithms, the Primary Language, bypassing their possible superficial connection to the modern human symbolic languages and sciences [49]. In addition, it should be easier to understand the structure of the Primary Language by investigating such algorithms together with the Primary Language as those texts carved on the Rosetta stone. We suggested that the list of those helpful algorithms includes, at least, Linguistic Geometry (LG), the theory of efficient warfighting [39, 55, 56], and the Algorithm of Discovery (AD). However, out of these three, only LG is well developed (to a certain degree) [38]. Thus, we investigate all three, the Primary Language, the AD and LG, together. The preliminary results of this investigation were published in [40–50]. Additional boost to this research comes from investigating the link between the Primary Language and the conventional sciences and human languages, as it should have been according to [60], see also Sects. 6 and 7. This new direction has just started.

In the following sections, we consider the details of the structure of the LVS. We assume that visual streams are uniform. However, they can be divided into classes and combined into groups by their purpose. The main groups include internal and communication streams. The internal streams communicate with other streams only while the communication streams pass information between the internal streams and the outer world. The other division of streams could be based on the information they are dealing with. These include streams that operate with mundane and scientific information. The streams involved with the AD can be divided according to their purpose. They include observation, construction and validation streams; streams utilizing various types of reasoning, such as proximity reasoning, mosaic reasoning, etc. In addition, visual streams can be grouped into parallel and nested structures (i.e., running concurrently or calling each other); groups of streams that have common themes, etc.

A very important member of the group of communication streams is an *expression* stream. This stream is concerned with passing information from other streams to the outside world, Sect. 6. There are at least two types of expression streams, *pictorial* and *symbolic*. The pictorial expression stream makes snapshots of the internal visual stream in the form of illustrations. The symbolic expression stream tags some of the objects, actions, and events shown in the internal stream and creates the so-called symbolic shell around the main visual stream. This shell eventually becomes a string of symbols, i.e., a novel, a speech, a scientific theory, etc., that can be communicated to others. Those strings may also serve as captions to the illustrations provided by the pictorial stream. It is likely that the symbolic expression stream employs the universal generating grammar and the tags as the terminal symbols of this grammar as suggested by Chomsky [6] and confirmed recently by Ding et al. [11]. Moreover, in Sects. 15–22, we demonstrate how a combination of expression streams can construct and utilize persistently a generating grammar that links the primary (visual) science to the symbolic conventional one. In Sects. 4 and 5, we describe the ability of a stream to reason about itself, for example, the ability to distinguish and understand the purpose of various entities involved in the stream. In particular, this includes the ability to tag those entities for using them as terminal symbols of the generating grammar. More details about the communications languages are provided in Sects. 6–10.

4 Purpose of Visibility

Visual streams include two major components, the ability to "see" inside the stream (the visual component) and the ability to logical reasoning (the reasoning component). We named this approach *visual reasoning* [44]. Accordingly, we assumed that the visual component (including pattern recognition) is sophisticated, while the reasoning component is relatively simple. We also assumed that the full-scale mental visibility (with pattern recognition) is rarely used, while the limited visibility can be simulated with a reasonable effort. Simulating the reasoning component should be within the scope of the modern software development. Our assumption of the need for "standard" visibility was based on the following. When we introduced mental visual streams, we assumed that we mentally observe events happening within those streams, i.e., we see what is going on. Moreover, for all of us observing or seeing something means recognizing it. Thus, as part of our original hypothesis, we assumed that pattern recognition should be an inherent component of the streams' visibility. In other words, we assumed that in order to simulate and animate the natural or artificial reality in the form of visual streams we have to understand it, and, certainly, recognize it in detail as a necessary component of this understanding. We also assumed that the same understanding and recognition should happen dynamically within the visual streams.

However, deeper consideration shows that utilization of the visual streams does not require pattern recognition at all. We utilize our brain's innate ability to simulate outside world, i.e., to "photocopy" the nature or invent and mentally construct artificial reality (which is, essentially, a modified nature) [18]. This ability goes further because we can animate simulation in the form of visual streams. In case of simulating (observing) the nature, we periodically verify simulation, employing information coming through our sensors, i.e., images, sounds, etc., bypassing conscious understanding of this information. In case of artificial reality, verification is not needed. In all cases simulation of reality within visual streams happens by construction, i.e., via internal subconscious understanding of the simulated reality and, consequently, without pattern recognition. We do not have to recognize it because we already know it. Specifically, because this simulation is driven by the internal engine (visual stream) the "mechanics" of this engine is used to understand the simulated reality. Roughly, we do not have to look at this simulation from outside, even mentally, (and employ pattern recognition). The visual stream is a self-aware entity: it can analyze its characters from inside because the stream is the one who created them. To realize similarity of two different visual streams or objects involved in those streams, we should recall that those streams are two data structures generated by the same algorithm, and these structures are available to the streams engine. Thus, to realize similarity of the streams we have to realize similarity between the data structures that implement them, which in its turn can be done by their comparison (or mutual analysis). We conclude that visual reasoning is essentially a construction-based reasoning, which does not involve pattern recognition. Roughly, the stream already knows who is who.

5 Analogy as Advanced Visibility

Visibility within the visual streams is more advanced than physical visibility of the real world objects. In particular, it includes realization of the analogy of the seemingly unrelated objects and processes in different visual streams. Our investigation shows that those analogies may be the keys to starting and focusing visual streams. To illustrate this idea we will introduce two examples, the first—from our experience of developing the theory of LG [38] and the second—from the theory of Evolution [8].

When attempting to represent mathematically chess pieces moving over the chessboard as well as real world moving entities we were running mental movies of morphing pictorial structures of the planning paths, the so-called trajectories. The key analogy appeared in another visual stream generating and reading sentences of the natural language. When a piece moves along a trajectory, this trajectory shortens—the part of the route left behind disappears; so does the respective sentence by dropping words from the front when they have been read. To realize the analogy, the cast of characters of the natural language stream was replaced. Indeed, every word was replaced by a square of the chessboard along the trajectory, and the "reader's cursor"—by a moving chess piece, and the grammar that generates sentences was replaced by the grammar for generating trajectories. This analogy led eventually to construction of the Hierarchy of Formal Languages, the foundation of the Linguistic LG [27–30, 38, 43].

When attempting to explain the variations in species, especially, their massive disappearance that C. Darwin had seen on his voyage, he should have run a stream of accelerated life on Earth [19]. The key analogy came from the stream of human life on Earth (introduced by Malthus [21]) where populations grow until they exceed the food supply and then the members of the population compete with each other. To realize this analogy, the cast of human populations was replaced by the cast of animal species competing for the natural resources (over much longer periods). In this competition, random variations of the species that created an advantage would be more likely to survive, breed, and transfer their traits to offspring, while others would be lost. This analogy led eventually to development of the theory of Evolution driven by natural selection [8].

6 The Communication Experiments

Every operation that involves the Primary Language is a thought experiment. There are two groups of experiments. The first group, the *internal* thought experiments, operates solely within the Primary Language while the second group, the *communication* thought experiments, provides link between the Primary Language and the external symbolic languages. These experiments are based on the internal and communication visual streams, respectively. The communication streams implement two mutually inverse operations of *expression* and *impression*. The first operation employs the expression visual streams and links the internal streams with external languages while the second one operates in reverse direction.

The expression streams drive generation for all the human symbolic languages including sound, pictorial, and written languages. This generation happens roughly as real time narration to running internal visual streams by tagging entities and actions involved in those streams. When talking, drawing, and writing, we describe what we "see" in an internal visual stream using symbolic summaries or shells. For this purpose, an internal stream can play real or imaginary world scenarios in real time or faster than real time. It can also branch into other streams, morph itself, zoom in and out, etc. Though these activities are controlled to support narration, the stream (or streams) has a lot of freedom of direction. It is likely that this freedom is not just a random choice but influenced by external features such as emotions. Essentially, talking, writing, and drawing are controlled by a pair of visual streams, the internal stream that is responsible for the contents of a message and the expression stream that converts this message into symbols (or pictures). Those two streams provide mental animation of the presentation and generate sequences of symbols (vocal, written), or pictures.

The communication experiment is the key for the conventional translation from one external language to another. A translation of a talk or a text to a different language is an expression of the respective internal stream employing this language. In general, understanding a symbolic human language, whether it results from reading, translating, or listening, does not happen directly. It happens via initiating and running a respective internal visual stream matching the original one that caused generation of this language by the same (or another) person. The degree of this matching defines the degree of understanding and the quality of translation. Essentially, the internal streams represent the semantic components of all the symbolic human languages (and their grammars).

Every symbolic expression stream includes generation of the strings of symbols, written, typed, or vocal. These strings of external symbols could be passed to a different person and received via human senses, vision, hearing, or touching. The symbolic impression visual streams operate in a different direction. These streams convert external strings of symbols into the internal streams. This is achieved by creating new streams or by altering the existing streams via morphing. The key operation in converting a string of symbols into the visual stream is conversion of terminal symbols (words of this string) into the cast of characters and actions of the stream.

Summarizing the above, we could divide the communication experiments into three classes: *presentation*, *conversation*, and *translation*.

- The presentation experiment includes a pair of streams, the internal stream and the expression stream. As we discussed above the first one is the animated presentation while the second one is responsible for presenting it to the outer world.
- The conversation experiment requires involvement of the two entities, humans or robots. Each of them is running internal, expression, and impression streams. The entities exchange information (strings of symbols) generated by their expression streams. For each person the reception of symbols is accomplished by their impression streams. The actual conversation is performed by the respective internal visual streams. However, those streams cannot communicate directly. For this purpose, they employ expression and impression streams. All the streams are running concurrently with possible alternations of the expression and impression streams between the entities (to provide information exchange).
- The translation experiment also requires two entities, a writer and an interpreter. Note that they could be combined in one. The writer producing written text is always involved in the presentation experiment. As such, he is running internal and expression streams. The interpreter receives the text employing his impression stream. So far, it is similar to the conversation experiment. However, his impression stream is highly constrained with the intent to reproduce the internal stream of the writer. In addition, the interpreter activates his own expression stream to produce text reflecting the essence of his/her internal stream employing the symbolic language different from the one utilized by the writer for the original text. The better match of the internal streams of the writer and the interpreter the better the translation. The actual evaluation of the quality of translation cannot be reduced to the comparison of texts—it still requires evaluation of the degree of match of the respective internal streams, which is certainly a subjective procedure.

All three communication experiments considered above could be supported by the modern communication based on computers.

7 The Science Experiments

Another classification of the thought experiments is related to the type of information they deal with. The major division involves experiments that operate with mundane and scientific information.

The science thought experiments serve as a foundation for all the sciences as was predicted by J. von Neumann. Indeed, all the sciences are based on algorithms. In many, if not all the cases, execution of those algorithms could be represented in the dynamic pictorial form, which could be animated as a visual stream. Let us reverse our reasoning and consider a special subset of internal visual streams as a universal form of sciences. Then, the science that we are accustomed to could be considered a result of applying the expression visual streams that convert internal science streams into the conventional scientific language. This language usually employs a human symbolic language and a scientific notation specific for different branches of science. For example, the Computer Science notation includes various types of algorithmic languages. Note that in this case the expression streams utilize different constraints than those involved in the communication experiments considered above, Sect. 6. In particular, the translation thought experiments translating one human (non-scientific) language to another one permit many options including synonyms, epithets, exaggerations, etc. while translations into the scientific language do not permit them. It is likely that the impression visual stream that maps texts written in the scientific language into the internal science stream provides "one-to-one" mapping. This means that there should be a perfect match between the primary and secondary sciences.

Note that the expression visual stream that generates strings in the scientific language still involves persistent tagging of the entities (and actions) participating in the respective internal science stream. As we already mentioned this generation is controlled by the universal grammar for which those entities serve as the terminal symbols (nouns, verbs, adjectives, etc.). On the other hand, these terminal symbols should serve as pointers to the entities and actions involved in the internal stream to be created or morphed by the inverse mapping employing the impression stream.

The science thought experiments are all belong to the primary science. They are divided naturally into the *established primary science* and the *discovery science*.

- The established science experiments involve a variety of the primary science streams, a subset of the LVS. In a number of papers, e.g., [40–50], we considered various examples of the visual streams that should be attributed initially to the discovery science and, later, converted into the established primary science. These streams were constructed in the process of revealing the nature of the AD. This was done by applying the AD manually to the past discoveries (Sect. 9). In addition, these papers include multiple figures, which are not just illustrations but snapshots of the respective visual streams. The *secondary science* is the conventional science reflected, as usual, in the scientific language. The expression streams convert the primary science streams, including both established and discovery streams, into the secondary science. This conversion process stays often unnoticed. The impression streams perform the inverse operation, by converting the secondary science into the primary one. These experiments generate visual streams from the primary science that animate the secondary (conventional) science. The impression streams are necessary because they generate input for the discovery streams. These streams may be generated a new, morphed from the existing stream or retrieved completely from the mental archive.

- The discovery thought experiments produce the primary science. This type of experiments was the first source of our original exposure to visual streams [40]. According to the results of our research [40–50], the primary science is the place and the means utilized for all the discoveries. The discovery streams operate exclusively within the primary science. At some point of time, when the AD decides that validation of the results is sufficient, the discovery streams turn into the established science streams, still, within the primary science. Only then, employing the expression experiments the established primary science is mapped into the secondary science, the conventional science reflected in the scientific language. Unfortunately, in many cases this mapping is rather sophisticated and may require involvement of the AD (see Sects. 15–22).

Our research on application of the AD to replaying various discoveries provided numerous examples of visual streams, especially, the discovery streams. Multiple visual streams representing LG, the AD, and applications of the AD to the discoveries in LG, Molecular Biology and sketches of discoveries in other branches of science are considered in [40–50].

8 The Computer Based Experiments

When J. von Neumann introduced the hypothesis of the Primary Language, he suggested that all the human symbolic languages as well as all the sciences are Secondary with respect to their original form based on the Primary Language (Sect. 1). Probably, unintentionally, or due to its underdeveloped status in 1957, he missed the computer-based form of expression.

The computer-based language is still a Secondary language, although, it is a powerful mix of pictorial and symbolic languages. The Primary Language does not have means to control directly the computer based expression, as is the case with the communication experiments considered in Sect. 6. The various internal and expression streams are involved in the multiple stages of software development, from development of requirements to programming to debugging to maintenance. Quite often, several programmers are involved. However, it appears that the power of computer expression by far exceeds the power of the human symbolic languages. Potentially, it is comparable with the visual streams of the Primary Language.

As we already mentioned, currently, there are no means for the internal streams of the Primary Language to control directly the computer generated visual streams. This control requires various expression streams involving symbolic or pictorial languages including those for the keyboard control (for computer games). However, the reverse communication could be direct (via visual contact). The impression stream following the computer visual stream could activate and direct an internal visual stream, which would duplicate the computer stream. In other words, a computer-generated movie can be duplicated as a mental movie. While the development of communication means between the Primary Language and the computer language is important, our goal is different. It is our intent to elevate the power of the computer visual streams to the level of those of the Primary Language.

A step in this direction is using computer visual streams for modeling the primary science streams, the established science and the discovery science. Success of such modeling would make internal visual streams and, in particular, the primary science streams interactive publicly. For example, interaction with the computer model of the primary science of LG would be a powerful tool for a student as well as an advanced scientist. Moreover, the computer models of the visual streams of the primary science should make them easily available and, possibly, more useful than publications of conventional secondary science.

9 The Discovery Streams

What would happen if discoveries were produced routinely as an output of computer programs? What a leap this would mean for humanity? Approaching this time of making discoveries on demand could be a byproduct of our efforts.

We have been developing a hypothesis that there is a universal AD driving all the innovations and, certainly, the advances in all sciences [40–50]. All the human discoveries from mastering fire more than a million years ago to understanding the structure of our Solar System to inventing airplane to revealing the structure of DNA to mastering nuclear power utilized this algorithm. The AD should be a major ancient item “recorded” in the Primary Language due to its key role in the development of humanity. This line of research involved investigating past discoveries and experiences of construction of various new algorithms, beginning from those, which we were personally involved [2, 3, 26–56]. So far, investigation of the AD made the greatest contribution to our knowledge of the Primary Language. We certainly investigate the AD in its natural environment, i.e., as a component of the primary science while branching to the conventional science for presenting results. In particular, our hypothesis that the Primary Language is the LVS owes on several years of persistent investigation of the structure of the AD.

The AD operates as a series of thought experiments, which are performed via the visual streams. These streams may or may not reflect the reality. In addition, those streams serve as the only interface of the AD. This interface is constructive, i.e., visual streams could be morphed in the desired direction. As all the streams of the primary science, the discovery streams may employ the communication streams to communicate with the secondary science.

The input to the AD is also a visual stream, which includes several visual instances of the object whose structure has to be understood or whose algorithm of construction has to be developed. Sometimes, the object is dynamic, i.e., its structure is changing in time. Then the input visual stream includes this visual dynamics. As a rule, neither the structure of the object nor the details of the dynamics are present in the stream. It simply replicates (mimics) the natural or imaginary phenomenon. The task of the AD is to reveal its structure including dynamics and/or develop an algorithm for reconstructing this object including its changes in time. This understanding happens in several stages. Importantly, it always ends up with the process of actual reconstruction of the object employing the construction set developed by the AD on the previous stages. If the AD investigates a real life object this imaginary reconstruction may be totally unrelated to the construction (replication) utilized by the nature. Usually, this reconstruction process is artificially developed by the AD with the only purpose to reveal the structure of the object. An example of artificial reconstruction of the structure of the DNA molecule, the Double Helix, employed by the AD, is the algorithm of multiple transformations of the DNA generator, which involves shift and rotation [45]. However, if the algorithm of natural replication is the goal of discovery than the AD will employ a set of different visual streams (or different themes, Sect. 14) to reveal the relevant components utilized by the nature [45].

All the discovery streams are divided into three classes, *observation*, *construction*, and *validation*. They usually follow each other. However, these streams may be nested hierarchically, with several levels of depth.

The AD initiates the observation stream, which must carefully examine the object. It has to morph the input visual stream and run it several times to observe (mentally) various instances of the object from several directions. Often, for understanding the object, it has to observe the whole class of objects considered analogous. If the object is dynamic (a process), it has to be observed in action. For this purpose, the observation stream runs the process under different conditions to observe it in different situations. The purpose of all those observations is erasing the particulars to reveal the general relations behind them.

A good example of multiple observations is related to developing the theory of LG [38]. It involves experiments generalizing movement of various chess pieces as well as real world entities like men, cars, ships, airplanes, missiles, etc. These experiments permitted to reveal the fundamental notion of LG, the general relations of reachability. For the construction experiments, like trajectory generation, the general relations are often reduced (visually) to the reachability of the chess King.

Another example of observation of processes is related to the thought experiments with various objects with respect to the inertial reference frames when discovering the theory of Special Relativity [12]. This includes thought experiments with uniformly moving ships, trains, entities inside the ship, etc. Experiments with inertial frames reveal that they are equivalent or indistinguishable in terms of the laws of Physics. Yet another series of thought experiments with water and sound waves as traveling disturbances in a medium, with the traveler trying to catch the beam of light, etc. revealed special nature of light or, more precisely, electromagnetic waves.

Once the relations have been revealed, a construction set and a visual prototype have to be constructed by the observation stream. They are not necessarily real world objects. They are not even objects from the problem statement. They are created by the observation stream out of various multiple instances of the real world objects by abstraction, i.e., by erasing the particulars. However, they may visually represent an abstract concept, usually, a class of objects or processes, whose structure is being investigated. For this purpose, a construction set and a prototype are reduced to specific visual objects representing abstract entities. In a sense those objects serve as "symbols" representing those entities.

For construction, the observation stream utilizes the construction stream with auxiliary purpose (which differs from its prime purpose—see below). Note that the prototype construction is different from the subsequent reconstruction of the object intended to reveal its structure. This prototype may differ substantially from the real object or class of objects that are investigated. Its purpose is to serve as a manual to be used for references during reconstruction. Various discoveries may involve a series of prototypes.

When the prototype and the construction set are ready, the AD initiates the construction stream with its main purpose. This purpose is to construct the object (or stage the process) by selecting appropriate construction parts of the set and putting them together. Several types of streams (different themes, Sect. 14) could be utilized to focusing the construction. In particular, they include the visual manual, the Ghost, the proximity and mosaic reasoning, etc.

If an object has a sequential nature, the construction also takes place sequentially, by repetition of similar steps. If multiple prototypes have been produced, the final object construction can also be considered yet another prototype construction. The construction stream operates similarly to a child playing a construction set. A child needs a manual to construct an object by looking constantly at its pictures included in this manual. This manual comes from the observation stream. The construction stream utilizes a construction set and a visual prototype (to be referenced during construction). This is similar to a model pictured in a manual (or a visual guide) enclosed to every commercial construction set. For example, all the thought experiments in LG related to construction investigated so far, utilized those manuals. A final version of the object constructed by the construction stream should be validated by the validation stream.

Prototypes and construction sets may vary significantly for different problems. Construction of the model begins from creation of the construction set and the relations between its components. Both items should be visually convenient for construction. The AD may utilize a different model for the same object if the purpose of development is different. Such a different model is produced by a different visual stream.

In many cases, the AD employs "a slave" to perform simple tasks for all types of visual streams. For example, this slave may be employed by the construction stream to move around, to observe construction parts, and put them together. More precisely, we introduced personality, a Ghost. This Ghost has very limited skills, knowledge and, even, limited visibility. The observation stream may utilize the Ghost to familiarize itself with the optional construction set, to investigate its properties. Next, the construction stream may use the Ghost to perform the actual construction employing those properties. Eventually, the validation stream may use the Ghost to verify visually, if properties of the constructed object match those revealed by the observation stream. In all cases, the Ghost is guided by the respective visual streams.

10 From Discovery to Conventional Science

At some point of execution of the AD, there may be a need to turn final or preliminary results from the primary to the secondary (conventional) science. This could happen after approval of the results by the validation stream or in the middle of construction in order to publish these results to be reviewed by the scientific community.

To achieve this goal, the AD initiates the expression stream (Sects. 7 and 9) to be executed concurrently with the construction stream. The purpose of the expression stream is to map the primary science model into the secondary science, e.g., to develop a model based on the conventional scientific notation. Essentially, every component of the original visual model carries an abstract class of components behind it. This way, visual reasoning about the model drives reasoning about abstract classes, which is turned eventually into the standard formal reasoning. This happens as follows. The expression stream drives construction of the symbolic model by tagging the key items in the visual model during its construction. Those tags will eventually represent the conventional symbolic model. To achieve this the expression stream employs a grammar. This could be a universal grammar [6] for a scientific text, a formal grammar for a formal language (including algorithms), etc., where the tagged entities serve as terminal symbols. At first, the symbolic model is incomplete. However, at some stage, the running expression stream generates a comprehensive *symbolic shell*. Running the expression stream (together with the shell) means doing a presentation thought experiment analogous the one introduced in Sect. 6. However, in this case the outcome is a text that employs scientific notation including formal algorithm, derivation, proof, etc., synchronized with a respective construction stream. While the shell is incomplete, the construction stream drives execution of the shell (via the expression stream), not the other way around. For example, a development of the formal algorithm is driven by the animated events within the respective visual stream. The AD, usually, runs the construction of the object several times. During those runs, persistent tagging leads to completion of the symbolic shell. Multiple runs of both streams may lead to additional morphing of the visual model and/or adjusting symbolic derivation if they initially mismatch. Eventually, the construction and expression streams change their roles. The construction stream loses its driving force while the expression stream turns to the inverse one, the impression stream. In the end, the construction stream becomes the animated set of illustrations, a movie, driven by the impression stream mapping the running symbolic shell back into the construction stream. For example, during the final runs (and only then), the visual streams, presented in [40–46], were driven by the constraints of the abstract board game, the abstract set theory and/or the productions of the controlled grammars. At this point, all the visual streams and the symbolic shell (the scientific text and/or formal algorithm) can be completely separated, and the streams could be dropped and even forgotten. A transparent example of the gradual construction of the scientific expression stream and the generating grammar for this stream employing the AD are considered in Sects. 15–22.

There are several means to focusing the AD visual streams in the desired direction. Some of them were described in Sects. 9 and 10. Additional powerful means include *proximity* and *mosaic* reasoning.

11 Proximity Reasoning

Proximity reasoning as a type of visual reasoning is utilized due to the need for approaching optimum for many discoveries. It is likely that all the technological inventions and discoveries of the laws of nature include "optimal construction" or, at least, have optimization components [46]. Thus, various construction steps performed by the AD require optimization, which, certainly, makes construction more difficult. As the appearance of the AD is lost in millennia, for its main purpose, it could not certainly utilize differential calculus even for the problems where it would be most convenient. For the same reason, it could not utilize any approximations based on the notion of a limit of function. Those components of differential calculus could certainly serve as auxiliary tools. In that sense, in order to reveal the main optimization components, the most interesting problems to be investigated should lack continuity compelling the AD to employ explicitly those components. Based on several case studies [46], we suggested that this optimization is performed by the imaginary movement via approaching a location (or area) in the appropriate imaginary space. Having such space and means, the AD should employ an agent to catch sight of this location, pave the way, and approach it. Contrary to the function-based approach, which is static by its nature, the AD operates with dynamic processes, the visual streams. Some of those streams approach optimum (in a small number of steps); other streams show dynamically wrong directions that do not lead to the optimum and prevent the AD from pursuing those directions. Both types of streams represent proximity reasoning. We suggested that proximity reasoning plays a special role for the AD as the main means for optimization. Proximity reasoning is a type of visual reasoning. This implies that the AD should reason about the space where distances are "analogous" to the 3D Euclidian distances. Roughly, when we approach something, the distance must be visually reduced, and this should happen gradually. The space for proximity reasoning should provide means to evaluate visually if the animated images representing various abstract objects approach each other or specific locations [46]. Construction of those spaces is the key component of the AD. There could be situations when an object approaching the desired location (of an expected optimal state) hits an obstacle, i.e., another object, but the time constraints do not permit to remove this obstacle. This way, the desired location and the optimal state appear to be unreachable [33, 34]. In such cases, the AD may consider a total retreat to one of the initial states where time constraints are conducive for removal of the obstacle. After removal, the AD may attempt to return objects to the originally desired locations, which may lead to reaching the optimum.

12 Mosaic Reasoning

The name of mosaic reasoning was introduced due to the analogy of the construction stream operation with assembling a mosaic picture of small colorful *tiles* [45]. Another analogy that is even more transparent is known as a jigsaw puzzle when a picture is drawn on a sheet of paper and then this paper is cut into small pieces that should be mixed up to be assembled later into the original picture. As Thompson [57] pointed "… the progress of science is a little like making a jigsaw puzzle. One makes collections of pieces which certainly fit together, though at first it is not clear where each group should come in the picture as a whole, and if at first one makes a mistake in placing it, this can be corrected later without dismantling the whole group". Both analogies, the pictorial mosaic and the jigsaw puzzle, represent well the key feature of the AD, the construction set. However, we prefer the former because the jigsaw puzzle looks more like an assignment in reassembling a construct, a picture, which has already been created and, certainly, well known. In that sense, a tile mosaic is created from scratch, including choosing or even creating necessary tiles. In addition, a jigsaw puzzle is reassembled out of pieces based on random cuts. On the contrary, in pictorial mosaic, in many cases, every tile should have unique properties; it should be shaped and colored to match its neighbors precisely. A similar specificity is related to a group of adjacent tiles, the *aggregate*.

Returning to the AD, for many discoveries, the components of the construction set should be developed with absolute precision, in the way that every part should be placed to its unique position matching its neighbors. We will use the same name, the tiles, for those construction parts. If precision is violated, the final mosaic will be ruined and the discovery will not happen. Though a group of tiles, an aggregate, may be configured properly, its correct placement in the mosaic may be unclear and requires further investigation. Moreover, a tile itself may have complex structure, which may require tailoring after placement in the mosaic. In some cases, a tile is a network of rigid nodes with soft, stretchable links.

Mosaic reasoning stretches through the observation, construction, and validation steps of the AD operating with tiles and aggregates. Overall, mosaic reasoning requires tedious analysis of the proper tiles and their *matching rules*. Investigation of the matching rules is the essential task of the observation stream. Multiplicity of those rules and their specificity with respect to the classes of construction tiles make the actual construction very complex. Selecting a wrong tile, wrong tailoring, choosing a wrong place, or incompatible neighbors may ruin the whole mosaic. The matching rules are the necessary constraints that control the right placement of the tiles. Missing one of them, usually, leads to the wrong outcome because the AD is pointed in the wrong direction.

We will illustrate various matching rules on the example of the discovery of the 3D structure of the DNA molecule considered in [45].

Some of the matching rules impact mosaic locally while other rules provide global constraints. The global matching rules include the requirement of the top-down analysis and construction, the global complementarity rule, certain statistical rules, the transformation rules, etc. For many if not all natural objects and processes, their structure is not reducible to a combination of the components. Large groups of tiles, i.e., large aggregates, may obey the rules, which are hardly reducible to the rules guiding placement of singular tiles. This matching rule must be understood globally first, implemented in the mosaic skeleton construction, and, only then, reduced to the placement of the specific tiles. An example of the global matching rule for the discovery of the structure of DNA is the choice of the helical structure of the DNA molecule including the number of strands [7, 45, 61–63]. The rule of the global complementarity means that placement of one aggregate may determine precisely the adjacent aggregate. In case of DNA, one strand of the helix with the sequence of the base tiles attached to it determines the unique complementary second strand with the corresponding sequence of the base tiles. The global statistical rules related to the whole mosaic may reflect the relationship between major structural components, the large aggregates. If understood and taken into account by the observation stream, they may focus the construction stream and lead to a quick discovery. In the case of DNA, the so-called Chargaff rules reflect the structural relationship between the base tiles of the complementary strands of the double helix [5, 45, 61]. Yet another class of global matching rules is called transformation rules [45]. This is an algorithm for reconstructing an aggregate out of another aggregate and placing this aggregate in the proper location. Applied sequentially, such a rule permits to turn an aggregate, the so-called *generator*, into the set of adjacent aggregates. This way the whole mosaic could be constructed. For example, the whole mosaic of the DNA molecule could be constructed if the generator and the singular transformation are defined. Over the course of four experiments, the double helix generator was constructed. It includes a pair of nucleotides with sugar-phosphate backbone and purine-pyrimidine base. The transformation is a combination of translation and rotation. Interestingly, this type of construction may be utilized by the AD as a convenient procedure to reveal the structure of an object, e.g., the DNA molecule, while the nature may have used a completely different algorithm for producing the same object.

The local matching rules include the local complementarity rule, the interchangeability rule, etc. The local complementarity means, roughly, that a protrusion of one tile corresponds to the cavity of the complementary adjacent tile. For the DNA molecule, a hydrogen bond of a base tile (a protrusion) corresponds to a negatively charged atom of the adjacent tile (a cavity). The local complementarity often expresses itself in the requirement of various kinds of symmetry within the pairs of matching construction tiles. The whole class of the local matching rules is based on interchangeability. In simple terms, if two aggregates that include several tiles are not identical but interchangeable, their internal structure may be unimportant. There are several levels of interchangeability. Two aggregates could be essentially the same, i.e., their skeletons coincide. Importantly, those skeletons must include nodes, which serve as the attaching points of the aggregates to the rest of the mosaic. The notion of an internal skeleton depends on the problem domain and is specific for different types of mosaic. For example, two different aggregates for the DNA mosaic may have identical ring structures but the atoms and respective bonds that do not belong to those structures may be different. Another lower level of interchangeability of the aggregates does not require their skeletons to coincide. The only requirement is that the attaching points of those aggregates are identical. In all cases, interchangeability means that the stream can take one aggregate off the mosaic and replace it with another. This will certainly change the picture but the whole structure will stand. We named those aggregates *plug-ins*. It appears that plug-ins played crucial role in the discovery of the structure of DNA because such a plug-in was the key component of the helical generator, a purine-pyrimidine base [45, 61].

Besides mosaic structural components that include tiles, aggregates, global and local matching rules, there is an unstructured component that we named a mosaic environment. Such environment may affect the structure of tiles, aggregates, application of matching rules, and the whole mosaic while being relatively unstructured itself. In case of DNA, this was the water content whose lack or abundance could seriously affect the structure of the whole mosaic.

13 Programming Discoveries

The process of discovery is a sequence of thought experiments. This sequence could be considered as a *program of discovery* as follows. This program may include subprograms. The actual creators of those programs and subprograms are visual streams. A stream may schedule other streams by creating a program with *experiment calls*. It may schedule a thought experiment to be executed immediately. After completion of this experiment based on its outcome, the stream can schedule and invoke another thought experiment, etc. We suggest that such sequence of executed experiments recorded visually (as sequence of movies) is a program of discovery. However, this is not a program in conventional sense.

The program of discovery may be executed several times, each time with different outcome. This might happen due to the different input received by each of the experiments included in the program, beginning from the first one. Moreover, this program may change during execution. Indeed, the output of an experiment usually includes request for the next experiment. Different output may include request for the experiment that differs from the one recorded previously.

Besides experiments created, executed, and recorded sequentially, one after another, a stream can create a sequence of thought experiments to be executed in the future. These experiments could, in their turn, initiate new visual streams (with new experiments). The purpose, the nature, and the general outcome of those future experiments should be known to the stream created this sequence. This sequence is different from the list of procedure calls in conventional procedural (or imperative) programming. The algorithms of those procedures, i.e., the algorithms to be utilized by the respective thought experiments are generally unknown. The experiments are not programmed—they are staged. The actual algorithm should be developed as a result of execution of such experiment. In a sense, this is analogous to the notion of declarative programming when a function is invoked by a problem statement while the function's body does not include an algorithm for solving this problem. The ability of a visual stream to schedule a sequence of thought experiments (with delayed execution) permits to create a nested top-down structure of visual streams with several levels of depth. However, we suspect that the actual depth of the nested programmed experiments never exceeds two or three.

The programming system of the AD reminds remotely a conventional compiler. The output of this compiler is analogous to the executable code, i.e., the program of discovery. Different inputs of this compiler (different "source code") produce different outputs, i.e., different programs of discovery. Almost all of them except for one or two, are incorrect and do not lead to the discovery. These incorrect inputs reflect our current knowledge, e.g., incorrect experimental data, incorrect theoretical principles, etc. The failure of making a discovery, verified by the validation stream, usually causes the AD to rerun several parts of the discovery program with different inputs. The program does not have to be rerun from the very beginning. Different parts of the program may be restarted with new inputs. As we already mentioned, different inputs to the program of discovery may change this program completely. These partial reruns with new inputs lead to the effect that, eventually, the discovery program looks like a small tree with branches at different levels. Usually, only one branch leads to discovery. This tree records the history of making a discovery including failures. With correct input the program of discovery will produce and reproduce the right branch of the tree and this way it will produce and reproduce the discovery as many times as it will be executed.

14 The Themes of Streams

Every visual stream is based on a *theme*, which controls its scope and focuses its morphing. The theme is the environment visualized and animated by the stream. For a stream to morph beyond the current theme, it has to change themes by switching from one visual environment to the other. Changing the theme is equivalent to switching streams, which requires generating a request for changing within the current stream. Such request means that reasoning utilizing the current theme should be suspended or terminated. Several themes could be utilized concurrently by the same stream and by different concurrent streams. The theme without constraints would have permitted a stream to morph in unlimited fashion, from following dragons flying and spitting fire to watching the universe big bang. A theme imposes a set of constraints on the stream, which cannot be violated. These constraints create a visual environment and focus stream in a desired direction. The streams utilize different types of themes. All the different streams considered in this paper are instances of the same universal stream with different themes.

The communication streams utilize the elaborate themes intended to generate strings of symbols. For example, the internal visual stream employed by a novel writer can use a theme representing a country and a city where the characters of his novel live. However, for the actual writing the same writer should initiate the expression stream to map his/her internal stream into the textual form. For this purpose, the expression stream may use a theme representing the grammar of the writer's native language. The themes employed by the communication streams of the scientific discovery are even more elaborate. The expression stream mapping the internal discovery stream into the conventional scientific language, in addition to the English grammar, has to use the grammar of the language of the specific branch of science like Math, Physics, or Computer Science. Both the English grammar and the science grammar make up the theme of this expression stream.

The discovery streams themselves utilize many generic and problem specific themes to focus their search for discovery. The generic discovery themes include observation, construction, and validation streams as well as proximity and mosaic reasoning (Sects. 9, 11 and 12). All of those themes are utilized by the streams in various branches of science. In particular, examples considered in [40–50] are related to Computer Science, Physics, and Molecular Biology. On the other hand, the problem specific themes like those of Pictorial LG, Mountain-triangle, State Space Chart, and Tree World [42, 45, 46] were used by the streams for discovering the No-Search Approach and solving other problems in LG. A combination of themes related to Pictorial LG, various representations of the abstract board (in ABG) and symbolic reasoning are utilized in the comprehensive example of discovering the Grammar of Zones, Sects. 15–22.

A discovery, i.e., a development of the final algorithm for the object construction is based usually on constructing a series of themes of streams. A theme may be utilized in multiple experiments. A theme may be used in multiple streams including observation, construction, and validation streams. Interestingly, those themes may represent the same object, though, be completely different. The purpose of these themes is to look at the object from different prospective to reveal different properties. The themes do not appear at once. Experiments with one theme may lead to demonstrating the need for the next one. After being invoked, several themes could be utilized simultaneously. The themes construction is based on the wide use of the principle of erasing the particulars. For each theme, some of the particulars of an object under investigation are erased while other particulars are emphasized.

All the thought experiments of a discovery program could be broken into several groups. Experiments that operate with the same theme are related to each other while they differ from those groups that operate with another theme. Once the theme has been invoked, it can usually schedule experiments within itself to be executed in the future. As a rule, after being invoked the theme is used in several experiments. Then another theme could be invoked. At that time, previous theme is turned into the idle mode, which means that the stream is suspended to pass control to the stream of a new theme. The experiments with different themes cannot be scheduled until those themes have actually been invoked. The problem is that until that moment, the stream is not aware of their existence and even the need for them, unless this theme has already been invoked, used, and currently preserved in an idle mode. Consequently, scheduling of experiments with delayed execution is possible with the active and idle visual themes, only.

It appears that the most sophisticated steps of execution of the AD, i.e., generation and execution of the discovery program, are those related to invoking new themes of visual streams. While reasoning within the active and idle themes requires just logic, activating a new theme requires constructing it from scratch or retrieving it from the archive of the past themes and morphing it to match current needs.

A theme can apply the ultimate set of constraints to the stream by restraining its flow down to the only sequence of events. For example, the stream of established primary science is obtained by "recording" the sequence of streams that has led to discovery, i.e., the specific branch of the discovery program, and dropping all other options (Sect. 13).

Various examples of themes were considered in [39–50] without explicitly using this term.

15 From Primary to Conventional Science

In Sects. 15–22, we demonstrate a series of thought experiments that turn a piece of the primary science to the secondary one. In presenting these experiments, we follow the general framework of relationship between the primary (internal) science and its external mapping, the secondary science (Sects. 6, 7, and 10). According to the [60] hypothesis "... when we talk mathematics, we may be discussing a secondary language, built on the primary language truly used by the central nervous system." The following sections demonstrate our first sketchy vision of the relationship between the primary and secondary mathematics. The N. Chomsky universal grammar that is possibly utilized for conversion of the internal streams, the components of Primary Language, into the secondary languages is still unknown. However, the example presented below is highly instructive for future research because in this case the AD developed a complete tool kit for conversion of the internal animated pictorial representation into the external formal language of mathematics. Both representations (including snapshots of the internal streams) have been used extensively in our research and applications. This tool kit includes the formal language with syntactic and semantic components as well as the formal grammar generating this language. This tool kit employs persistent tagging of the events (and their components) animated in the Primary Language. It converts those events into the strings of symbols.

More precisely, we introduce several internal visual streams that operate with the theme of Pictorial LG as well as the streams of Linguistic LG (Sect. 17). Then, employing the AD, we construct a series of expression streams that operate with a mix of the Pictorial and Linguistic LG themes, gradually approaching the pure Linguistic LG stream. The AD executes its construction stream three times, each time in the form of an expression stream. The first expression stream is intended to develop the visual time distribution algorithm to control semantics of the future Grammar of Zones. The second expression stream is intended to sketch the syntactic structure of the Grammar of Zones, the list of productions (with due attention to the formal semantics). The third expression stream is intended to complete construction of the Grammar of Zones by completing development of all its components, functions and predicates. The result of this gradual construction is the expression stream with a piece of the formal mathematical knowledge, the Grammar of Zones. This grammar can generate an arbitrary zone for an arbitrary ABG given the proper input. As a typical abstract mathematical object, it does not have a pictorial representation. However, various instances of the generated objects could be represented as pictures.

16 Abstract Board Games

To describe the Pictorial and Linguistic LG themes we have to introduce class of problems that LG is intended to solve. Abstract Board Games (ABG) are defined as follows (see complete definition in Stilman [38]):

$$\langle X, P, R_p, SPACE, val, S_0, S_t, TR \rangle,$$

where

- $X = \{x_i\}$ is a finite set of points (or locations of an abstract board);
- $P = \{p_i\}$ is a finite set of pieces (or mobile entities); $P = P_1 \bigcup P_2$, $P_1 \bigcap P_2 = \varnothing$, called the opposing sides;
- $R_p(x, y)$ is a set of binary relations of reachability in X ($x \in X$, $y \in X$, and $p \in P$) representing mobility rules of the pieces;
- *val* is a function on P representing the values of pieces;
- SPACE is the state space. A state $S \in$ SPACE consists of a partial function of *placement* ON: $P \rightarrow X$ and additional parameters.
- $ON(p) = x$ means that element $p \in P$ occupies location x at state S. Thus, to describe function ON at state S, we write equalities $ON(p) = x$ for all elements p, which are present at S. We use the same symbol ON for each such partial function, though the interpretation of ON may be different at different states. Every state S from SPACE is described by a list of formulas $\{ON(p_j) = x_k\}$ in the language of the first order predicate calculus, which matches with each relation a certain formula.

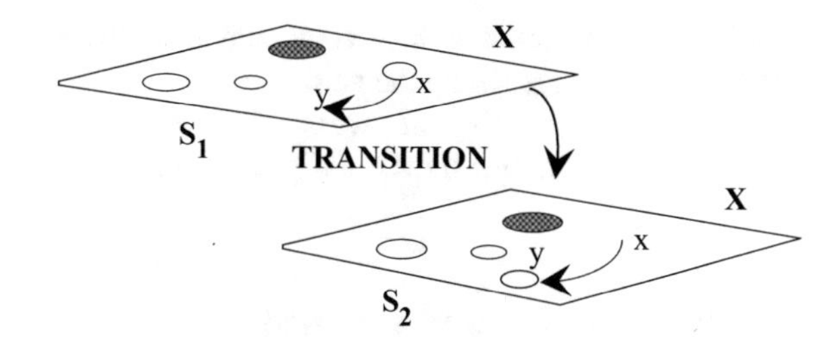

Fig. 1 Interpretation of ABG: TRANSITION(p, x, y)

- S_0 and S_t are the sets of *start* and *target* states. Thus, each state from S_0 and S_t is described by a certain list of formulas $\{ON(p_j) = x_k\}$. S_0 and S_t are the sets of Start and target states. $S_t = S_t^1 \cup S_t^2 \cup S_t^3$, where all three are disjoint. S_t^1 and S_t^2 are the subsets of *win* target states for the opposing sides P_1 and P_2. S_t^3 is the subset of the *draw* target states.
- TR is a set of *transitions* (moves), TRANSITION(p, x, y), of the ABG from one state to another (Fig. 1). These transitions are described in terms of the lists of formulas (to be removed from and added to the description of the state) and a list of formulas of applicability of the transition. These three lists for state S ∈ SPACE are as follows:

  $$\textbf{Applicabilitylist}: \ (ON(p) = x) \wedge R_p(x, y);$$
  $$\textbf{Remove list}: \ ON(p) = x;$$
  $$\textbf{Add list}: \ ON(p) = y,$$

  where p ∈ P. The transitions are defined and carried out by means of a number of elements p from P_1, P_2, or both. This means that each of the lists may include a number of items shown above. Transitions may be of *two types*. A transition of the *first type* (shown above) occurs when element p moves from x to y without removing an opposing element. In this case, point y is not occupied by an opposing element. A transition of the *second type* occurs if element p moves from x to y and does remove an opposing element q. Typically, the opposing element has to occupy y before the move of p has commenced. In the latter case, the **Applicability list** and the **Remove list** include additional formula ON(q) = y. For the concurrent systems [38], this is not necessary: element q may arrive at y simultaneously with p and be removed.

The goal of each side is to reach a state from its subset of the win target states, S_t^1 or S_t^2, respectively, or, at least, a draw state from S_t^3. The problem of the optimal operation of ABG is considered as a problem of finding a sequence of transitions leading from a Start State of S_0 to a target state of S_t assuming that each side makes only the best moves (if known), i.e., such moves that could lead ABG to the respective subset of target states. To solve ABG means to find a *strategy* (an algorithm to select moves) for one side, if it exists, that guarantees that the proper subset of target states will be reached assuming that the other side could make arbitrary moves to prevent the first side from reaching the goal.

For simplicity, in this paper, we consider *serial alternating* ABG, i.e., ABG where only one piece at a time moves and pieces from P_1 and P_2 alternate moves.

17 Pictorial LG and Linguistic LG

In this section, we provide minimal information required for explaining the Pictorial LG and the Linguistic LG. In the Pictorial LG, a zone is a trajectory network—a set of interconnected trajectories with one singled out trajectory (called the main trajectory). An example of an *attack* zone is shown in Fig. 2. Below, we utilize this image extensively to explain semantics of the zone. This image represents an idealized snapshot of the situation representing a local skirmish, a subset of a state of the ABG. This situation is shown with pieces located at various points of the abstract board. The board, a finite set of points, is not shown explicitly—it is represented by the lines and circles on the 2D plane. The opposing pieces are shown with black and white circles, respectively. The straight lines represent trajectories, the planning paths over the abstract board. The small circles represent stop locations along those trajectories. The straight lines do not reflect possible sophisticated shapes of those trajectories, which may exist in case of complex relations of reachability, the rules of movement of pieces. When the Pictorial LG theme was originally constructed all those complexities were removed (encapsulated in the form of straight lines) according to the principle of erasing the particulars. Note that the best way to understand the Pictorial LG is to assume that it represents positions from the game of chess (as an instance of the ABG).

The basic idea behind a zone is as follows. Piece p_0 should move along the so-called *main* trajectory $\boldsymbol{a}(1)\boldsymbol{a}(2)\boldsymbol{a}(3)\boldsymbol{a}(4)\boldsymbol{a}(5)$ to reach the end point 5 and remove the target q_4 (an opposing piece marked with a dark circle). Numbers 1, 2, 3, 4 and 5

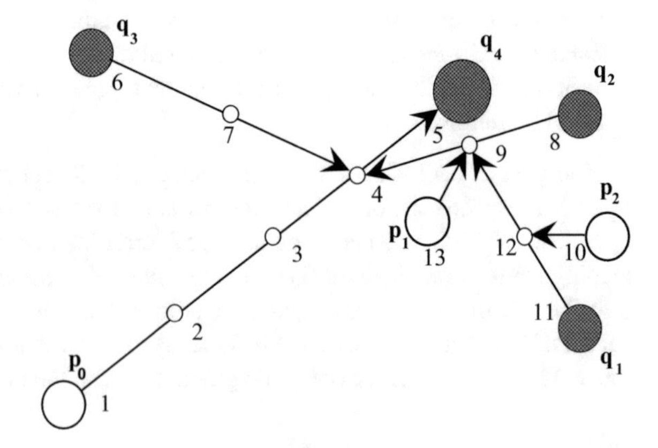

Fig. 2 Pictorial LG image of a sample attack zone

represent the identifiers of the respective points on the abstract board. Naturally, the opposing pieces should try to obstruct this movement by *dominating* the intermediate stop points of the main trajectory. They should come closer to these points (to point 4 in Fig. 2) and remove piece p_0 after its arrival (at point 4). For this purpose, pieces q_3 or q_2 should move along the trajectories $\boldsymbol{a}(6)\boldsymbol{a}(7)\boldsymbol{a}(4)$ and $\boldsymbol{a}(8)\boldsymbol{a}(9)\boldsymbol{a}(4)$, respectively, and wait (if necessary) on the next to last point (7 or 9) for the arrival of the piece p_0 at point 4. These trajectories are called the *first negation trajectories*. Similarly, piece p_1 of the same side as p_0 (i.e., White) might try to obstruct the movement of q_2 by dominating point 9 along the trajectory $\boldsymbol{a}(13)\boldsymbol{a}(9)$. It makes sense for the opposing side to employ the trajectory $\boldsymbol{a}(11)\boldsymbol{a}(12)\boldsymbol{a}(9)$ of piece q_1 to prevent this domination.

Consider a complete Linguistic LG representation (Table 1) of a sample zone depicted in Fig. 2.

Below, in analyzing this example, we describe the movements of pieces in the zone (Table 2). Note that the zone itself is represented in the Linguistic LG as a string of symbols $\boldsymbol{t}$ with parameters $\boldsymbol{t}(\ldots)\boldsymbol{t}(\ldots s)\boldsymbol{t}(\ldots)\boldsymbol{t}(\ldots)$, a static (*not dynamic*) object that represents a network of trajectories. It represents a subsystem of the ABG in a certain state. With every change of a state, a new zone should be constructed. Some of these zones are considered successors of the original zone. For example, zones with the main trajectories that start at locations 2, 3 and 4 are the successors of the zone shown in Fig. 2. Zones shown in Table 2 after the successive moves $\mathbf{q_1 11\text{-}12}$, $\mathbf{p_0 1\text{-}2}$ are the successors of the same zone. For simplicity, movements of pieces considered below are guided by the pictorial image of the same zone, and the zone itself is considered isolated from the rest of the ABG.

Let us reconsider these movements to explain reasoning behind the construction of this zone. Assume that the goal of White (Fig. 2) is to remove target q_4, while the goal of Black is to protect it. Also, assume that Black and White alternate turns. Black starts first to move pieces q_2 or q_3 to intercept p_0. White follows, in its turn, by moving piece p_0 to the target. Only those Black trajectories are to be included in the zone where movement of the piece makes sense, i.e., the *length of the trajectory is less than (or equal) the amount of time allotted to it*. For example, the movement along the trajectories $\boldsymbol{a}(6)\boldsymbol{a}(7)\boldsymbol{a}(4)$ and $\boldsymbol{a}(8)\boldsymbol{a}(9)\boldsymbol{a}(4)$ makes sense, because they are of length 2 and time allotted equals 4. All the above information is reflected in the values of parameters of the symbols of the string Z, $\boldsymbol{t}(q_3, \boldsymbol{a}(6)\boldsymbol{a}(7)\boldsymbol{a}(4), 4)$ and $\boldsymbol{t}(q_2, \boldsymbol{a}(8)\boldsymbol{a}(9)\boldsymbol{a}(4), 4)$, respectively. Indeed, each of the pieces, q_2 and q_3, has 4 time intervals to reach point 4 to intercept piece p_0, assuming that Black starts first and p_0 would go along the main trajectory without delay. By definition of a zone, the trajectories of White pieces (except p_0) could only be of the length 1, e.g., $\boldsymbol{a}(13)\boldsymbol{a}(9)$ or $\boldsymbol{a}(10)\boldsymbol{a}(12)$. Because piece p_1 can intercept piece q_2 at point 9, Black trajectory $\boldsymbol{a}(11)\boldsymbol{a}(12)\boldsymbol{a}(9)$ of the piece q_1 should be included, i.e., q_1 has enough time for movement to prevent this interception. The total amount of time allotted to the whole bundle of the Black trajectories connected (directly or indirectly) with a given point of the main trajectory is determined by the ordinal number of that point. For example, for point 4, it is equal to 4 time intervals. This number gives us the value of time bound for the first negation trajectories connected to the main trajectory at point 4. However, the length of the trajectory $\boldsymbol{a}(8)\boldsymbol{a}(9)\boldsymbol{a}(4)$ is 2, thus there is enough

Table 1 Linguistic representation of a sample zone shown in Fig. 2

$$Z = \boldsymbol{t}\left(p_0, \boldsymbol{a}(1)\,\boldsymbol{a}(2)\,\boldsymbol{a}(3)\,\boldsymbol{a}(4)\,\boldsymbol{a}(5), 5\right) \boldsymbol{t}\left(q_3, \boldsymbol{a}(6)\,\boldsymbol{a}(7)\,\boldsymbol{a}(4), 4\right)$$
$$\boldsymbol{t}\left(q_2, \boldsymbol{a}(8)\,\boldsymbol{a}(9)\,\boldsymbol{a}(4), 4\right) \boldsymbol{t}\left(p_1, \boldsymbol{a}(13)\,\boldsymbol{a}(9), 3\right)$$
$$\boldsymbol{t}\left(q_1, \boldsymbol{a}(11)\,\boldsymbol{a}(12)\,\boldsymbol{a}(9), 3\right) \boldsymbol{t}\left(p_2, \boldsymbol{a}(10)\,\boldsymbol{a}(12), 2\right)$$

time for the second negation trajectories such as $\boldsymbol{a}(11)\boldsymbol{a}(12)\boldsymbol{a}(9)$ linked to the first negation.

A set of zones generated in every state of a problem is a unique representation of this state. A piece may be involved in several zones and in several trajectories in the same zone. All the trajectories and zones are evaluated with respect to their quality [38]. Only the highest quality trajectories are considered for generating strategies. The quality function is based on the prediction of the "rate of difficulty" for a piece for moving along the trajectory. For example, for the attack zone (Fig. 2) piece p_0 has to pass four locations 2, 3 and 4 to reach destination and destroy its target at 5. This passage may be free or it may be obstructed by the enemy pieces if they approach these locations. In particular, piece p_0 can be captured at location 4 by q_2 or q_3 if they approach location 4. The notion of passage through location for the game of chess is based on the values of pieces (surrounding this location, i.e., one step away) and on the result of optimal exchange of these pieces [38]. For the military operations employing trajectories of physical entities (missiles, planes, single soldiers) and shooting, the notion of passage through location is based on the notion of probability-of-kill, which is defined for all the entity-weapon pairs [51–54]. These probabilities permit calculating qualities of trajectories and zones based on the integrated probabilities of successful passage. In all cases, the less "difficulties" a piece would experience in passing along a trajectory the higher quality of this trajectory is. Every location along a trajectory, where a piece can be intercepted (for the game of chess) or destroyed with high probability (for modern military operations) reduces quality of this trajectory. A trajectory, which includes at least one such location, is called a trajectory with closed location or a *closed trajectory*. A trajectory without such locations is called an *open trajectory*. An example of an open main trajectory of an attack zone, trajectory for p_0, is shown in Fig. 2. An example of the closed first negation trajectory is a trajectory $\boldsymbol{a}(8)\boldsymbol{a}(9)\boldsymbol{a}(4)$ for q_2.

The grammars constructed in the formal theory of LG generate all kinds of trajectories and zones. For example, zone Z (Fig. 2 and Table 1) for a specific ABG could be generated by the Grammar of Zones with the input of just the current state of the ABG, the coordinates of two locations 1 and 5, and the length of the main trajectory, which is 4 steps, Tables 6 and 7.

In this explanation, for simplicity, we described only one type of zones called *attack* zones. There are several different types of zones. For example, *domination* zone looks similar to the attack zone (Fig. 2) but it would not have a target at location 5. The purpose of this zone would be for p_0 to reach location 4 and guard location 5 (which might belong to other zones).

Table 2 Pictorial and Linguistic LG: Movement of pieces inside the zone shown in Fig. 2

Moves	Snapshots of the Primary Science Stream: **Pictorial LG**	Conventional Science Language: **Linguistic LG**
$q_1$11-12	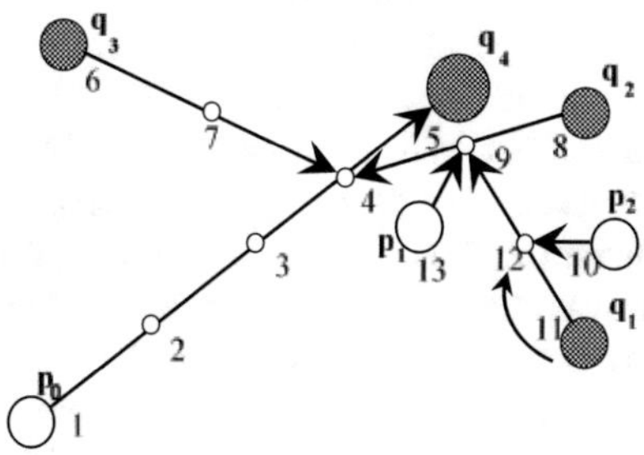	$t(p_0, a(1)a(2)a(3)a(4)a(5), 5)$ $t(q_3, a(6)a(7)a(4), 4)$ $t(q_2, a(8)a(9)a(4), 4)$ $t(p_1, a(13)a(9), 3)$ $t(q_1, a(11)a(12)a(9), 3)$ $t(p_2, a(10)a(12), 2)$
$p_0$1-2	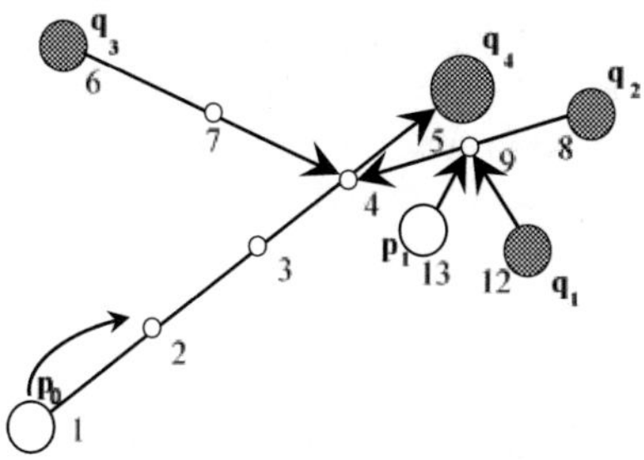	$t(p_0, a(1)a(2)a(3)a(4)a(5), 5)$ $t(q_3, a(6)a(7)a(4), 4)$ $t(q_2, a(8)a(9)a(4), 4)$ $t(p_1, a(13)a(9), 3)$ $t(q_1, a(12)a(9), 3)$
$q_2$8-9	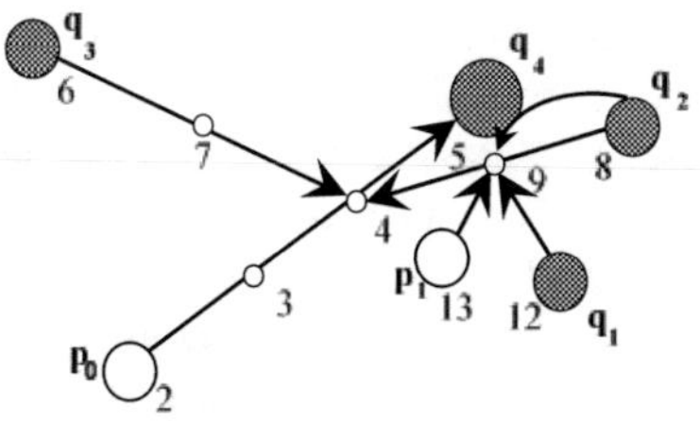	$t(p_0, a(2)a(3)a(4)a(5), 4)$ $t(q_3, a(6)a(7)a(4), 3)$ $t(q_2, a(8)a(9)a(4), 3)$ $t(p_1, a(13)a(9), 2)$ $t(q_1, a(12)a(9), 2)$

The visual streams of the Pictorial LG permit natural representation of the pieces moving along trajectories within the ABG (like those moving over the chessboard). An example of those is shown Table 2, middle section. However, representing movement within the Linguistic LG is not trivial, Table 2, right.

It is obvious that movement along a trajectory could be implemented by shortening the respective string of symbols, i.e., by dropping symbols one by one (one symbol per move), from the front of this string. It was much harder to understand what is happening to the string representing a zone when pieces are moving inside this zone. Historically, the formal representations of the trajectories and zones, the strings of symbols, were mentally and often visually (on paper) attached as tags to the respective visual images. The purpose was to represent those morphing pictorial structures of trajectories and zones via changes to their formal representations. The key analogy for representing morphing was as follows. When a piece moves along a trajectory, this trajectory shortens—the part of the route left behind disappears; so does the respective string of symbols by dropping symbols from the front when a piece moves from one location to another. When pieces move inside a zone this zone shrinks; so does the string, representing this zone, by dropping some of the symbols, though, not necessarily from the front, Table 2, middle and right (gray highlights). The part of a zone that was dropped on the way forward was called a freezing part, to unfreeze on the way backward. To reveal formally the structure of the freezing part required substantial analysis employing formal representation. In the end, the entire movement process was formalized as a sequence of Translations of Languages of Trajectories and Zones [38]. All the above advancements were based on the development of the Grammar of Zones to link the internal streams of Pictorial LG with the external formal representation of the Linguistic LG. This grammar provided not only one-to-one mapping between the Primary and Secondary science of LG; it laid the algorithmic foundation for the self-awareness of the Pictorial LG streams, Sects. 4 and 8.

18 Controlled Grammars: From Images to Strings

The observation stream explored various classes of languages and grammars while searching for the languages that include powerful means for handling formal semantics in addition to syntactic structures. The class of appropriate languages was found in [59]. These were strings of symbols with sophisticated parameters generated by the so-called BUPPG grammars [38]. Below we introduce the reader to the controlled grammars, which is a slight modification of those BUPPG grammars. This introduction is informal (Table 3) because this type of informal example-based introduction guided the observation stream. For a complete formal definition, see [38]. Controlled grammars generate strings of symbols with parameters. The lists of parameters incorporate semantics of the string. This semantics is determined by the problem domain, e.g., by squares of the chessboard, military aircraft, type of the robot or obstacle, etc. The values of actual parameters are calculated and assigned in the process of generation. Moreover, the generation itself is controlled by the state of the problem domain. This is achieved by providing the grammar with a control mechanism. This mechanism includes two major additions to the standard Chomsky structure [6], lists of productions admitted for application at each step of the generation and conditions of applicability of the productions, i.e., certain formulas of the predicate calculus. During generation, this control mechanism, in turn, is controlled by the problem domain through the values of formulas and actual parameters of the substring generated so far. Let us consider a rough schema of operation of such grammars.

Table 3 Informal schema of controlled grammars

Label	Condition	Kernel, π_k	π_n	F_T	F_F
l	$Q(,,)$	$A(,,) \rightarrow B(,,)$	$C(,,) := D(,,)$	L_T	L_F

V_T is the alphabet of terminal symbols
V_N is the alphabet of nonterminal symbols
V_{PR} is the alphabet of the first order predicate calculus (including ***Pred***, the list of predicates Q)
E is the subject domain
Parm is the finite set of parameters
L is the finite set of labels L referencing the productions

Parameters (variables and functions) are shown in parenthesis
– If Q is true and the current string contains nonterminal A, production with label L is applied and we go to the production with label from the list L
– If Q is not true or the current string does not contain nonterminal A, production L does not apply and we go to production with label from the list L
Values of parameters are changed when we apply productions

Every production $A(,,) \rightarrow B(,,)$ of a controlled grammar is equipped with the label l, predicate $Q(,,)$, separate formulas $C(,,) := D(,,)$, i.e., assignment operators, and the two sets of labels of transition in case of success and failure, F_T and F_F, respectively. Expressions shown in parenthesis may include variables and functional formulas. The grammar operates as follows. The initial permissible set of productions consists of the production with label 1. It should be applied first. Let us describe the application of an arbitrary production in a controlled grammar. Suppose that we attempt to apply production with label l to rewrite the symbol A. We have to choose the leftmost entry of symbol A in the current string and compute the value of predicate Q for the subject domain E (the condition of applicability of the production). There are two options for applying this production.

- If the current string *does not* contain A, or if $Q = F$ (*false*), then the application of the production is ended, and the next production is chosen from the failure section of labels F_F, i.e., F_F becomes the current permissible set.
- If the current string *does* contain A and $Q = T$ (*true*), A is replaced by the string in the right hand side of the production, $B(,,)$. In addition, we carry out computation of the values of all formulas corresponding to the parameters of the symbols in the right hand side of the production $B(,,)$—from section π_k and those standing separately $D(,,)$—from section π_n, and the parameters assume their new values. After application of the production is ended, the next production is chosen from the success section F_T, which is now the current permissible set. If the permissible set is empty, the derivation halts.

19 Observation Stream: Preparing for Construction

The observation stream investigating application of the controlled grammars to generating zones made several important conclusions.

A terminal symbol of the future grammar of zones could be associated with a whole trajectory via a parameter list. Thus, every trajectory of a zone could be reflected by the same terminal symbol with a parameter evaluated as a string generated for this trajectory by the grammar of trajectories. In such case, the final string to be generated would include the number of terminal symbols equal to the number of different trajectories in this zone. In addition, the rigid zone's network structure could be preserved by marking the location of the abstract board where one trajectory has to be connected to another and storing the coordinate of this location. (Note that an abstract board is a finite set, so the locations can be marked, enumerated or assigned a coordinate.) This type of connection based on the coordinates should have provided precision and respective rigidness to the resulting structure.

Besides absence of the actual zones' generating grammar, the main issue remained was the need for an algorithm for the time distribution. The notion of time distribution was obtained by the observation stream in the form of the permission rule. This rule means that only those opposing (to the main piece) trajectories are to be included in the zone where the movement of the piece makes sense, i.e., the length of the trajectory is less than (or equal) the amount of time allotted to it. Thus, during zone construction, every trajectory has to be assigned a number, the amount of time allotted to it. The observation stream concluded that for every trajectory this number could be generated employing simple algorithm utilized already for creating examples of zones via the Pictorial LG stream. Time allotted to the second and higher negation trajectories is equal to $TIME(\mathrm{y}) - l + 1$; $TIME(\mathrm{y})$ is the time already allotted to the lower negation trajectory, to which the current one is attached (at the location y); l is the length of the current trajectory. For the first negation trajectories, time allotted is just $TIME(\mathrm{y})$, where $TIME(\mathrm{y})$ is the ordinal number (beginning from the start) of the location y of the main trajectory to which the current trajectory is attached. This algorithm is recursive and requires step-by-step zone construction, i.e., beginning from the main trajectory, then, first negation trajectories, following with all the subsequent higher negations sequentially. The current negation should be started when the previous (lower) negation is finished. Every inclusion of a trajectory in a zone should be accompanied by assigning time allotted to it according to the algorithm considered above. The main trajectory is the special case in a way that such assignment should be performed for every stop location along this trajectory (except start)—they should be assigned just their ordinal numbers. As long as the trajectories are superimposed over the abstract board via the coordinates of their locations, the value of time allotted to a trajectory in a zone could also be assigned to the appropriate location of the board, specifically, to the end location of the trajectory. For the main trajectory, the time assignment should be made for all the stop locations along the trajectory (except for the start location). The time distribution for the sample zone (Fig. 2), is shown in Fig. 3. This visual theme was introduced by the

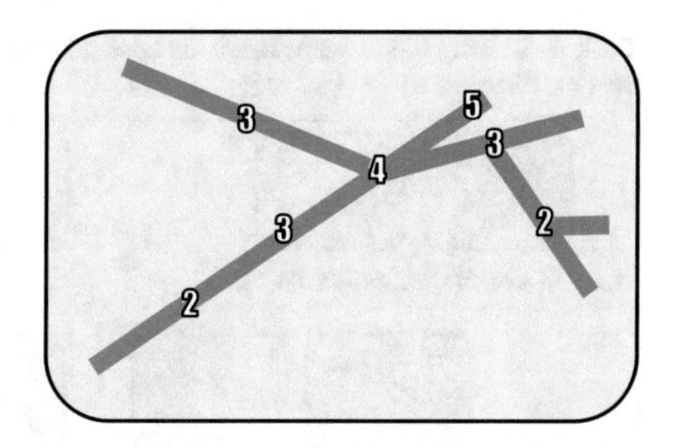

Fig. 3 Construction stream sketching time distribution for a zone generation grammar

observation stream to show values of time assigned to the respective board locations. It was named *TIME*, a *TIME*-board of the abstract board. In Fig. 3, *TIME* is shown as a gray rectangle with rounded edges meaning that it represents a screen showing the locations of the abstract board where the time distribution numbers are assigned. The dark bold lines are present just for the reader to understand which trajectories these numbers represent. Otherwise, we could use a clear gray screen (as an area representing abstract board) with the time distribution numbers (shown white).

20 The First Expression Stream: Sketching Zone Generation Without a Grammar

The observation stream passed all its conclusions to the construction stream. In addition, it passed a prototype, the sample zone shown in Fig. 2. The construction stream was invoked to generate this Pictorial LG zone as a string of symbols (shown in Table 1) without a grammar. It assumed that a grammar would be developed by the following streams based on this sketch.

The stream started from introducing a dummy start symbol $\boldsymbol{S}(u, v, w)$ to be replaced with the nonterminal symbol $\boldsymbol{A}(u, v, w)$. The second and third parameters of both nonterminal symbols are shown as gray screens (Table 4). These gray screens are the models of the abstract board of the ABG shown in Fig. 3, v-board and w-board. In addition, two more copies of the abstract board were introduced, *TIME*-board and *NEXTTIME*-board, to accompany every step of the generation (Table 4). All the four copies are to be updated in the course of generation. At the beginning of the generation all the four copies of the abstract board are unmarked—clear gray screens. In this respect, this first run of the construction stream could be considered as a combination of the five expression streams. The symbolic/pictorial stream whose snapshots are shown in Table 4 includes the major steps of the zone generation. The other four auxiliary pictorial streams shown in Table 4 as four screens were intended to support time distribution.

To generate all the required trajectories the construction stream assumed that it could employ a grammar of trajectories. The observation stream had information

Table 4 Construction stream sketching zone generation including the trajectories generation and the time distribution

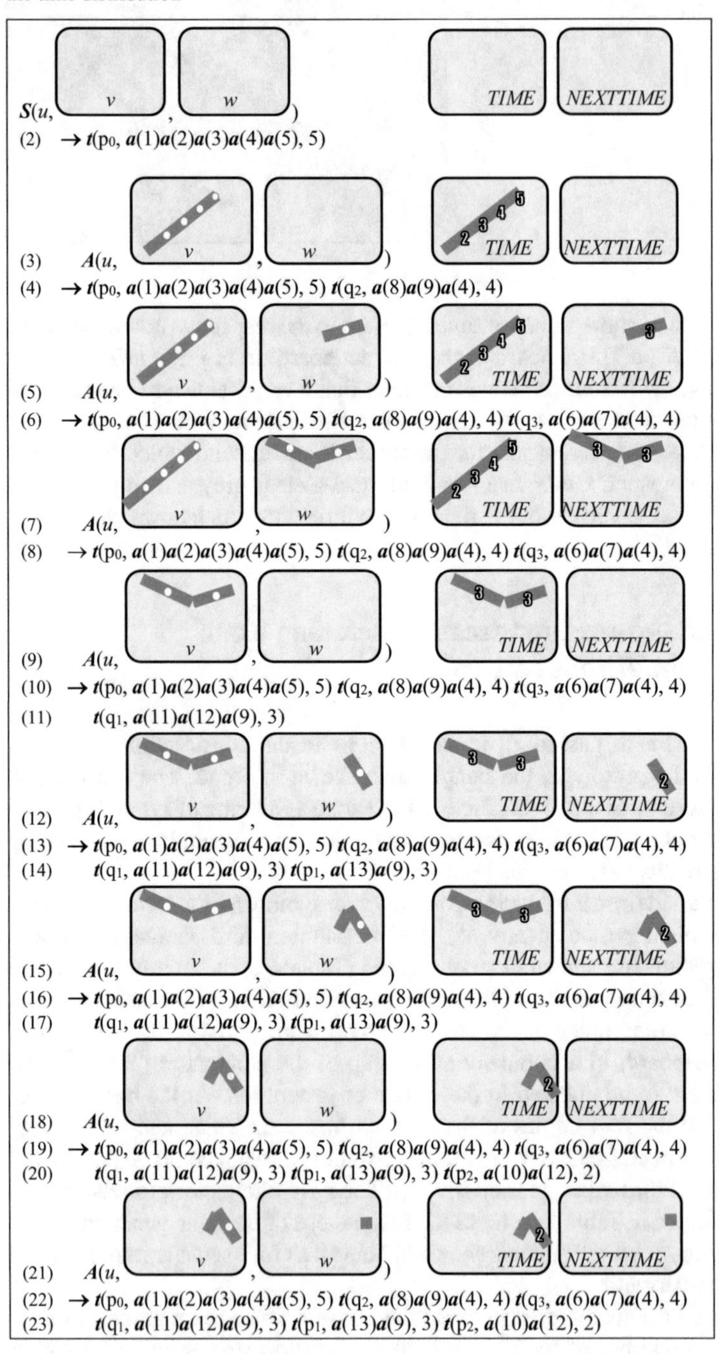

S(*u*, *v*, *w*) *TIME* *NEXTTIME*

(2) → $t(p_0, a(1)a(2)a(3)a(4)a(5), 5)$

(3) *A*(*u*, *v*, *w*) *TIME* *NEXTTIME*

(4) → $t(p_0, a(1)a(2)a(3)a(4)a(5), 5)\ t(q_2, a(8)a(9)a(4), 4)$

(5) *A*(*u*, *v*, *w*) *TIME* *NEXTTIME*

(6) → $t(p_0, a(1)a(2)a(3)a(4)a(5), 5)\ t(q_2, a(8)a(9)a(4), 4)\ t(q_3, a(6)a(7)a(4), 4)$

(7) *A*(*u*, *v*, *w*) *TIME* *NEXTTIME*

(8) → $t(p_0, a(1)a(2)a(3)a(4)a(5), 5)\ t(q_2, a(8)a(9)a(4), 4)\ t(q_3, a(6)a(7)a(4), 4)$

(9) *A*(*u*, *v*, *w*) *TIME* *NEXTTIME*

(10) → $t(p_0, a(1)a(2)a(3)a(4)a(5), 5)\ t(q_2, a(8)a(9)a(4), 4)\ t(q_3, a(6)a(7)a(4), 4)$

(11) $t(q_1, a(11)a(12)a(9), 3)$

(12) *A*(*u*, *v*, *w*) *TIME* *NEXTTIME*

(13) → $t(p_0, a(1)a(2)a(3)a(4)a(5), 5)\ t(q_2, a(8)a(9)a(4), 4)\ t(q_3, a(6)a(7)a(4), 4)$

(14) $t(q_1, a(11)a(12)a(9), 3)\ t(p_1, a(13)a(9), 3)$

(15) *A*(*u*, *v*, *w*) *TIME* *NEXTTIME*

(16) → $t(p_0, a(1)a(2)a(3)a(4)a(5), 5)\ t(q_2, a(8)a(9)a(4), 4)\ t(q_3, a(6)a(7)a(4), 4)$

(17) $t(q_1, a(11)a(12)a(9), 3)\ t(p_1, a(13)a(9), 3)$

(18) *A*(*u*, *v*, *w*) *TIME* *NEXTTIME*

(19) → $t(p_0, a(1)a(2)a(3)a(4)a(5), 5)\ t(q_2, a(8)a(9)a(4), 4)\ t(q_3, a(6)a(7)a(4), 4)$

(20) $t(q_1, a(11)a(12)a(9), 3)\ t(p_1, a(13)a(9), 3)\ t(p_2, a(10)a(12), 2)$

(21) *A*(*u*, *v*, *w*) *TIME* *NEXTTIME*

(22) → $t(p_0, a(1)a(2)a(3)a(4)a(5), 5)\ t(q_2, a(8)a(9)a(4), 4)\ t(q_3, a(6)a(7)a(4), 4)$

(23) $t(q_1, a(11)a(12)a(9), 3)\ t(p_1, a(13)a(9), 3)\ t(p_2, a(10)a(12), 2)$

about two such grammars developed earlier. These were the Grammar of Shortest Trajectories $\mathbf{G_t^{(1)}}$ and the more advanced Grammar of Shortest and Admissible Trajectories $\mathbf{G_t^{(2)}}$ [38, 43–46]. The first parameter u of the start symbol $\boldsymbol{S}(u, v, w)$ in such case would carry the values of input parameters for the grammar of trajectories $u = (\mathrm{x}, \mathrm{y}, l)$. According to Fig. 3, the initial value of $u =$ (coordinate of location 1, coordinate of location 5, 4), the start, the end, and the length of the main trajectory of the zone to be generated.

Additional input parameter for the grammar of trajectories as well as the grammar of zones (being sketched) is yet another copy of the abstract board. However, unlike the four copies of the abstract board introduced above this copy would not be shown explicitly. It is used all the way through generation as a depository of pieces of the ABG with their proper locations at the current state. A sample copy could be visualized using Fig. 2, if we would remove all the trajectories and leave only pieces, Black and White circles, with their locations, 6, 5, 8, 11 and 1, 13, 10, respectively. Here we assume that every piece carries a reachability tag in such a way that if a grammar checks location of a piece, e.g., location 1, it would know that piece p_0 is located at 1 and its reachability would be available.

The construction stream assumed that it will produce one terminal symbol at a time employing right-linear productions, $\boldsymbol{A} \to \boldsymbol{tA}$. The stream produced the first terminal symbol $\boldsymbol{t}(p_0, \boldsymbol{a}(1)\boldsymbol{a}(2)\boldsymbol{a}(3)\boldsymbol{a}(4)\boldsymbol{a}(5), 5)$ with three parameters. The main trajectory of the zone $\boldsymbol{a}(1)\boldsymbol{a}(2)\boldsymbol{a}(3)\boldsymbol{a}(4)\boldsymbol{a}(5)$ is the value of the second parameter, Table 4, line (2). The key part of this generation step is the nonterminal $\boldsymbol{A}(u, v, w)$ as well as the current values of *TIME* and *NEXTTIME*, Table 4, line (3).

The v-board includes markings (small white circles) at the locations 2, 3, 4 and 5 of the main trajectory; the *TIME*-board includes values of the allotted time assigned to the same locations. These locations should be considered as potential end points for the first negation trajectories. This means that v and *TIME* store information for generating first negation trajectories on the following generation steps. Indeed, on the next step, the stream produced another terminal symbol $\boldsymbol{t}(q_2, \boldsymbol{a}(8)\boldsymbol{a}(9)\boldsymbol{a}(4), 4)$ with the first negation trajectory, Table 4, line (4). The nonterminal symbol $\boldsymbol{A}(u, v, w)$ produced on this step includes the w-board where location 9 of the trajectory $\boldsymbol{a}(8)\boldsymbol{a}(9)\boldsymbol{a}(4)$ is marked (with small white circle), Table 4, line (5). The *NEXTTIME*-board includes value of the allotted time at the same location—it is 3. This means that w and *NEXTTIME* store information for generating the second negation trajectories on the following generation steps. The v-board and the *TIME*-board are left unchanged. Note, that this generation step should not be considered as the second step as it is shown in Table 4. The construction stream produced a sketch of the generation, i.e., the major steps, while several intermediate steps could have been skipped. The lines (6) and (7), Table 4, show generation of the terminal symbol $\boldsymbol{t}(q_3, \boldsymbol{a}(6)\boldsymbol{a}(7)\boldsymbol{a}(4), 4)$, which completes generation of the first negation trajectories. The next generation step should pass information about the end points and the allotted time for generating the second negation trajectories. This is shown in lines (8) and (9), Table 4, where no new terminal symbols were generated. However, the previous values of the screens w-board and *NEXTTIME*-board were assigned to v-board and

TIME-board, respectively, while the *w*-board and *NEXTTIME*-board were cleaned after that. This way, the same productions, which were utilized for generating first negation trajectories, could be reused for generating second negations. After generating second negations in lines (10) through (15), Table 4, the same procedure was utilized in lines (16)–(18) to pass information about potential third negation trajectories to *v*-board and *TIME*-board for reusing the same productions for the third negation. This process stopped at the third negation because generation of those third negation trajectories did not produce information for generating higher negation trajectories. This happened because the third negation trajectories were too short (just $\boldsymbol{a}(10)\boldsymbol{a}(12)$ of length 1, Fig. 3) to leave any marks in *w*-board and *NEXTTIME*-board, Table 4, line (21). Emptiness of *w* after completion of generating the third negation trajectories was considered as a stop signal for completion of the whole generation sketch. This sketch demonstrated a possibility of generating a zone as a rigid network of trajectories with allotted time distribution.

21 The Second Expression Stream: A Sketch of the Grammar of Zones

The AD invoked the construction stream again. The purpose was to refine the example of zone generation (Sect. 20) and sketch the general Grammar of Zones (Table 5) employing this refinement. The construction stream was tasked to sketch the Grammar at the top level, i.e., to construct productions with terminals and nonterminals including required parameter lists and functions. The sketch should have also included conditions of applicability of the productions as well as lists of transitions in case of success and failure, Table 3. The intent was to introduce the major grammar components in the form of scheduled thought experiments defined as declarative statements without algorithms implementing them, Sect. 13. The AD assumed that those algorithms should be developed during the future invocations of the construction stream via execution of the postponed experiments.

Production 1 of the Grammar (introduced by the stream) is intended to replace the start symbol $\boldsymbol{S}$ by the nonterminal $\boldsymbol{A}$ with the same list of parameters, Table 5. The boards *v*, *w*, *TIME* and *NEXTTIME* are implemented as n-vectors, where n is the number of locations of the abstract board X.

The set of productions 2_i (Table 5) is intended to implement lines (2) and (3) of the original generation sketch, Table 4. Function $h_i^o(u)$ generates the bundle of main trajectories from location 1 to location 5 for the piece p_0 and picks trajectory number *i* of this bundle. Function *g* is introduced as yet another copy of the abstract board, a *g*-board. When applied to the $h_i^o(u)$ main trajectory this trajectory is superimposed over the *g*-board, and the *g*-board locations that match the trajectory are marked with 1. Evaluating the second parameter of nonterminal $\boldsymbol{A}$ as $g(h_i^o(u), w)$ means assignment of marks of the *g*-board to the *v*-board, Table 4, line (3), and Table 5. The locations of the *TIME*-board are modified by direct assignment of new values.

Table 5 Sketch of the Grammar of Zones $\mathbf{G_Z}$

L	Q	Kernel, π_k ($\forall z \in X$)		π_n ($\forall z \in X$)	F_T	F_F
1	Q_1	$S(u, v, w) \rightarrow$	$A(u, v, w)$		*two*	∅
2_i	Q_2	$A(u, v, w) \rightarrow$	$t(h_i^o(u), l_o+1)$ $A((0, 0, 0), g(h_i^o(u), w), zero)$	$TIME(z) = DIST(z, h_i^o(u))$	3	∅
3	Q_3	$A(u, v, w) \rightarrow$	$A(f(u, v), v, w)$	$NEXTTIME(z) = init(u, NEXTTIME(z))$	*four*	5
4_j	Q_4	$A(u, v, w) \rightarrow$	$t(h_j(u), TIME(y)))$ $A(u, v, g(h_j(u), w))$	$NEXTTIME(z) = ALPHA(z, h_j(u), TIME(y) - l + 1)$	3	3
5	Q_5	$A(u, v, w) \rightarrow$	$A((0, 0, 0), w, zero)$	$TIME(z) = NEXTTIME(z)$	3	6
6	Q_6	$A(u, v, w) \rightarrow e$			∅	∅

$f(u, v)$, a 3-vector (f_x, f_y, f_l), together with Q_3, Q_4 and Q_5 form a board scanner. Repetitive applications of productions 3 and 4_j lead to scanning the entire abstract board in search for the potential trajectories of certain negation. When the scan is finished, Q_5 of production 5 checks if the new scan for the higher negation trajectories should start. If not, generation is finished, and the grammar transitions to production 6.

$h_i^o(u)$ and $h_j(u)$ generate a bundle of trajectories with input of u and pick one trajectory of this bundle. $h_i^o(u)$ generates a main trajectory, $h_j(u)$ generates a negation trajectory.

g is a copy of the abstract board, the g-board. When function g is applied to the trajectory $h_i^o(u)$ (or $h_j(u)$) this trajectory is superimposed over the g-board, and those locations are marked with 1.

TIME is a copy of the abstract board, the *TIME*-board, whose locations (that match the locations of the newly generated main trajectory) are modified with time distribution values. This modification is performed by the function *DIST*.

NEXTTIME is a copy of the abstract board, the *NEXTTIME*-board, whose locations (that match the locations of the newly generated negation trajectory) are modified with time distribution values, $TIME(y) - l + 1$. This modification is performed by the function *ALPHA*. *NEXTTIME* is reinitialized employing function *init*.

At the beginning of generation:

$u = (x_o, y_o, l_o)$, $w = zero$ (n-vector of zeros), $v = zero$, $x_o \in X$, $y_o \in X$, $l_o \in \mathbf{Z}_+$, $p_o \in P$, $TIME(z) = 2n$, $NEXTTIME(z) = 2n$ for all z from X, n is the number of locations of the abstract board X, $|X| = n$.

This assignment modifies only those locations that match the locations of the newly generated main trajectory (excluding start). This modification is performed by the function *DIST*. The development of the algorithms implementing those functions was postponed until the future runs of the construction stream.

Production 3 (Table 5) replaces $\boldsymbol{A}(u, v, w)$ with $\boldsymbol{A}(f(u, v), v, w)$ in the way that each application of this production increments f_x or f_y of the 3-vector function $f(u, v)$ to find the start x and the end y of a potential negation trajectory. The length of such trajectory is limited by the value of $l = TIME(\mathrm{y})$. Function f, combined with multiple cyclic applications of productions 3 and 4_j, implements the scan through all locations of the abstract board X to register those, which could serve as the start and the end for potential negation trajectories. This loop should be terminated when the board is scanned and all the trajectories of certain negation were generated. Predicate $\boldsymbol{Q}_3$ should implement this termination by failing production 3 and transitioning the grammar to production 5 (Table 5).

Every production 4_j (Table 5) operates analogously to productions 2_i. First successful application of this production implements lines (4) and (5), second—(6) and (7), Table 4. This completes the first board scan and, thus, completes generation of the first negation trajectories. On the second scan, the first successful application of production 4_j implements lines (10), (11), and (12), the second application implements lines (13), (14), and (15), Table 4, which completes the second scan and generation of the second negation trajectories for this zone. The only successful application of production 4_j for generating one third negation trajectory $\boldsymbol{a}(10)\boldsymbol{a}(12)$ during the third scan is demonstrated in lines (19), (20), and (21).

Every successful application of production 4_j is separated by a series of pairs, successful application of production 3 and failed attempt to apply production 4_j. This way the grammar scans the abstract board while searching for the start and the end of the next trajectory of certain negation. Function $h_j(u)$ generates the bundle of the first negation trajectories. In case of the original sketch, these are trajectories from location 8 to 4 for the piece q_2 and picks trajectory $\boldsymbol{a}(8)\boldsymbol{a}(9)\boldsymbol{a}(4)$, which is number j of this bundle, Table 4, line (4). Next, it generates $\boldsymbol{a}(6)\boldsymbol{a}(7)\boldsymbol{a}(4)$, Table 4, line (6), and so on. Function g is the same as in production 2_i. When it is applied to the $h_j(u)$ trajectory this trajectory is superimposed over the g-board, and the g-board locations that match the trajectory are marked with 1. Evaluating the third parameter of nonterminal $\boldsymbol{A}$ as $g(h_j(u), w)$ means assignment of the g-board to the w-board, Table 4, line (5), and Table 5. The locations of the *NEXTTIME*-board are modified by direct assignment of the time distribution values, $TIME(\mathrm{y}) - l + 1$. This assignment modifies only those locations that match the locations of the newly generated trajectory (excluding its start and end). This modification is performed by the function *ALPHA*. The development of the algorithms implementing those functions was postponed until the future runs of the construction stream (Sect. 22).

Production 5 (Table 5) allows the grammar to switch to the next negation, to generate symbols related to trajectories of the second negation. The second parameter, the v-board, of the nonterminal $\boldsymbol{A}((0, 0, 0), w, zero)$ assumes the value of the w-board, while the *NEXTTIME*-board is explicitly assigned to the *TIME*-board in the section $\boldsymbol{\pi}_n$ of this production, Table 5. These two assignments pass information about the end points and the allotted time for generating second negation trajectories stored in w and *NEXTTIME* on the previous steps. Production 5 performs generation step shown in the sketch in lines (8) and (9), Table 4. This way, the second negation trajectories are generated employing the same series of productions 3, 4_j, and 5 as the first negation trajectories. All the trajectories of higher negations are generated employing the same series. Generation is stopped by applying production 6 after production 5 fails and predicate $\boldsymbol{Q}_5$ shows that no more trajectories of higher negation exist.

Functions $f, g, h, h^o, init, ALPHA, DIST$ were introduced in the form of scheduled thought experiments, i.e., as declarative statements without algorithms implementing them, Table 5, bottom. The AD assumed that those algorithms should be developed by the next invocation of the construction stream, Sect. 22.

22 The Third Expression Stream: Complete Grammar of Zones

The construction stream initiated execution of the sketch of the grammar for the zone shown in Fig. 3, and Table 4, with the purpose of complete execution of the experiments scheduled earlier, Sect. 21. It was expected that those experiments would result in the complete Grammar of Zones, Tables 6 and 7.

We will demonstrate the experiment that led to construction of the function $f(u, v)$ assuming the functions and predicates involved in productions 1 and 2_i have already been constructed. The stream expects that the first scan of the abstract board will lead to two successful applications of productions 4_j that generated two first negation trajectories $\boldsymbol{a}(8)\boldsymbol{a}(9)\boldsymbol{a}(4)$ and $\boldsymbol{a}(6)\boldsymbol{a}(7)\boldsymbol{a}(4)$, Table 4. For convenient visualization, the stream assumed that the board has rectangular shape and the board locations are enumerated similarly to the chessboard but in a linear fashion, each row from left to right. Consequently, the scan will start from location in the bottom left corner and continue to the right through the end of this row, then it will return to the left and continue through the end of the second row, etc. To achieve successive movement along the v-board, one location at a time, the stream made the first step in constructing function $f(u, v)$ as follows:

$$f\ (u, v) = (1, \mathrm{y} + 1, l)\ .$$

Note that values 1 and y + 1 here are ordinal numbers of the locations of the abstract board, i.e., v-board, *TIME*-board, etc. They should not be confused with symbols 1, 2, 3, 4, etc. in Fig. 3 and Table 4, which are the identifiers of the specific locations of trajectories of the same board. The stream found that location marked as location 2 (in Fig. 3) is the first location (hit by the scan) of the v-board with non-zero value. The ordinal number of this location is y + 1, thus, $v_{\mathrm{y}+1} = 1$, Table 4, line (3). The stream considered this location as the end of potential first negation trajectories. The third component l of function $f\ (u, v)$ should yield the length of those potential

Table 6 Grammar of Zones $\mathbf{G_Z}$

L	Q	Kernel, π_k ($\forall z \in X$)	π_n ($\forall z \in X$)	F_T	F_F
1	Q_1	$S(u, v, w) \to A(u, v, w)$		*two*	$\varnothing$
2_i	Q_2	$A(u, v, w) \to t(h_i^o(u), l_o+1)\ A((0, 0, 0), g(h_i^o(u), w), zero)$	$TIME(z) = DIST(z, h_i^o(u))$	3	$\varnothing$
3	Q_3	$A(u, v, w) \to A(f(u, v), v, w)$	$NEXTTIME(z) = init(u, NEXTTIME(z))$	*four*	5
4_j	Q_4	$A(u, v, w) \to t(h_j(u), TIME(y)))\ A(u, v, g(h_j(u), w))$	$NEXTTIME(z) = ALPHA(z, h_j(u), TIME(y) - l + 1)$	3	3
5	Q_5	$A(u, v, w) \to A((0, 0, 0), w, zero)$	$TIME(z) = NEXTTIME(z)$	3	6
6	Q_6	$A(u, v, w) \to e$		$\varnothing$	$\varnothing$

$V_T = \{t\}$, $V_N = \{S, A\}$,

V_{PR}

$Pred = \{Q_1, Q_2, Q_3, Q_4, Q_5, Q_6\}$

$Q_1(u) = (ON(p_o) = x) \wedge (MAP_{x,p_o}(y) \leq l \leq l_o) \wedge (\exists q\ ((ON(q) = y) \wedge (OPPOSE(p_o, q))))$

$Q_2(u) = T$

$Q_3(u) = (x \neq n) \vee (y \neq n)$

$Q_4(u) = (\exists p\ ((ON(p) = x) \wedge (l > 0) \wedge (x \neq x_o) \wedge (x \neq y_o)) \wedge ((\neg OPPOSE(p_o, p) \wedge (MAP_{x,p}(y) = 1)) \vee (OPPOSE(p_o, p) \wedge (MAP_{x,p}(y) \leq l)))$

$Q_5(w) = (w \neq zero)$

$Q_6 = T$

$Var = \{x, y, l, \tau, \theta, v_1, v_2, ..., v_n, w_1, w_2, ..., w_n\}$;

for the sake of brevity: $u = (x, y, l)$, $v = (v_1, v_2, ..., v_n)$, $w = (w_1, w_2, ..., w_n)$, $zero = (0, 0,..., 0)$;

$Con = \{x_o, y_o, l_o, p_o\}$;

$Func = Fcon \cup Fvar$;

$Fcon = \{f_x, f_y, f_l, g_1, g_2, ..., g_n, h_1, h_2, ..., h_M, h_1^o, h_2^o, ..., h_M^o, DIST, init, ALPHA\}$,

$f = (f_x, f_y, f_l)$, $g = (g_{x_1}, g_{x_2}, ..., g_{x_n})$,

$M = |\mathbf{L}_t^{l_o}(S)|$ is the number of trajectories in $\mathbf{L}_t^{l_o}(S)$;

$Fvar = \{x_o, y_o, l_o, p_o, TIME, NEXTTIME\}$

$E = \mathbf{Z}_+ \cup X \cup P \cup \mathbf{L}_t^{l_o}(S)$ is the subject domain;

Parm: $S \to Var$, $A \to \{u, v, w\}$, $t \to \{p, \tau, \theta\}$;

$L = \{1, 3, 5, 6\} \cup two \cup four$, $two = \{2_1, 2_2, ..., 2_M\}$, $four = \{4_1, 4_2, ..., 4_M\}$.

At the beginning of generation:

$u = (x_o, y_o, l_o)$, $w = zero$, $v = zero$, $x_o \in X$, $y_o \in X$, $l_o \in \mathbf{Z}_+$, $p_o \in P$,

$TIME(z) = 2n$, $NEXTTIME(z) = 2n$ for all z from X.

Table 7 Definition of functions of the Grammar of Zones $\mathbf{G_Z}$

$D(init) = X \times X \times \mathbf{Z_+} \times \mathbf{Z_+}$

$$init(u, r) = \begin{cases} 2n, & \text{if } u = (0,0,0), \\ r, & \text{if } u \neq (0,0,0). \end{cases}$$

$D(f) = (X \times X \times \mathbf{Z_+} \cup \{0, 0, 0\}) \times \mathbf{Z_+}^n$

$$f(u, v) = \begin{cases} (x+1, y, l), & \text{if}((x \neq n) \wedge (l > 0)) \vee ((y = n) \wedge (l \leq 0)) \\ (1, y+1, TIME(y+1) \times v_{y+1}), & \text{if } (x = n) \vee ((l \leq 0) \wedge (y \neq n)). \end{cases}$$

$D(DIST) = X \times P \times \mathbf{L_t}^{lo}(S)$.

Let $t_o \in L_t^{lo}(S)$, $t_o = \boldsymbol{a}(z_o)\boldsymbol{a}(z_1)\ldots\boldsymbol{a}(z_m)$, $t_o \in t_{p_o}(z_o, z_m, m)$;

If $((z_m = y_o) \wedge (p = p_o) \wedge (\exists\, k\, (1 \leq k \leq m) \wedge (x = z_k))) \vee$

$(((z_m \neq y_o) \vee (p \neq p_o)) \wedge (\exists\, k\, (1 \leq k \leq m - 1) \wedge (x = z_k)))$

then $DIST(x, p_o, t_o) = k+1$

else $DIST(x, p_o, t_o) = 2n$

$D(ALPHA) = X \times P \times \mathbf{L_t}^{lo}(S) \times \mathbf{Z_+}$

$$ALPHA(x, p_o, t_o, k) = \begin{cases} max\,(NEXTTIME(x), k), & \text{if}(DIST(x, p_o, t_o) \neq 2n) \\ & \wedge (NEXTTIME(x) \neq 2n); \\ k, & \text{if } DIST(x, p_o, t_o) \neq 2n) \\ & \wedge (NEXTTIME(x) = 2n); \\ NEXTTIME(x), & \text{if } DIST(x, p_o, t_o) = 2n). \end{cases}$$

$D(g_r) = P \times \mathbf{L_t}^{lo}(S) \times \mathbf{Z_+}^n$, $r \in X$.

$$g_r(p_o, t_o, w) = \begin{cases} 1, & \text{if } DIST(r, p_o, t_o) < 2n, \\ w_r, & \text{if } DIST(r, p_o, t_o) = 2n. \end{cases}$$

$D\,(h_i{}^o) = X \times X \times \mathbf{Z_+}$; Let $TRACKS_{p_o} = \{p_o\} \times (\bigcup_{1 \leq k \leq l} L[\mathbf{G_t}^{(2)}(x, y, k, p_o)]$

If $TRACKS_{p_o} = \varnothing$

then $h_i^o(u) = \varepsilon$

else $TRACKS_{p_o} = \{(p_o, t_1), (p_o, t_2), \ldots, (p_o, t_b)\}, (b \leq M)$ **and**

$$h_i^o(u) = \begin{cases} (p_o, t_i), & \textbf{if } i \leq b, \\ (p_o, t_b), & \textbf{if } i > b. \end{cases}$$

$D(h_i) = X \times X \times \mathbf{Z_+}$; Let $TRACKS_p = \{p\} \times (\bigcup_{1 \leq k \leq l} L[\mathbf{G_t}^{(2)}(x, y, k, p)]$

If $TRACKS_p = \varnothing$

then $h_i(u) = \varepsilon$

else $TRACKS_p = \{(p_1, t_1), (p_1, t_2), \ldots, (p_m, t_m)\}$, $(m \leq M)$ **and**

$$h_i(u) = \begin{cases} (p_i, t_i), & \textbf{if } i \leq m, \\ (p_m, t_m), & \textbf{if } i > m. \end{cases}$$

Trajectories should not be embedded (as sub-trajectories) in the trajectories of the same negation.

trajectories. This value could be obtained from the same location with ordinal number y + 1 of the *TIME*-board, $TIME(y + 1) = 2$, Table 4, line (3). Based on this simple reasoning the stream refined further function $f(u, v)$ by constructing its first branch as follows (Table 7):

$$f\ (u, v) = (1, y + 1, TIME\ (y + 1) \times v_{y+1})$$

During the scan before hitting location 2 of the *v*-board while incrementing the second parameter y + 1 and having the third parameter $TIME(y + 1) \times v_{y+1} = 0$ (because $v_{y+1} = 0$), the grammar considers application of production 3 successful. Consequently, after every such application, it transitions to production 4_j, Table 5. This production should generate a trajectory ending at location with symbol 2 of the *v*-board employing $f(u, v) = (1, y + 1, 0)$ as an input for the grammar of trajectories. The stream introduced $\boldsymbol{Q_4}$ (Table 6) as a necessary condition for applying the grammar of trajectories inside a zone. These are standard conditions for a zone construction discussed in Sects. 17 and 19. In particular, they require presence of a piece p at the start location x of the trajectory, i.e., $ON(p) = x$, which is currently location with ordinal number 1 (not symbol 1), and it is empty. In addition, these conditions require that the length of the trajectory $l > 0$, which is currently zero. This all means that production 4_j fails. It will continue failing until the first and the second components of $f(u, v)$ hit locations with symbols 8 and 4, the start and the end of the first negation trajectory, Table 4. The stream has already constructed $f(u, v)$ to increment y in a search (scan) for the end of potential first negation trajectories. Now it has to make further refinement to start incrementing x (when the end location has been found). The stream refined further function $f(u, v)$ by constructing its second branch as follows (Table 7):

$$f\ (u, v) = (x + 1, y, l)$$

The complete refinement is shown in Table 7. It includes conditions of applicability, the domains, of each of the branches of this function, i.e., when to increment y, when to increment x, and when to initialize one or both components.

The second scan led to two successful applications of productions 4_j that generated two second negation trajectories $\boldsymbol{a}(11)\boldsymbol{a}(12)\boldsymbol{a}(9)$ and $\boldsymbol{a}(13)\boldsymbol{a}(9)$, Table 4, etc. The construction stream completed execution of all other thought experiments and this way completed construction of the Grammar of Zones, Tables 6 and 7.

23 Concluding Remarks

The Grammar of Zones was originally developed in 1981 by the author with intent to provide a scientific foundation for the performance of the chess program PIONEER [27, 38]. The results demonstrated by this program in solving chess endgames almost without search were unbelievable to the scientific community [3, 38]. Until 1981, the only means for explaining the algorithm of the program PIONEER were snapshots of the Pictorial LG. Most importantly, those snapshots were used consistently in the development of PIONEER [2, 3]. In spite of their convenience, they certainly could not be considered as an acceptable scientific notation. They were suggested by Professor Botvinnik and, obviously, they reflected well our thinking about the algorithm being developed. We had no idea that those snapshots reflected the Pictorial LG visual stream based directly on the Primary Language. When developing the Linguistic LG and the Grammar of Zones we could not imagine that we were reflecting the expression stream mapping the primary science to the conventional one. At the same time, we understood that the Pictorial LG snapshots were completely rejected by the scientific community. This kind of rejection of the reflections of the primary science and, specifically, the reflections of the visual streams utilized by the AD is very typical for all the discoveries. They are considered as non-rigorous toys that do not deserve to be mentioned in the scientific publications. It appears that those toys are the only means to discoveries.

In Sects. 20–22, the construction stream of the AD utilized the expression stream three times. The construction stream morphed gradually the expression stream until this morphing led to constructing the complete Grammar of Zones and completion of formal generation of the string shown in Table 1. As was the case in all the discoveries, we investigated so far mosaic reasoning (Sect. 12) played the crucial role in this one. Out of multiple matching rules used by the observation and construction streams for this discovery, we would emphasize the transformation rule and the generator. In here, the object to be constructed was a string of symbols as is typical for an expression stream. In addition, the AD constructed a mathematical tool, the grammar, for generating those strings. It appears that the major part of this construction was the construction of the transformation rule. The generator here was a trajectory reflected as a line in the Pictorial LG, Fig. 2. The transformation rule here consisted of several scans of the abstract board for generating negation trajectories, one negation per scan. Discovery of the details of this rule and explicit reflection of those details in the Grammar of Zones, productions 3, 4 and 5 (Table 6), were the most difficult parts of the construction. The grammar itself is the implementation of the transformation rule for this discovery. It is likely that constructing transformation rule for the mosaic reasoning has universal nature. It works not only for the internal discovery streams for constructing visual models of natural and artificial objects but for the expression streams as well, for constructing strings of symbols reflecting abstract mathematical objects.

The next remark is related to the value of this very first construction of the expression stream that maps a piece of the primary science to the conventional one. In this case, the mapping is sufficiently precise. It establishes one-to-one correspondence between the "congruent" classes of zones in the Pictorial LG and their formal representations in the Linguistic LG. A simple reverse impression stream maps those representations back into the Pictorial LG "movies." It would be interesting to construct those impression streams for other discoveries in order to reveal further details of the primary science. Maybe, this is the main road ahead.

In Sects. 3 and 4, we emphasized the issues of visibility and self-awareness of the visual streams. We pointed that a stream knows about itself and does not require visual pattern recognition. This observation was around for several years. Interestingly, the first construction of the expression stream brought the first explicit confirmation of this idea. It was necessary to construct a generating grammar, the Grammar of Zones, in order to complete construction of the expression stream that maps the Pictorial LG to the Linguistic LG. The Grammar of Zones, in this case, serves as the tool, which implements self-awareness of this expression stream.

References

1. Botvinnik, M.: Chess, Computers, and Long-Range Planning. Springer, New York (1970)
2. Botvinnik, M.: Blok-skema algorithma igry v shahmaty (A Flow-Chart of the Algorithm for Playing Chess). Sovetskoe Radio (1972) (in Russian)
3. Botvinnik, M.: Computers in Chess: Solving Inexact Search Problems. Springer (1984)
4. Brown, J.: The Laboratory of the Mind: Thought Experiments in the Natural Sciences, 2nd edn. Taylor & Francis Group, Routledge, New York (2011)
5. Chargaff, E.: Structure and function of nucleic acids as cell constituents. Fed. Proc. **10**, 654–659 (1951)
6. Chomsky, N.: Syntactic Structures. Mouton de Gruyter (1957)
7. Crick, F.H.C., Watson, J.D.: The complementary structure of deoxyribonucleic acid. Proc. R. Soc. Lond. A **223**, 80–96 (1954)
8. Darwin, C.: The Origin of Species by Means of Natural Selection. Penguin, London (1968)
9. Deheaene, S.: A few steps toward a science of mental life. Mind Brain Educ. **1**(1), 28–47 (2007)
10. Deheaene, S.: Edge In Paris, Talk at the Reality Club: Signatures of Consciousness (2009). http://edge.org/3rd_culture/dehaene09/dehaene09_index.html
11. Ding, N., Melloni, L., Zhang, H., Tian, X., Poeppel, D.: Cortical tracking of hierarchical linguistic structures in connected speech. Nat. Neurosci. **19**, 158–164 (2016). https://doi.org/10.1038/nn.4186
12. Einstein, A.: On the electrodynamics of moving bodies. Annalen der Physik **17**, 891 (1905) (in German)
13. Einstein, A.: Autobiographical Notes. Open Court, La Salle, IL (1991)
14. Foppl, A.: Introduction to Maxwell's Theory of Electricity. B. G. Teubner, Leipzig, in German (1894)
15. Galilei, G.: Dialogue Concerning the Two Chief World Systems (trans.: Stillman Drake) (1632)
16. Gleick, J.: Genius: The Life and Science of Richard Feynman. Pantheon Books, a division of Random House, New York (1992)
17. Hadamard, J.: The Mathematician's Mind: The Psychology of Invention in the Mathematical Field. Princeton University Press, Princeton, NJ (1996)
18. Iacoboni, M.: Mirroring People: The New Science of How We Connect with Others. Farrar, Straus and Giroux, New York (2008)
19. Klein, G.: Seeing What Others Don't: The Remarkable Ways to Gain Insights. PublicAffairs, a Member of the Perseus Books Group, New York (2013)
20. Kosslyn, S., Thompson, W., Kim, I., Alpert, N.: Representations of mental images in primary visual cortex. Nature **378**, 496–498 (1995)
21. Malthus, T.: An Essay on the Principle of Population. London, Printed for J. Johnson, in St. Paul's Church-Yard (1798)
22. Miller, A.: Insights of Genius: Imagery and Creativity in Science and Art. Copernicus, an imprint of Springer (1996)

23. Nasar, S.: A Beautiful Mind. Touchstone, New York, NY (2001)
24. Nersessian, N.: Conceptual change: creativity, cognition, and culture. In: Meheus, J., Nicles, T. (eds.) Models of Discovery and Creativity, pp. 127–166. Springer (2009)
25. Pauling, L., Corey, R.B.: A proposed structure for the nucleic acids. Nature **171**, 346–359 (1953); Proc. US Nat. Acad. Sci. **39**, 84–97 (1953)
26. Stilman, B.: Formation of the set of trajectory bundles. In: Botvinnik, M. M. (ed.) Appendix 1 to: On the Cybernetic Goal of Games, pp. 70–77. Soviet Radio, Moscow (1975) (in Russian)
27. Stilman, B.: Ierarhia formalnikh grammatik dla reshenia prebornikh zadach (Hierarchy of Formal Grammars for Solving Search Problems). Technical Report, 105 pp., VNIIE, Moscow (1981) (in Russian)
28. Stilman, B.: A formal language for hierarchical systems control. Int. J. Lang. Des. **1**(4), 333–356 (1993)
29. Stilman, B.: A linguistic approach to geometric reasoning. Int. J. Comput. Math. Appl. **26**(7), 29–58 (1993)
30. Stilman, B.: Network languages for complex systems. Int. J. Comput. Math. Appl. **26**(8), 51–80 (1993)
31. Stilman, B.: Linguistic geometry for control systems design. Int. J. Comput. Appl. **1**(2), 89–110 (1994)
32. Stilman, B.: Translations of network languages. Int. J. Comput. Math. Appl. **27**(2), 65–98 (1994)
33. Stilman, B.: Deep search in linguistic geometry. In: Symposium on Linguistic Geometry and Semantic Control, Proceedings of the First World Congress on Intelligent Manufacturing: Processes and Systems, Mayaguez, Puerto Rico, pp. 868–879 (1995)
34. Stilman, B.: A linguistic geometry for 3D strategic planning. In: Proceedings of the 1995 Goddard Conference on Space Applications of Artificial Intelligence and Emerging Information Technologies, NASA Goddard Space Flight Center, Greenbelt, MD, USA, pp. 279–295 (1995)
35. Stilman, B.: Linguistic geometry tools generate optimal solutions. In: Proceedings of the 4th International Conference on Conceptual Structures—ICCS'96, Sydney, Australia, pp. 75–99 (1996)
36. Stilman, B.: Managing search complexity in linguistic geometry. IEEE Trans. Syst. Man Cybern. **27**(6), 978–998 (1997)
37. Stilman, B.: Network languages for concurrent multi-agent systems. Int. J. Comput. Math. Appl. **34**(1), 103–136 (1997)
38. Stilman, B.: Linguistic Geometry: From Search to Construction, 416 pp. Kluwer Academic Publishers (now Springer) (2000)
39. Stilman, B.: linguistic geometry and evolution of intelligence. ISAST Trans. Comput. Intell. Syst. **3**(2), 23–37 (2011)
40. Stilman, B.: Thought experiments in linguistic geometry. In: Proceedings of the 3rd International Conference on Advanced Cognitive Technologies and Applications—COGNITIVE'2011, pp. 77–83, Rome, Italy (2011)
41. Stilman, B.: Discovering the discovery of linguistic geometry. Int. J. Mach. Learn. Cybern. **4**(6), 575–594 (2012). https://doi.org/10.1007/s13042-012-0114-8
42. Stilman, B.: Discovering the discovery of the no-search approach. Int. J. Mach. Learn. Cybern. 27 pp. (2012). https://doi.org/10.1007/s13042-012-0127-3
43. Stilman, B.: Discovering the discovery of the hierarchy of formal languages. Int. J. Mach. Learn. Cybern. 25 pp. (2012). https://doi.org/10.1007/s13042-012-0146-0
44. Stilman, B.: Visual reasoning for discoveries. Int. J. Mach. Learn. Cybern. 23 pp. (2013). https://doi.org/10.1007/s13042-013-0189-x
45. Stilman, B.: Mosaic reasoning for discoveries. J. Artif. Intell. Soft Comput. Res. **3**(3), 147–173 (2013)
46. Stilman, B.: Proximity reasoning for discoveries. Int. J. Mach. Learn. Cybern. 31 pp. (2014). https://doi.org/10.1007/s13042-014-0249-x
47. Stilman, B.: The algorithm of discovery: making discoveries on demand. In: Kunifuji, S., et al. (eds.) Advances in Intelligent Systems and Computing, pp. 1–16. Springer International Publishing, Switzerland (2016). https://doi.org/10.1007/978-3-319-27478-2_1

48. Stilman, B.: Discoveries on demand. Int. J. Des. Nat. Ecodyn. **11**(4), 495–507 (2016)
49. Stilman, B., Aldossary, M., The algorithm of discovery: programming the mosaic of the shortest trajectories. In: Proceedings of the IEEE International Conference on Intelligent Systems, Sofia, Bulgaria, pp. 346–351 (2016)
50. Stilman, B., Alharbi, N.: Towards the algorithm of discovery: mosaic reasoning for the discovery of the genetic code. In: Proceedings of the IEEE International Conference on Intelligent Systems, Sofia, Bulgaria, pp. 353–363 (2016)
51. Stilman, B., Yakhnis, V., Umanskiy, O.: Winning strategies for robotic wars: defense applications of linguistic geometry. Artif. Life Robot. **4**(3) (2000)
52. Stilman, B., Yakhnis, V., Umanskiy, O.: Knowledge acquisition and strategy generation with LG wargaming tools. Int. J. Comput. Intell. Appl. **2**(4), 385–409 (2002)
53. Stilman, B., Yakhnis, V., Umanskiy, O.: Strategies in Large Scale Problems, Chapter 3.3. In: Kott and McEneaney (eds.) Adversarial Reasoning: Computational Approaches to Reading the Opponent's Mind, pp. 251–285 (2007)
54. Stilman, B., Yakhnis, V., Umanskiy, O.: Linguistic geometry: the age of maturity. J. Adv. Comput. Intell. Intell. Inform. **14**(6), 684–699 (2010)
55. Stilman, B., Yakhnis, V., Umanskiy, O.: Revisiting history with linguistic geometry. ISAST Trans. Comput. Intell. Syst. **2**(2), 22–38 (2010)
56. Stilman, B., Yakhnis, V., Umanskiy, O.: The primary language of ancient battles. Int. J. Mach. Learn. Cybern. **2**(3), 157–176 (2011)
57. Thomson, G.: The Inspiration of Science. Oxford University Press, London (1961)
58. Ulam, S.: Adventures of a Mathematician. Charles Scribner's Sons, New York, NY (1976)
59. Volchenkov, N.: Interpreter of context-free controlled programmed grammars with parameters in Voprosy Kibernetiky: Intellektualnie banky dannikh (Proceedings on Cybernetics: Intelligent Data Banks), pp. 147–157. USSR Academy of Sciences, Scientific Board on Complex Problem Cybernetics (1979) (in Russian)
60. Von Neumann, J.: The Computer and the Brain. Yale University Press (1958)
61. Watson, J.D.: The Double Helix: A Personal Account of the Discovery of the Structure of DNA, Scribner Classics Edition (1996). Atheneum, New York (1968)
62. Watson, J.D., Crick, F.H.C.: A structure for deoxyribose nucleic acid. Nature **171**, 737–738 (1953)
63. Watson, J.D., Crick, F.H.C.: The structure of DNA. In: Cold Spring Harbor Symposia on Quantitative Biology, vol. 18, pp. 123–131 (1953)

Intelligent Two-Level Optimization and Model Predictive Control of Degrading Plants

Mincho Hadjiski, Alexandra Grancharova and Kosta Boshnakov

1 Introduction

Model predictive control (MPC) has become the accepted methodology to solve complex control problems related to process industries [1, 2]. MPC involves the solution at each sampling instant of a finite horizon optimal control problem subject to the system dynamics, and state and input constraints. The MPC methodology is especially very powerful for the design of hierarchical multilayer systems, i.e. control systems made by a number of control algorithms working at different time scales (see [3] for a review of hierarchical MPC approaches). Multilayer structures are useful either to control plants characterized by significantly different dynamics or to use different models of the same plant with the aim to optimize a number of criteria. In [4, 5], a two-level systematic approach for integrating model based dynamic real-time optimization and control of industrial processes is developed. The rationality behind it is that in the presence of disturbances and changing plant parameters, a re-optimization at the higher level may be necessary for optimal operation. The two-level strategy distinguishes economic and control objectives and solves the problem on different time scales. The multilayer control structure can be particularly very useful to increase the long-term operational effectiveness of technological plants with degrading performance.

M. Hadjiski
Institute of Information and Communication Technologies,
Bulgarian Academy of Sciences, Acad. G. Bonchev Str., Bl. 2, 1113 Sofia, Bulgaria
e-mail: hadjiski@uctm.edu

A. Grancharova (✉) · K. Boshnakov
Department of Industrial Automation, University of Chemical Technology
and Metallurgy, Bul. Kl. Ohridski 8, 1756 Sofia, Bulgaria
e-mail: alexandra.grancharova@abv.bg

K. Boshnakov
e-mail: kosta.boshnakov@gmail.com

V. Sgurev et al. (eds.), *Learning Systems: From Theory to Practice*, Studies in Computational Intelligence 756, https://doi.org/10.1007/978-3-319-75181-8_6

Here, a two-level structure for control and optimization of degradation type of plants is proposed, which incorporates case-based reasoning for an appropriate modification of the lower level criteria and constraints. An associated two-level MPC problem is formulated, which uses plant models in the "fast" and in the "slow" time scales. In particular, the plantwide optimization of a Peirce-Smith converter is considered, which is a typical example of a plant with degrading performance. The purpose is to optimize the number of blocked tuyeres for an efficient long-term operation.

2 Case-Based Reasoning

The Case-Based Reasoning (CBR) method allows to use the retrospective experience (precedents) in the decision making process when choosing alternative solutions. At the same time, the decision support is based on structured data and formal procedures (unlike some other decision making methods with significant degree of subjectivity). The CBR approach applies the principle "similar situation/similar solution". It is widely used because of the following reasons:

- The theory of the CBR approach is well developed [6–10] and its practical applications confirm its efficacy.
- The industrial process systems usually implement distributed control systems and SCADA systems, which contain a significant amount of information. This information can be used to build the base of precedents (the case base).
- The precedents (the cases) may include all necessary indicators of the quality of the accepted solutions.
- The practical experience of many engineers-operators is embedded in the solution part of the precedent, which guarantees their participation in the decision making process.
- The CBR method does not require mathematical models, which is an advantage when applied to complex technological systems, where the mechanism of the undergoing processes is not well known.
- There exists freely available software (jColibri, myCBR) for implementing the CBR approach [11–17].

The CBR method allows the use of the old "problems-decisions" to directly solve new situations, unlike the Rule-Based Reasoning approach that require building a rule base out of the known precedents.

With the CBR approach, the cases are represented in the standard form of "problem (P) – solution (S)":

$$C = \langle P, S \rangle \tag{1}$$

The problem P is defined by the occurred situation H with its attributes h_i and their values v_{ij}:

$$P \equiv H\,(h(v)) \tag{2}$$

$$H = (h_1, h_2, .., h_n) \tag{3}$$

$$v_i = (v_{i1}, v_{i2}, \ldots, v_{in_i}), \tag{4}$$

The attribute values v_{ij} can be scalars or vectors. For the purpose of optimization and control of degrading plants it is appropriate to use the following attributes:

- h_1—**States**
- h_2—**Constraints**
- h_3—**Criteria**

The second component of each case is the solution *S*, which can be represented in the form:

$$S = \langle D, E, I \rangle \tag{5}$$

Here, *D* is the structured content of the decision, *E* is the estimate of its usefulness, and *I* contains additional information. Similar to the problem *P*, the solution *S* is defined by its attributes and their values.

For industrial processes, the decision *D* is usually represented by the attributes:

$$D = (d_1, d_2, d_3, d_4) \tag{6}$$

where d_i, $i = 1, \ldots, 4$ are:

- d_1—**Business decisions**
- d_2—**Technological decisions**
- d_3—**Decisions for the technical maintenance**
- d_4—**Decision making method being used**

The estimate *E* of decision usefulness is often characterized by six attributes:

$$E = (e_1, e_2, e_3, e_4, e_5, e_6) \tag{7}$$

where e_i, $i = 1, \ldots, 6$ contain the following information:

- e_1—**Economy of resources for current, mid-term and ling-term maintenance of the converter**
- e_2—**Profit increase due to the increased productivity**
- e_3—**Bonus for adherence to the production schedule**
- e_4—**Economy related to the decrease of unforeseen downtimes**
- e_5—**Risk estimate**
- e_6—**Additional costs for technical maintenance**
  - e_{61}—*Equipment purchase and qualification courses for the personnel*
  - e_{62}—*Use of external specialists to perform specific measurements*

The additional information I about the decision is defined as:

$$I = (i_1, i_2, i_3) \tag{8}$$

where the attributes i_1, i_2 and i_3 have the following meaning:

- i_1—**Interpretation of the decision**
- i_2—**Argumentation for taking the decision**
- i_3—**Clarifications of different issues related to the decision**

The case-based reasoning (CBR) method is based on the principle of local-global similarity [6, 8, 9, 18], where the similarity measure is defined as:

$$\text{sim}\left(H_0, H_j\right) = \sum_{j=1}^{N} w_j \,\text{sim}_j\left(h_0, h_j\right), \quad \sum_{j=1}^{N} w_j = 1 \tag{9}$$

Here, $H_0(k)$ and $h_0(k)$ are the new situation and its attributes (2)–(4), H_j and h_j are j-th known situation and its attributes in the extended case base (CB), limited to the N-number of nearest neighbours. The expression $\text{sim}_j\left(h_0, h_j\right)$ characterizes the local similarity between the attributes h_0 and h_j, and it contains the specific knowledge for the pair $(0, j)$. The weighting coefficients w_j represent the relative importance of these attributes on the global similarity between the situations H_0 and H_j. They are specified according to the empirical knowledge of the operators and can be updated by learning in the course of accumulating precedents in the case base and improving the quality of the decisions.

3 Hierarchical Control Structure for Plantwide Optimization of Degrading Plants

In the process industry it is common to design the overall control system according to the hierarchical structure shown in Fig. 1 [3].

At the higher layer, real time optimization is performed to compute the optimal operating conditions with respect to a performance index representing an economic criterion. At this stage a detailed, although static, physical nonlinear model of the system is used. At the lower layer a simpler linear dynamic model of the same system, often derived by means of identification experiments, is used to design a regulator with MPC, guaranteeing that the target values transmitted from the higher layer are attained. Also in this case, the lower level can transmit bottom-up information on constraints and performance.

Here, a two-level structure for control and optimization of degrading plants is proposed, which incorporates case-based reasoning (Fig. 2). In Fig. 2, x_1 and x_2 are the "fast" and "slow" state variables (assumed to be measurable), and u_1 and u_2 are the "fast" and "slow" control inputs. It is supposed that the dynamics of the degrading

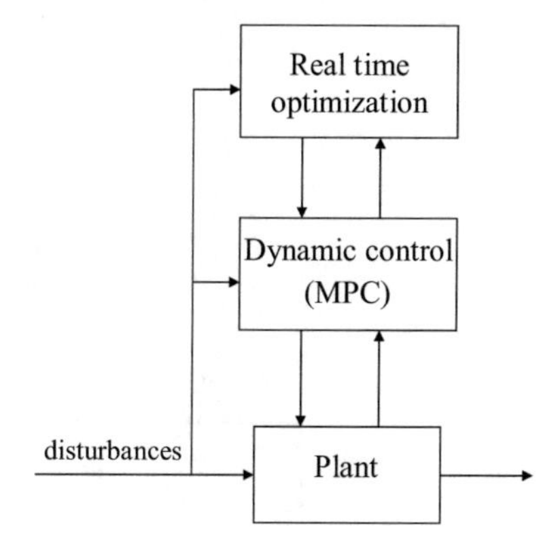

Fig. 1 Hierarchical structure for plantwide control and optimization [3]

plant is described in two time scales. The plant model M1 represents the plant behavior at the "fast" time scale and it has slowly varying parameters due to the degradation of the plant operation. The change of these parameters over the "slow" time scale is captured by the model M2 (the details are given in the next section). The high level objectives depend on the economic goals and are associated to the plant management policy. The lower level criteria and constraints are determined by a case-based reasoning (CBR) system that relies on the experts knowledge about the functioning of the particular plant (the reader is referred to [19] for the basis of the CBR systems). The CBR systems can modify the criteria and the constraints in accordance with the high level objectives and the data about plant performance/degradation. Further, the weighting coefficients for the individual criteria in the single criterion "fast" MPC problem can be determined by a rule-based reasoning system (the foundations of the rule-based systems can be found in [20]).

4 Hierarchical MPC Problem for Degrading Plants

Here, we consider a class of technological plants whose long-term functioning is considered in a multiple time scale. For simplicity, it is assumed that plant behavior is described by a discrete-time model in two time scales (two levels model):

Level 2 (Model M2):

$$\begin{aligned}
&x_2((j+1)T_2) = f_2(x_2(jT_2), u_2(jT_2), p(jT_2), x_1(jT_2))\\
&p((j+1)T_2) = g(x_2(jT_2), u_2(jT_2), p(jT_2))\\
&x_1(jT_2) = x_1(N_1T_1)\\
&j \geq 0\\
&x_2(0) = x_{2,0}
\end{aligned} \tag{10a}$$

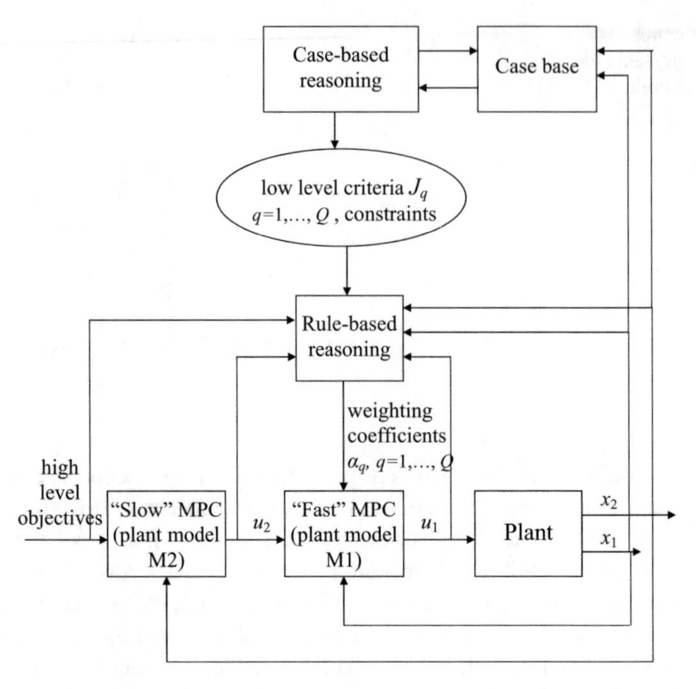

Fig. 2 Hierarchical structure for plantwide control and optimization incorporating case-based and rule-based reasonings

Level 1 (Model M1):

$$
\begin{aligned}
&x_1((i+1)T_1) = f_1(x_1(iT_1), u_1(iT_1), p(jT_2)) \\
&i = 0, 1, \ldots, N_1 \\
&x_1(0) = x_1(jT_2), \;\; j \geq 1 \\
&x_1(0) = x_{1,0}, \;\; j = 0
\end{aligned} \tag{10b}
$$

where $T_1 \in \mathbb{R}$ and $T_2 \in \mathbb{R}$ are sampling times with ratio $T_2/T_1 = r \in \mathbb{N}$, $x_1 \in \mathbb{R}^{n_1}$, $u_1 \in \mathbb{R}^{m_1}$ and $y_1 \in \mathbb{R}^{r_1}$ are the state, control input and output vectors corresponding to level 1, $x_2 \in \mathbb{R}^{n_2}$, $u_2 \in \mathbb{R}^{m_2}$ and $y_2 \in \mathbb{R}^{r_2}$ are the state, control input and output vectors corresponding to level 2, and $p \in \mathbb{R}^{n_p}$ are relatively slowly varying plant parameters that indicate its performance degrading. It is assumed that $r \geq 2$ and that the initial states $x_{1,0}$ and $x_{2,0}$ are known. From the relation between the two sampling times it follows that the model in level 1 describes the fast plant dynamics, while the model in level 2 describes the degradation in plant functioning. By omitting the sampling times, the model (10a, 10b) obtains the form:

Level 2 (Model M2):

$$\begin{aligned} &x_2(j+1) = f_2(x_2(j), u_2(j), p(j), x_1(j)) \\ &p(j+1) = g(x_2(j), u_2(j), p(j)) \\ &x_1(j) = x_1(N_1) \\ &j \geq 0 \\ &x_2(0) = x_{2,0} \end{aligned} \tag{11a}$$

Level 1 (Model M1):

$$\begin{aligned} &x_1(i+1) = f_1(x_1(i), u_1(i), p(j)) \\ &i = 0, 1, \ldots, N_1 \\ &x_1(0) = x_1(j),\ \ j \geq 1 \\ &x_1(0) = x_{1,0},\ \ j = 0 \end{aligned} \tag{11b}$$

The following constraints are imposed on the plant variables:

Level 2:

$$\begin{aligned} &x_{2,\min} \leq x_2(j) \leq x_{2,\max} \\ &u_{2,\min} \leq u_2(j) \leq u_{2,\max},\ \ j \geq 0 \end{aligned} \tag{12a}$$

Level 1:

$$\begin{aligned} &x_{1,\min}(u_2(j)) \leq x_1(i) \leq x_{1,\max}(u_2(j)) \\ &u_{1,\min}(u_2(j)) \leq u_1(i) \leq u_{1,\max}(u_2(j)),\ \ i = 0, 1, \ldots, N_1 \end{aligned} \tag{12b}$$

In this constraints definition it is assumed that the decision taken at level 2 determines the bounds in the constraints that have to be respected at level 1.

The formulated two level (cascade) MPC problem is:

Level 2 MPC ("Slow" MPC):

$$J^*_{\text{level }2}(\bar{x}_2) = \min_{U_2}\ J_{\text{level }2}(U_2, x_{2,0}) \tag{13a}$$

subject to $x_{2,\,j\,|\,j} = x_{2,0}$ and:

$$u_{2,\min} \leq u_{2,\,j+k} \leq u_{2,\max},\ \ k = 0, 1, \ldots, N_2 - 1 \tag{13b}$$

$$x_{2,\min} \leq x_{2,\,j+k\,|\,j} \leq x_{2,\max},\ \ k = 1, 2, \ldots, N_2 \tag{13c}$$

$$x_{2,\,j+k+1\,|\,j} = f_2(x_{2,\,j+k\,|\,j}, u_{2,\,j+k}, p_{j+k}, x^*_{1,\,j+k}),\ \ k = 0, 1, \ldots, N_2 - 1 \tag{13d}$$

$$p_{j+k+1} = g(x_{2,\,j+k\,|\,j}, u_{2,\,j+k}, p_{j+k}),\ \ k = 0, 1, \ldots, N_2 - 1 \tag{13e}$$

$$x^*_{1,i+s+1|i} = f_1(x^*_{1,i+s|i}, u^*_{1,i+s}, p_{j+k}),\ \ i = (j+k)r,\ \ s = 0, 1, \ldots, N_1 \tag{13f}$$

$$x^*_{1,\,j+k} = x^*_{1\,i+N_1|\,i},\ k = 1\,2, \ldots, N_2 \tag{13g}$$

$$U_1^* = [u^*_{1,i}, u^*_{1,i+1}, \ldots, u^*_{1,i+N_1}] \tag{13h}$$

Level 1 MPC ("Fast MPC"):

$$U_1^* = \arg\min_{U_1} J_{\text{level }1}(U_1, \bar{x}_1) \tag{14a}$$

$$\bar{x}_1 = x^*_{1,\,j+k},\ k \geq 1 \tag{14b}$$

$$\bar{x}_1 = x_{1,0},\ k = 0 \tag{14c}$$

$$x_{1,i|i} = \bar{x}_1,\ i = (j+k)r \tag{14d}$$

$$u_{1,\min}(U_2) \leq u_{1,i+s} \leq u_{1,\max}(U_2),\ s = 0, 1, \ldots, N_1 - 1 \tag{14e}$$

$$x_{1,\min}(U_2) \leq x_{1,i+s|i} \leq x_{1,\max}(U_2),\ s = 1, 2, \ldots, N_1 \tag{14f}$$

$$x_{1,i+s+1|i} = f_1(x_{1,i+s|i}, u_{1,i+s}, p_{j+k}),\ s = 0, 1, \ldots, N_1 \tag{14g}$$

The cost function $J_{\text{level }2}(U_2, x_{2,0})$ is defined in accordance with the high level objectives (see Fig. 2), while $J_{\text{level }1}(U_1, \bar{x}_1)$ is a low level criterion that usually has the form:

$$J_{\text{level }1}(U_1, \bar{x}_1) = \sum_{q=1}^{Q} \alpha_q J_q \tag{14h}$$

In (14h), $J_q,\ q = 1, 2, \ldots, Q$ are the individual optimality criteria determined by the CBR system (see Fig. 2) and $\alpha_q,\ q = 1, 2, \ldots, Q$ are the weighting coefficients specified by the rule-based system.

The two-level MPC problem (4)–(5) is a bilevel optimization problem, which can be solved by applying the methods for bilevel optimization [21, 22]. There are two possibilities for real-time implementation of the two-level MPC strategy:

- For plants with slow dynamics, the bilevel optimization problem (4)–(5) can be solved by PC-based online optimization;
- For plants with fast dynamics, either methods for explicit MPC may be applied [23] (for small-scale plants) or approaches to embedded MPC [24, 25] may be used. Both methods would lead to high-performance predictive controllers on a low-cost embedded hardware with limited computational resources.

5 Case Study: Case-Based Reasoning and MPC for Plantwide Optimization of a Peirce-Smith Converter

Here, it is described how the case-based reasoning approach and the MPC methodology are combined for plantwide optimization of a Peirce-Smith converter (a typical degrading plant).

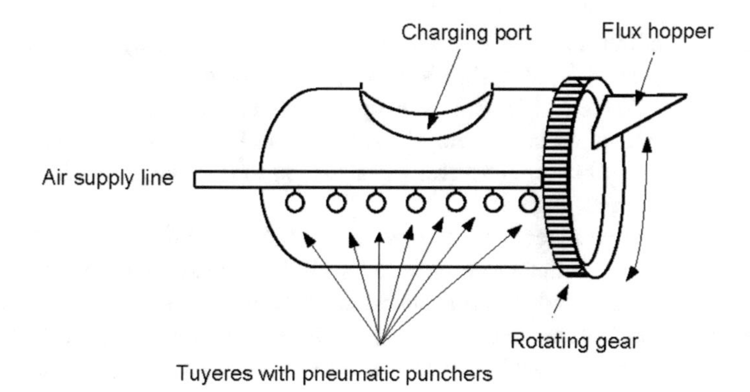

Fig. 3 Peirce-Smith convertor

5.1 Peculiarities of the Peirce-Smith Converter

The Peirce-Smith converter (PSC) is a pyrometallurgical plant working in a periodic mode [26], where each cycle duration is 6.5–8.5 h. The inlet matte with 58–65% copper content is blown by air that comes through 50–60 horizontal tuyeres (see Fig. 3 [27]). At the end of the cycle, a blister copper with 96–98% Cu is obtained. The converting process is strongly exothermic and includes removing the undesirable components *Fe*, *S*, *Pl*, *Si*.

The air blowing is divided in three sequential stages. The autothermic oxidation process passes as interactions between three phases—melt, slag and gas. To intensify the converting process, an enriched with O_2 air is injected at the beginning of *Fe*-removal stage. To reach an appropriate slag process a silica flux SiO_2 is added during the process and a corresponding slag volume is skimming. The melt/slag temperature is controlled by adding cold reverts. The discrete operations (supplement of matte and $Si\,O_2$ slag skimming) are fulfilled after the first and second blowing sub-stages. The end of the cycle is determined via analysis of *PbO* and *PbS* into the outlet gas. A series of 240–340 cycles are consolidated into a campaign. Between two campaigns the PCS is shut down for and repairing of the refractory and/or tuyeres is performed.

Detailed description of the PSC is given in [26, 28, 29]. From control point of view, it has the following main features:

(i) Cu—Converting exothermic process at 1100–1250 °C realized due to the air blowing through 54 horizontal submerged tuyeres;
(ii) The PSC is a degradation type plant. Two degradation processes run in parallel: wear and erosion of blow tuyeres, and converter refractory lining corrosion. The deterioration of the tuyeres is about four times faster. The degradation of individual tuyeres is strongly non-uniform with most intensity into the middle zone (from 20-th to the 35-th tuyere). To cope with this, maintenance actions have to be adopted in the form of particular tuyeres blocking. The purpose is to optimize these actions.

The length of the tuyeres is inspected regularly after each 10 cycles. Small maintenance works are fulfilled by blocking and/or punching the tuyeres in the same periods of time. At the end of each campaign a partial repair is carried out including all tuyeres replacing. After the end of the series (4 campaigns) a overhaul is executed (all refractory bricking, tuyeres replacing, converter mantle repair). The cost of maintenance is very high.

Some degradation models of the tuyeres and the conventional refractory lining are published [30, 31], but they are derived for idealized conditions (one dimensional thermal, aerodynamic and slag concentration environment, ideal mixing, deterministic and stationary boundary conditions), and thus have limited practical usability. In [28, 29], data driven models are received. In PSC the tuyer's degradation strongly influences not only the reliability of the converting process, but also its total operational state.

The optimization problems related to the PSC functioning, considered in [30, 32] are focused only on technological or design decisions without considering the plant operation change due to degradation.

5.2 Case-Based Reasoning Approach for the Peirce-Smith Converter

In this paper, the CBR approach is applied to the optimization and control of a Peirce-Smith converter. For this particular plant, the three attributes related to the problem definition (see (2)–(4)) contain the following information:

- **States** h_1:
  - h_{11}—*Industrial situation*
    h_{111}—Campaign number since the beginning of the series
    h_{112}—Cycle number in the campaign
    h_{113}—Cu content of the stein
    h_{114}—Stein temperature
    h_{115}—Average duration of the last 3 cycles
    h_{116}—Average production of the last 3 cycles
  - h_{12}—*Technical condition of the converter*
    h_{121}—Length of all tuyeres
    h_{122}—Average residual length of the tuyeres
    h_{123}—Standard deviation of the length of the tuyeres
    h_{124}—The smallest residual tuyer length
    h_{125}—The number of the shortest tuyer
    h_{126}—Estimate of the minimal thickness of the converter heat insulation wall
    h_{127}—Estimate of the average thickness of the converter heat insulation wall
  - h_{13}—*Condition of the tuyeres zone*
    h_{131}—Total number of the blocked tuyeres

h_{132}—Positions of all blocked tuyeres

- h_{14}—*Main technological parameters*

  h_{141}—Average temperature of the gas phase in the converter
  h_{142}—Total amount of the inlet air
  h_{143}—Total amount of the inlet oxygen
  h_{144}—Total amount of the silica flux (SiO_2)
  h_{145}—Total amount of the cooling materials
  h_{146}—Average pressure of the inlet air
  h_{147}—Cu content of the stein after the cycle

- **Constraints** h_2:

  - h_{21}—*Constraints on the acceptable wear and erosion of the tuyeres and the heat insulation wall*

    h_{211}—Minimal allowed residual length of tuyeres
    h_{212}—Minimal allowed residual thickness of the heat insulation wall

  - h_{22}—*Resources constraints*

    h_{221}—Constraints on the amount of the inlet air
    h_{222}—Constraints on the amount of the cooling materials

- **Criteria** h_3:

  - h_{31}—*Production requirements*

    h_{311}—Production of the cycle
    h_{312}—Production of the campaign

  - h_{32}—*Technological requirements*

    h_{321}—Cu content of the stein at the end of the cycle
    h_{322}—Allowed production costs related to the use of oxygen
    h_{323}—Allowed production costs related to the use of silica flux
    h_{324}—Allowed production costs related to the use of electricity

The attributes associated to the decision D are structured as follows:

- d_1**—Business decisions**

  - d_{11}—*Maximal allowed cycle duration*
  - d_{12}—*Maximal allowed duration of the campaign*
  - d_{13}—*Minimal allowed productivity of the cycle*

- d_2**—Technological decisions (change of the load of the Peirce-Smith converter)**

  - d_{21}—*Average air flowrates during the first and the second stage of the cycle*
  - d_{22}—*Average oxygen flow rates during the first and the second stage of the cycle*
  - d_{23}—*Amount of the cooling materials during the first and the second stage of the cycle*
  - d_{24}—*Frequency of measuring the residual length of the tuyeres*
  - d_{25}—*The weighting coefficients* $\alpha_q,\ q = 1, 2, \ldots, Q$ *for the MPC problem solved at the lower level (see* Fig. 2*)*

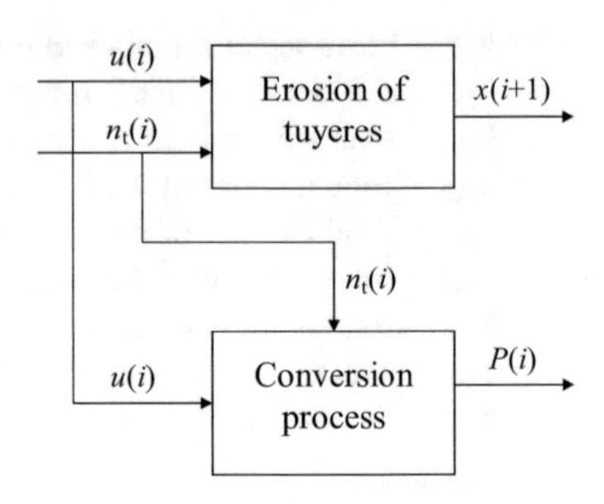

Fig. 4 Principal scheme of the converter [33]

- d_4—**Decision making method—case-based reasoning based on the similarity measure**

The estimate E of decision usefulness and the additional information I have the common structure as given in Sect. 2, where the principle of local-global similarity has been also described.

5.3 Plantwide Optimization of the Peirce-Smith Converter

- **Description of the converter operation from control point of view**

The copper converter is a semi-batch plant with time varying parameters (because of degradation), whose principal scheme from control pint of view is shown in Fig. 4 [33].

Here, $x(i) \in \mathbb{R}$ and $u(i) \in \mathbb{R}^6$ are the average length of tuyeres and the control input vector for the i-th cycle ($i = 1, 2, \ldots, 300$), $n_t(i)$ is the number of blocked tuyeres (it is constrained to be $0 \leq n_t(i) \leq 15$), and $P(i) \in \mathbb{R}$ is the productivity of the converting process (defined below). The vector of control inputs is:

$$u(i) = [G_{\text{air},1}(i), G_{\text{ox},1}(i), \theta_{\text{melt},1}(i), G_{\text{air},2}(i), G_{\text{ox},2}(i), \theta_{\text{melt},2}(i)] \tag{15}$$

where $G_{\text{air},1}(i)$, $G_{\text{ox},1}(i)$, $\theta_{\text{melt},1}(i)$ are the average flowrates (in nm^3/h) of the air and oxygen blown into the converter and the melt temperature (in °C) during the first conversion stage of the i-th cycle, and $G_{\text{air},2}(i)$, $G_{\text{ox},2}(i)$, $\theta_{\text{melt},2}(i)$ have the same meaning but related to the second stage of the cycle. The productivity $P(i)$ of the conversion process for the i-th cycle is defined as [29]:

$$P(i) = \frac{G_1(i) + G_2(i) - G_{\text{slag}}(i)}{\tau_{\text{cycle}}(i)}, \text{ tons/h} \tag{16}$$

In (16), $G_1(i)$ and $G_2(i)$ are the amounts of copper matte loaded into the converter in the first and in the second conversion stage (usually $G_1(i) = 90$ tons, $G_2(i) =$

60 tons), $G_{\text{slag}}(i)$ is the slag amount (normally $G_{\text{slag}}(i) = 40$ tons), and $\tau_{\text{cycle}}(i)$ is the cycle duration, which is [29]:

$$\tau_{\text{cycle}}(i) = \tau_0 + T_{\text{stage},1}(i) + T_{\text{stage},2}(i), \;\; \text{h} \tag{17}$$

with $T_{\text{stage},1}(i)$ and $T_{\text{stage},2}(i)$ being the durations (in h) of the first and the second stage of the conversion process in the cycle and τ_0 is the length of the preparative period ($\tau_0 = 1, h$). Thus, the expression for productivity becomes:

$$P(i) = \frac{G_1(i) + G_2(i) - G_{\text{slag}}(i)}{\tau_0 + T_{\text{stage},1}(i) + T_{\text{stage},2}(i)}, \;\; \text{tons/h} \tag{18}$$

- **Mathematical models**

Several mathematical models describing the operation of a particular Peirce-Smith converter are obtained based on experimental data for 5000 individual cycles [28, 29, 33]. The data were available from the existing SCADA system and the working shifts reports. The models for the time periods $\overline{T}_{\text{stage},1}$ and $\overline{T}_{\text{stage},2}$ (in min) of the first and the second stage of the conversion process are [29]:

$$\overline{T}_{\text{stage},1}(i) = 154 - 0.00167G_{\text{air},1}(i) + 0.16n_{\text{t}}(i) - 0.0282G_{\text{ox},1}(i) - 0.0098\theta_1(i) \tag{19}$$

$$\begin{aligned}&\overline{T}_{\text{stage},2}(i) = 259 - 0.00667G_{\text{air},2}(i) + 4.24n_{\text{t}}(i) + 0.0511G_{\text{ox},2}(i) + 0.155\theta_2(i)\\ &i = 1, 2, \ldots, 300\end{aligned} \tag{20}$$

where $\theta_1(i)$ and $\theta_2(i)$ are the average temperatures of the outlet gas for the two cycle periods of the i-th cycle. The cycle duration in min is then:

$$\bar{\tau}_{\text{cycle}}(i) = 60\tau_0 + \overline{T}_{\text{stage},1}(i) + \overline{T}_{\text{stage},2}(i), \;\; \text{min} \tag{21}$$

The dynamic model describing the degradation process along the series (each series contains 10 cycles), associated with the decreasing length of the tuyeres, is [33]:

$$\begin{aligned}&\bar{x}(j+1) = 10.19 + 0.7823\bar{x}(j) - 0.0178N_{\text{cycle}}(j) + 0.173\bar{n}_{\text{t}}(j)\\ &j = 1, 2, \ldots, 30\end{aligned} \tag{22}$$

where $\bar{x}(j) \in \mathbb{R}$ is the average length of tuyeres for the j-th series, $N_{\text{cycle}}(j) = 10j$ is the consecutive cycle number, and $\bar{n}_{\text{t}}(j)$ is the number of blocked tuyeres during the j-th series (this number remains constant during the 10 cycles in the series).

- **MPC problem**

Let the initial average length of tuyeres be $\bar{x}_0 = 55$ cm. Assume that at the end of each series (at every 10 cycles) the average length can be determined by measurements of the lengths of the individual tuyeres. Let the measured average length of tuyeres at the j-th series be $\bar{x}_{\mathrm{m}}$. The purpose is to determine the optimal profile of the number of blocked tuyeres along 30 series by optimizing a given optimality criterion. This is obtained by solving the following optimization problem at the current (the j-th series) corresponding to the "slow" MPC in the cascade control system (cf. Fig. 2):

MPC problem:

$$J^*(\bar{x}_{\mathrm{m}}) = \min_{\overline{N}_{\mathrm{t}}} J(\overline{N}_{\mathrm{t}}, \bar{x}_{\mathrm{m}}) \tag{23a}$$

subject to $\bar{x}_{j\,|\,j} = \bar{x}_{\mathrm{m}}$ and:

$$\bar{n}_{\mathrm{t},\, j+k} \in \left\{0, 1, 2, \ldots, \bar{n}_{\mathrm{t,max}}\right\}, \; k = 0, 1, \ldots, N_{\mathrm{p}} - 1 \tag{23b}$$

$$\bar{x}_{\min} \le \bar{x}_{j+k\,|\,j}, \; k = 1, 2, \ldots, N_{\mathrm{p}} \tag{23c}$$

$$\bar{\tau}_{\mathrm{cycle},\, j+k} \le \bar{\tau}_{\mathrm{cycle,max}}, \; k = 1, 2, \ldots, N_{\mathrm{p}} \tag{23d}$$

$$\begin{aligned} \bar{x}_{j+k+1\,|\,j} = {} & 10.19 + 0.7823\bar{x}_{j+k\,|\,j} - 0.0178 N_{\mathrm{cycle},\, j+k} \\ & + 0.173\bar{n}_{\mathrm{t},\, j+k}, \; k = 0, 1, \ldots, N_{\mathrm{p}} - 1 \end{aligned} \tag{23e}$$

$$N_{\mathrm{cycle},\, j+k} = 10(j + k), \; k = 0, 1, \ldots, N_{\mathrm{p}} - 1 \tag{23f}$$

$$\bar{\tau}_{\mathrm{cycle},\, j+k} = 60\tau_0 + \overline{T}_{\mathrm{stage,\, 1,}\, j+k} + \overline{T}_{\mathrm{stage,\, 2,}\, j+k}, \; k = 1, 2, \ldots, N_{\mathrm{p}} \tag{23g}$$

$$\begin{aligned} \overline{T}_{\mathrm{stage,\, 1,}\, j+k} = {} & 154 - 0.00167\widetilde{G}_{\mathrm{air,1}} + 0.16\bar{n}_{\mathrm{t},\, j+k} \\ & - 0.0282\widetilde{G}_{\mathrm{ox,1}} - 0.0098\tilde{\theta}_1, \; k = 1, 2, \ldots, N_{\mathrm{p}} \end{aligned} \tag{23h}$$

$$\begin{aligned} \overline{T}_{\mathrm{stage,\, 2,}\, j+k} = {} & 259 - 0.00667\widetilde{G}_{\mathrm{air,2}} + 4,24\bar{n}_{\mathrm{t},\, j+k} \\ & + 0.0511\widetilde{G}_{\mathrm{ox,2}} + 0.155\tilde{\theta}_2, \; k = 1, 2, \ldots, N_{\mathrm{p}} \end{aligned} \tag{23i}$$

with vector of optimization variables $\overline{N}_{\mathrm{t}} = [\bar{n}_{\mathrm{t},\, j}, \bar{n}_{\mathrm{t},\, j+1}, \ldots, \bar{n}_{\mathrm{t},\, j+N_{\mathrm{p}}-1}]$ and the cost function given by:

$$J(\overline{N}_{\mathrm{t}}, \bar{x}_{\mathrm{m}}) = \alpha_1 \underbrace{\sum_{k=0}^{N_{\mathrm{p}}} \left(\frac{\bar{x}_{j+k|j} - \bar{x}_0}{\bar{x}_0} \right)^2}_{J_1} + \alpha_2 \underbrace{\sum_{k=0}^{N_{\mathrm{p}}} \left(\frac{\bar{\tau}_{\mathrm{cycle},\, j+k}}{\bar{\tau}_{\mathrm{cycle,max}}} \right)^2}_{J_2} + \alpha_3 \underbrace{\sum_{k=0}^{N_{\mathrm{p}}-1} \left(\frac{\Delta\bar{n}_{\mathrm{t},\, j+k}}{\bar{n}_{\mathrm{t,max}}} \right)^2}_{J_3} \tag{23j}$$

Let the optimal solution to the optimization problem (23a–23j) be denoted $\overline{N}_{\mathrm{t}}^* = [\bar{n}_{\mathrm{t},\, j}^*, \bar{n}_{\mathrm{t},\, j+1}^*, \ldots, \bar{n}_{\mathrm{t},\, j+N_{\mathrm{p}}-1}^*]$. In (23j), $\Delta\bar{n}_{\mathrm{t},\, j+k}$, $k = 0, 1, 2, \ldots, N_{\mathrm{p}} - 1$ is determined as:

$$\Delta\bar{n}_{\mathrm{t},\, j+k} = \begin{cases} \bar{n}_{\mathrm{t},\, j} - \bar{n}_{\mathrm{t}}^*(j-1) & \text{for } k = 0 \\ \bar{n}_{\mathrm{t},\, j+k} - \bar{n}_{\mathrm{t},\, j+k-1} & \text{for } k = 1, 2, \ldots, N_{\mathrm{p}} \end{cases} \tag{24}$$

where $\bar{n}_{\mathrm{t}}^*(j-1)$ is the optimal number of tuyeres that are actually blocked at the $(j-1)$-th series (in accordance with the receding horizon strategy we have $\bar{n}_{\mathrm{t}}^*(j-1) =$

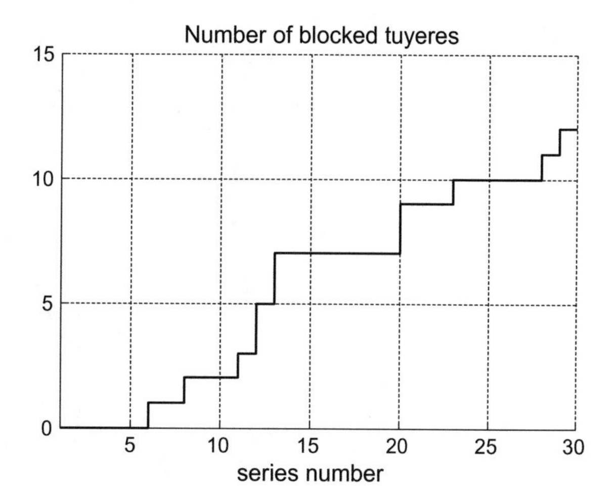

Fig. 5 Optimal profile of the number of blocked tuyeres

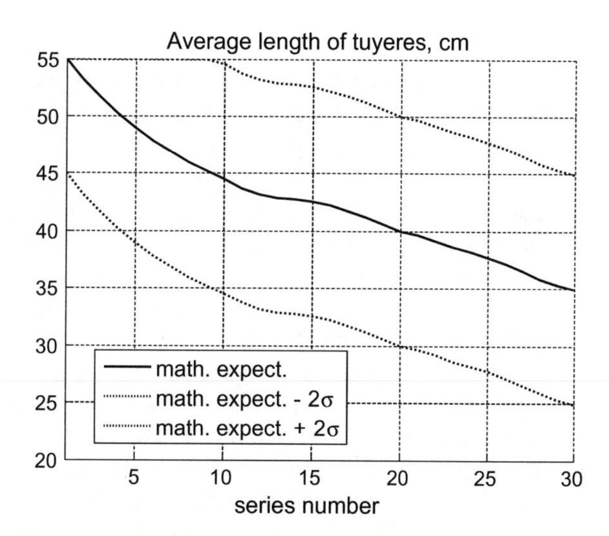

Fig. 6 The change of average length of tuyeres along the series

$\bar{n}^*_{t,\,j-1}$). Further in (23j), $N_p = 10$ is the prediction horizon and $\alpha_i > 0,\ i = 1, 2, 3$ are weighting coefficients determined by the case reasoning part of the control system. In the problem (23a–23j), $\bar{n}_{t,\max} = 15$ is the maximal allowed number of blocked tuyeres, $\bar{x}_{\min} = 8$ cm is the minimal allowed length of tuyeres, $\bar{\tau}_{\text{cycle},\max} = 480$ min is the maximal allowed cycle duration (determined by the CBR system). In (23h) and (23i), $\widetilde{G}_{\text{air},1} = 31472\,\text{nm}^3/\text{h}$, $\widetilde{G}_{\text{ox},1} = 132.2\,\text{nm}^3/\text{h}$, $\widetilde{G}_{\text{air},2} = 39750\,\text{nm}^3/\text{h}$, $\widetilde{G}_{\text{ox},2} = 132.2\,\text{nm}^3/\text{h}$ are the average values of the air flowrate and oxygen flowrate (part of the decision taken by the CBR system). The average temperatures of the outlet gas in the two stages of the cycle are $\tilde{\theta}_1 = 988$ °C and $\tilde{\theta}_2 = 1065$ °C. The values of the weighting coefficients in the cost function (23j) are $\alpha_1 = 1, \alpha_2 = 0.5, \alpha_3 = 0.5$ (determined by the CBR system).

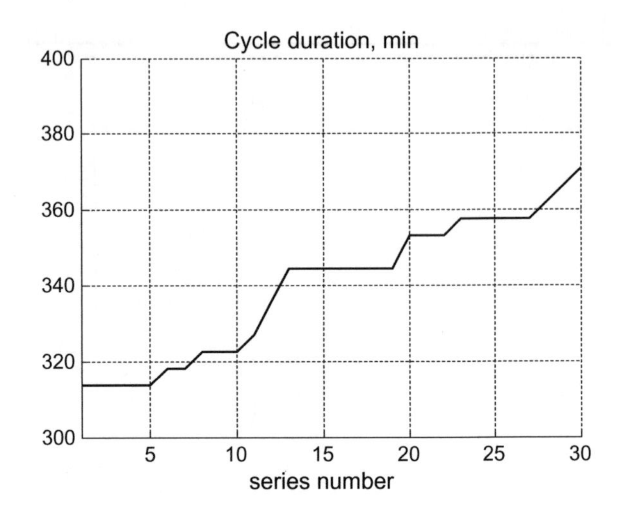

Fig. 7 The change of cycle duration along series

The optimization problem (23a–23j) represents a Quadratic Integer Programming (QIP) problem, because the optimization variables $\overline{N}_{\mathrm{t}} = [\bar{n}_{\mathrm{t},\,j}, \bar{n}_{\mathrm{t},\,j+1}, \ldots, \bar{n}_{\mathrm{t},\,j+N_{\mathrm{p}}-1}]$ take integer values (cf. (23b)), the cost function is quadratic in $\overline{N}_{\mathrm{t}}$ and the constraints are linear in $\overline{N}_{\mathrm{t}}$. It is solved by using the DIRECT algorithm [34], that handles problems with both nonlinear and integer constraints. The DIRECT algorithm (DIviding RECTangles) [34] is a deterministic sampling algorithm for searching for the global minimum of a multivariate function subject to constraints, using no derivative information. It is a modification of the standard Lipschitzian approach that eliminates the need to specify a Lipschitz constant.

The obtained optimal profile of the number of blocked tuyeres is shown in Fig. 5, where it can be seen that this number increases along the series. The associated changes of the average length of tuyeres and the cycle duration along the series are given in Figs. 6 and 7. In Fig. 6, both the mathematical expectation (computed from the model (22)) and the 95% confidence interval of the average length of tuyeres are shown, where the value of the standard deviation is $\sigma = 5\,\mathrm{cm}$ (estimated from experimental data [28, 29]).

It can be seen from Figs. 5 and 7 that along the series the number of the blocked tuyeres increases, which leads to an increase in the cycle duration.

6 Conclusions

In this paper, a two-level structure for control and optimization of degrading plants is proposed, which uses plant models in two time scales and incorporates case-based reasoning. A hierarchical MPC problem is formulated, where the change of the plant parameters over the "slow" time scale is captured by the model in the higher level. It is described how the case-based reasoning approach and the MPC methodology are combined to optimize the long-term performance of a Peirce-Smith converter. The optimal profile of the number of blocked tuyeres of the converter and the associated optimal changes of the average length of tuyeres and the cycle duration along series are determined by solving a Quadratic Integer Programming problem.

References

1. Mayne, D.Q., Rawlings, J.B., Rao, C.V., Scokaert, P.O.M.: Constrained model predictive control: Stability and optimality. Automatica **36**, 789–814 (2000)
2. Qin, S.J., Badgwell, T.A.: A survey of industrial model predictive control technology. Control Eng. Pract. **11**, 733–764 (2003)
3. Scattolini, R.: Architectures for distributed and hierarchical model predictive control—A review. J. Process Control **19**, 723–731 (2009)
4. Backx, T., Bosgra, O., Marquardt, W.: Integration of model predictive control and optimization of processes. In: Proceedings of IFAC Symposium on Advanced Control of Chemical Processes, Pisa, Italy, pp. 249–260 (2000)
5. Kadam, J.V., Marquardt, W., Schlegel, M., Backx, T., Bosgra, O.H., Brouwer, P.J., Dünnebier, G., van Hessem, D., Tiagounov, A., de Wolf, S.: Towards integrated dynamic real-time optimization and control of industrial processes. In: Proceedings of the Foundations of Computer-Aided Process Operations (FOCAPO2003), pp. 593–596 (2003)
6. Aamodt, A., Plaza, E.: Case-based reasoning: foundational issues, methodological variations and system approaches. Proc. AICOM **7**, 39–59 (1994)
7. Kaster, D., Medeiros, C., Rocha, H.: Supporting modeling and problem solving from precedent experiences: the role of workflows and case-based reasoning. Environ. Modell. Softw. **20**, 689–704 (2005)
8. Kolodner, J.: Case-Based Reasoning. Morgan Kaufmann Publishers, San Mateo, CA (1993)
9. Pal, S., Shin, S.: Foundation of Soft Case-Based Reasoning. Wiley (2004)
10. Recèo-Garcia, J.A., Diaz-Agudo, B., Sanches-Ruiz, A.A., Gonzales-Calero, P.A.: Lessons Learned in the Development of a CBR Framework, Expert Update, vol. 10, no. 1 (2010)
11. Case-Based Reasoning Framework jCOLIBRI. http://gaia.fdi.ucm.es/projects/jcolibri/
12. CBR*Tools. www-sop.inria.fr/axis/cbrtools/usermanual-eng/Introduction.html#OOFrame work
13. Component Architecture Technology for Developing CBR systems, CAT-CBR. http://www.iiia.csic.es/Projects/cbr/cat-cbr/
14. Zilles, L.: MyCBR Tutorial, myCBR Project (2009)
15. myCBR. http://www.mycbr-project.net/
16. Protege. http://protege.stanford.edu/
17. Sanchez-Ruiz-Granados, A.A., Recio-Garcia, J.A., Diaz-Agudo, B., Gonzalez-Calero, P.A.: Case Structures in jCOLIBRI. In: 24th SGAI International Conference on Inovative Technologies, UK (2005)
18. Mitra, R., Basak, J.: Methods of case adaptation: A survey. Int. J. Intell. Syst. **20**(6), 627–645 (2005)
19. Richter, M.M., Weber, R.: Case-Based Reasoning. Springer, Berlin, Heidelberg (2013)
20. Ligeza, A.: Logical Foundations for Rule-Based Systems. Series: Studies in Computational Intelligence, vol. 11. Springer, Berlin, Heidelberg (2006)
21. Bard, J.F.: Practical Bilevel Optimization: Applications and Algorithms, Series: Nonconvex Optimization and Its Applications, vol. 30. Springer, US (1998)
22. Colson, B., Marcotte, P., Savard, G.: An overview of bilevel optimization. Ann. Oper. Res. **153**, 235–256 (2007)

23. Grancharova, A., Johansen, T.A.: Explicit Nonlinear Model Predictive Control: Theory and Applications, LNCIS, vol. 429. Springer, Berlin, Heidelberg (2012)
24. Giselsson, P.: Improved fast dual gradient methods for embedded model predictive control. In: Proceedings of the 19th World Congress, Cape Town, South Africa, pp. 2303–2309 (2014)
25. Kufoalor, D.K.M., Aaker, V., Johansen, T.A., Imsland, L., Eikrem, G.O.: Automatically generated embedded model predictive control: Moving an industrial PC-based MPC to an embedded platform. Optim. Control Appl. Methods **36**, 705–727 (2015)
26. Ng, K.W., Kapusta, J.P.T., Harris, R., Wraich, A.E., Parra, R.: Modeling Peirce-Smith converter operating costs. J. Miner. Met. Mater. Soc., pp. 52–57 (2005)
27. Davenport, W.G., King, M., Schlesinger, M., Biswas, A.K.: Extractive Metallurgy of Copper, 4th edn. Pergamon, Elsevier, Oxford, UK (2002)
28. Hadjiski, M., Boshnakov, K.: Extended supervisory control of Peirce-Smith converter. Comptes rendus de l'Academie bulgare des Sciences **67**(5), 705–714 (2014)
29. Hadjiski, M., Boshnakov, K.: Nonlinear hybrid control of copper converter. Comptes rendus de l'Academie bulgare des Sciences **67**(6), 855–862 (2014)
30. Goni, C., Barbes, M., Bazan, V., Brandalez, E., Parra, R., Gonzales, L.: The mechanism of thermal spalling in the wear of the Peirce-Smith copper converter. J. Ceram. Soc. Jpn. **114**(8), 672–675 (2006)
31. Oprea, G., Lo, W., Trozynski, T., Rigby, J.: Corrosion of refractories in PSC. In: Kapusta, J., Warner, T.: International Peirce-Smith Converting Centenial. Wiley, California, USA (2010)
32. Song, Y., Peng, X., Dong, W., Hu, Z.: Data driven optimal decision making modeling for copper-matte converting process. J. Comput. Inf. Syst. **7**(3), 754–761 (2011)
33. Boshnakov, K., Ginchev, T., Petkov, V., Mihailov, E.: Strategy for predictive maintenance of Peirce-Smith converters. In: Proceedings of the Technical University of Sofia, vol. 62, Book 1, pp. 345–354 (2012) (in Bulgarian)
34. Jones, D.R.: The direct global optimization algorithm. In: Floudas, C.A., Pardalos, P.M. (eds.) Encyclopedia of Optimization, vol. 1, pp. 431–440. Kluwer, Dordrecht (2001)

Collaborative Clustering: New Perspective to Rank Factor Granules

Shihu Liu, Xiaozhou Chen and Patrick S. P. Wang

1 What is a Factor Granule?

Mathematically speaking, a factor f can be regarded as a mapping, which is based on the universe of discourse U. Withal, every pattern u in U can be mapped into a so-called state space, denoted by $U(f)$. Here, the state of $U(f)$ can be an attribute, a feature, or a characteristic, etc., they form a dimension with respect to the factor f. On these bases, a special name, **factor**, is applied to describe this dimension [37, 38]. Certainly, it is not the attribute appeared in rough set theory [26]. Because the concept "attribute" appeared in rough set theory is the connotation-description with respect to the categorized results, but if a series of attributes belongs to the common categorized results, then it can be described by the factor. In other words, the factor can be applied to describe these attributes. So, one can find that in essence, factor is the attribute of attributes.

In principle, a factor granule can be described as a triple $G = (U, F, I)$, where

- $U = \{u_i \mid i = 1, 2, \ldots, n\}$ is a nonempty finite set and $u_i \in U$ for $i = 1, 2, \ldots, n$ are patterns;
- $F = \{f_i \mid i = 1, 2, \ldots, k\}$ is the set of factors and $f_i \in F$ for $i = 1, 2, \ldots, k$ is a factor;

S. Liu (✉)
School of Mathematics and Computer Sciences, Yunnan Minzu University,
Kunming 650500, People's Republic of China
e-mail: liush02@126.com

X. Chen
Key Laboratory of IOT Application Technology of Universities in Yunnan Province,
Yunnan Minzu University, Kunming 650031, People's Republic of China
e-mail: ch_xiaozhou@163.com

P. S. P. Wang
College of Computer and Information Science, Northeastern University,
360 Huntington Avenue, Boston, MA 02115, USA
e-mail: patwang@ieee.org

V. Sgurev et al. (eds.), *Learning Systems: From Theory to Practice*, Studies
in Computational Intelligence 756, https://doi.org/10.1007/978-3-319-75181-8_7

- $I = \{f_i(u_s, u_t) \mid f_i \in F, u_s, u_t \in U\}$ is the collection of relationship information, $f_i(u_s, u_t)$ represents the relationship information between u_s and u_t about $f_i \in F$.

Obviously, $f_i(u_s, u_t)$ can be qualitative, or quantitative. It depends entirely, of course, on the property of the factor f_i. In practical problem solving, people seem to be fond of treating the quantitative values. For this, Yu and Luo [41] had proposed granule factors space to fulfill the transformation for certain problem.

Formally, if $u_s \in U$ is nothing but a symbolic representation of the sth pattern, then only the relationship information I in $G = (U, F, I)$ can be used. What is more, if the number of patterns is finite (i.e., $n < \infty$), then the relationship information I can be shown by a matrix

$$I = \begin{pmatrix} I_{11} & I_{12} & \dots & I_{1n} \\ I_{21} & I_{22} & \dots & I_{2n} \\ \vdots & \vdots & \ddots & \vdots \\ I_{n1} & I_{n2} & \dots & I_{nn} \end{pmatrix}, \quad (1)$$

where $I_{st} = (f_1(u_s, u_t), f_2(u_s, u_t), \dots, f_k(u_s, u_t))$.

Besides, if the pattern u_s is not a symbolic representation but is depicted by some factors $h_1, h_2, \dots, h_w$, then it can be expressed as $u_s = (h_1(u_s), h_2(u_s), \dots, h_w(u_s))$. Thereupon, one can find that for a factor granule $G = (U, F, I)$, it contains not only the relationship information I induced by factors in F, but also the information J induced by factors in $H = \{h_1, h_2, \dots, h_w\}$. Hereinto, $J = (h_1(U), h_2(U), \dots, h_w(U))$ and $h_j(U) = (h_j(u_1), h_j(u_2), \dots, h_j(u_n))^T$ for $j = 1, 2, \dots, w$.

Obviously, $f_i(u_s, u_t)$ and $h_j(u_s)$ would be qualitative or quantitative with respect to $i = 1, 2, \dots, k$, $j = 1, 2, \dots, w$ and $s, t = 1, 2, \dots, n$. In general, fuzzy number [12], interval [30], set [31], et al., would be applied to describe it. This is almost entirely dependent on the property of $f \in F$ and $h \in H$. Besides, for different factors $f_i, f_j \in F$, $f_i(u_s, u_t)$ and $f_j(u_s, u_t)$ can have the same/different data representation. But, the same factor must has the fixed representation format.

Without otherwise specifications, in what follows we adhere to the hypothesis that $G = (U, F + H, I + J)$ would be more appropriate for showing a factor granule. To exemplify this, next we introduce an example.

Example 1 Given that there have 20 staffs in one laboratory. If we regard each staff as a pattern, then the universe of discourse is $U = \{u_i \mid i = 1, 2, \dots, 20\}$.

Obviously,

f_1 : the number of coauthored articles;

f_2 : the number of co-cited articles;

f_3 : the scientific research ability.

can be regarded as three factors to describe the scientific cooperation relationship between any two staffs, i.e., $F = \{f_1, f_2, f_3\}$. Hereinto, factors f_1 and f_2 are quantitative factors, but factor f_3 is qualitative factor. Based on these, the relationship information $I = \{f_i(u_s, u_t) \mid i = 1, 2, 3; s, t = 1, 2, \dots, 20\}$ can be determined.

Certainly, each staff has his or her self-characteristics, such as sex, age, native place, stature, etc. All of these factors, i.e., $H = \{h_1, h_2, \ldots\}$, can be used to describe their individual characteristics. Naturally, the factor granule also obtain the information J induced by factors in H.

2 Why to Rank Factor Granules?

The purpose of ranking, for decision makers, is to discover a mechanism that can produce a linear sequence of the data to be ranked. Because of its practicality, the ranking problem has been studied widely in many fields, such as assignment problem [10, 27, 29], information retrieval [9, 21, 32], similarity measure [7, 40] and others [6, 15, 19, 22, 33]. To some extent, the available information participated in ranking process is vector-typed pattern's attribute values. For this, it becomes a multi-criteria decision making problem [1, 14].

In general, this is not enough. As for some problems, the available information is not the vector-typed pattern's attribute value, but the relational information between any two patterns, i.e., the graph becomes the main information provider. So far, a lot of studies, especially in the fields of machine learning [4, 8, 13], have been done. For example, Agarwal [2, 3] investigated the ranking problem on graph data; Mihalcea [23] presented an innovative unsupervised method for automatic sentence extraction using graph-based ranking algorithms; Lee et al. [18] focused on the flexibility of recommendation and proposed a graph-based multidimensional recommendation method, and others [4, 5].

Sometimes, the available information include not only the vector-typed patterns attribute but also the relationship information between patterns. In other words, two kinds of these information can be used at the same time. For instance, if we intend to make a sequence for the teachers of one school in aspect of their scientific research ability, each teacher can be described by many factors (such as the published papers, research projects, etc.). Meanwhile, the cooperation relations between/among other teachers should also be considered. Therefore, it is necessary to sort data with such information.

Whether the involved information is vector-typed pattern's attribute value, relationship between patterns or both of them, how to depict the ranking objects is very important. By considering the advantages of factor space theory [35, 36, 39, 41] in aspects of attribute or feature description, here the ranking objects are named as factor granules. For each factor granule, it is composed by at least two types of information: the factor-induced relational information, and factor-induced pattern's attributes. Given this situation, the to-be-ranked objects are not the simplified set of patterns any more, but the granules: the set of some patterns, with corresponding factor-induced information.

The remainder of this chapter is organized as follows. Section 1 proposes the mathematical description of factor granule, and introduced a synthetical example to illustrate its essence. Section 2 elaborates the reason why we discusses the problem of ranking factor granules. Section 3 develops the ranking algorithm for factor granules in terms of collaborative clustering. First, we briefly introduce the collaborative information which are provided by factor granules to be ranked. Then, we construct the detailed ranking algorithm. A synthetical example is given in Sect. 4 to demonstrate the validity of the proposed ranking approach. We describe some conclusions in Sect. 5.

3 How to Rank Factor Granules in Terms of Collaborative Clustering?

In practical, the available information of factor granule would not be applied freely and openly all the time. To retain the privacy of factor granules, the fuzzy collaborative clustering mechanism [11, 28] is introduced to execute the ranking process. But, the collaborative clustering could not provide the finial ranking result. Withal, inspired by the ideology of the famous TOPSIS method [16], a referential factor granule is pre-proposed to deal with the collaborated clustering results. Due to the practical concerns of data privacy, here the self-related information of each factor granule is completely confidential. The only available information (i.e., collaborative information) is the corresponding clustering centers of them, i.e., $V[ii]$ of $G^{(ii)}$ for $ii = 1, 2, \ldots, p$. Here, it should be pointed out that the patterns of the referential factor granule are the same as that of the factor granules $G^{(ii)}$ to be ranked.

3.1 Collaborative Information

Given that $G = (U, F + H, I + J)$ is a factor granule, one has that the factor $h \in H$ would be qualitative, or quantitative. For convenience, we suppose that $H = (H_1, H_2)$, where $h \in H_1$ is quantitative and $h \in H_2$ is qualitative. Therefore, the information J can be divided into two irrelevant parts: quantitative information J_1 induced by factors in H_1 and qualitative information J_2 induced by factors in H_2. For any $u_s, u_t \in U$ and $f_i \in F$, in this paper, we presume that all $f_i \in F$ are quantitative. In other words, I_{st} in Eq. (1) is a numerical vector for $s, t = 1, 2, \ldots, n$.

By considering what was discussed above, we have that the collaborative information, provided by factor granule $G^{(ii)}$ for $ii = 1, 2, \ldots, p$, would be originated from three resources: the relationship information I, the non-relationship information J_1 or J_2. Withal, we have that

- if the collaborative information $V[ii]$ is derived from the relationship information I, then we use $\{\breve{v}_1, \breve{v}_2, \ldots, \breve{v}_{c_1}\}$ to denote it, i.e.,

$$\breve{V}[ii] = \{\breve{v}_1, \breve{v}_2, \ldots, \breve{v}_{c_1}\}; \tag{2}$$

- if the collaborative information $V[ii]$ is derived from the non-relationship information J_1, then we use $\{\dot{v}_1, \dot{v}_2, \ldots, \dot{v}_{c_1}\}$ to denote it, i.e.,

$$\dot{V}[ii] = \{\dot{v}_1, \dot{v}_2, \ldots, \dot{v}_{c_2}\}; \tag{3}$$

- if the collaborative information $V[ii]$ is derived from the non-relationship information J_2, then we use $\{\ddot{v}_1, \ddot{v}_2, \ldots, \ddot{v}_{c_2}\}$ to denote it, i.e.,

$$\ddot{V}[ii] = \{\ddot{v}_1, \ddot{v}_2, \ldots, \ddot{v}_{c_3}\}; \tag{4}$$

where c_1, c_2 and c_3 represent the corresponding number of clusters.

Certainly, if the collaborative information is derived from two or more resources, then the concrete expression of it can be constructed naturally, in terms of Eqs. (2)–(4). For example, if it is generated from the non-relationship information J_1 and J_2, then

$$V[ii] = \{\dot{V}[ii], \ddot{V}[ii]\}. \tag{5}$$

With above discussion, the collaborative information $V[ii]$ can have as many as seven types, which are $\dot{V}[ii]$, $\ddot{V}[ii]$, $\breve{V}[ii]$, $\{\dot{V}[ii], \ddot{V}[ii]\}$, $\{\dot{V}[ii], \breve{V}[ii]\}$, $\{\ddot{V}[ii], \breve{V}[ii]\}$ and $\{\dot{V}[ii], \ddot{V}[ii], \breve{V}[ii]\}$. Different factor granules can provide different types of collaborative information, but for a specified factor granule $G^{(ii)}$, the type of collaborative information is fixed.

3.2 Rank Factor Granules in Terms of Collaborative Clustering

Besides the truth that each pattern is depicted by the same factors, one know that a linear sequence is what we want. Hence, in this paper we introduce the ideology of the famous TOPSIS [16] method, to help we rank the factor granules. Naturally, a referential factor granule $G^{(*)}$ with n patterns is prescribed and all of its information can be shared freely. Especially, all of its factors are the same as that of the factor granules to be ranked.

By considering the collaborative clustering mechanism [11, 17, 25, 28], if the collaborative information is $\breve{V}[ii]$, then the collaborative mechanism based objective function can be formulated as

$$Q(G^{(*)}, G^{(ii)}) = \sum_{i=1}^{c}\sum_{k=1}^{n} u_{ik}^2 d^2(u_k, v_i) + \beta[ii, c_1]\sum_{i=1}^{c}\sum_{k=1}^{n} u_{ik}^2 \sum_{j=1}^{c_1} d^2(v_i, \breve{v}_j[ii]), \tag{6}$$

where $d^2(u_k, v_i)$ represents the Euclidean distance between $u_k \in U$ and $v_i \in V$,[1] the same to that of $d^2(v_i, \breve{v}_j[ii])$; $\beta[ii, c_1]$ represents the corresponding collaboration degree, and can be determined by following equation

$$\beta[ii, c_1] = \begin{cases} 1 & \text{if } c = c_1 \\ \dfrac{c \wedge c_1}{c \vee c_1} & \text{if } c \neq c_1 \end{cases}. \tag{7}$$

The interpretation of Eq. (6) is obvious: the first part is the objective function used to search for the structure of the referential factor granule $G^{(*)}$, and the second part is to calculate the difference between the centers (weighted by the partition matrix of the referential factor granule).

The similar collaborative mechanism based objective function can be constructed if the collaborative information is $V[ii] = \dot{V}[ii]$ or $V[ii] = \ddot{V}[ii]$. Certainly, one can combine all three of these cases into one, then a unified objective function can be expressed as

$$Q(G^{(*)}, G^{(ii)}) = \sum_{i=1}^{c}\sum_{k=1}^{n} u_{ik}^2 d^2(u_k, v_i) + \beta[ii] \sum_{i=1}^{c}\sum_{k=1}^{n} u_{ik}^2 T[ii, i], \tag{8}$$

where $T[ii, i]$ represents the distance between $v_i \in V$ and the collaborative information $V[ii]$, and can be expressed as

$$T[ii, i] = \begin{pmatrix} \sum_{j=1}^{c_1} \|v_i - \breve{v}_j[ii]\|^2 \\ \sum_{j=1}^{c_2} \|v_i - \dot{v}_j[ii]\|^2 \\ \sum_{j=1}^{c_3} \|v_i - \ddot{v}_j[ii]\|^2 \end{pmatrix}; \tag{9}$$

the notation $\beta[ii] = (\beta[ii, c_1], \beta[ii, c_2], \beta[ii, c_3])$ is the label function with respect to the type of collaborative information and can be determined by some ways, such as the following approach:

- if $V[ii] = \breve{V}[ii]$, then $\beta[ii, c_2] = \beta[ii, c_3] = 0$, but $\beta[ii, c_1] \neq 0$;
- if $V[ii] = \dot{V}[ii]$, then $\beta[ii, c_1] = \beta[ii, c_3] = 0$, but $\beta[ii, c_2] \neq 0$;
- if $V[ii] = \ddot{V}[ii]$, then $\beta[ii, c_1] = \beta[ii, c_2] = 0$, but $\beta[ii, c_3] \neq 0$.

The optimization of Eq. (8) involves the determination of the partition matrix $U[ii]$ and the cluster centers $V[ii]$. In the same way, we select the Lagrange multiplier method to solve this optimization problem. For each pattern $u_k \in U$, the extended

[1] In equation (6), V represents the cluster centers of the referential factor granule $G^{(*)}$.

Lagrange function can be formulated as

$$Q_\lambda = \sum_{i=1}^{c} u_{ik}^2 d^2(u_k, v_i) + \beta[ii] \sum_{i=1}^{c} u_{ik}^2 T[ii, i] - \lambda(\sum_{i=1}^{c} u_{ik} - 1). \quad (10)$$

Taking the derivative of Q_λ with respect to u_{is} for $i = 1, 2, \ldots, c$ and making it 0, we have

$$\begin{aligned} \frac{\partial Q_\lambda}{\partial u_{is}} &= 2u_{is} d^2(u_s, v_i) + 2\beta[ii] u_{is} T[ii, i] - \lambda \\ &= 0. \end{aligned} \quad (11)$$

With above equation, we have that

$$u_{is} = \frac{\lambda}{2d^2(u_s, v_i) + 2\beta[ii]T[ii, i]}. \quad (12)$$

By considering the constraint $\sum_{i=1}^{c} u_{is} = 1$, we have that

$$\lambda = \left[\sum_{i=1}^{c} \frac{1}{2d^2(u_s, v_i) + 2\beta[ii]T[ii, i]} \right]^{-1}. \quad (13)$$

Taking Eq. (13) into Eq. (12), we can obtain

$$u_{is} = \left[\sum_{j=1}^{c} \frac{d^2(u_s, v_i) + \beta[ii]T[ii, i]}{d^2(u_s, v_j) + \beta[ii]T[ii, j]} \right]^{-1}. \quad (14)$$

For the calculation of cluster centers, we can make a derivation for Eq. (8) with respect to variant v_i for $i = 1, 2, \ldots, c$, and let the result equal to 0, then it comes

$$\begin{aligned} \frac{\partial Q(G^{(*)}, G^{(ii)})}{\partial v_i} &= \sum_{s=1}^{n} u_{is}^2 \frac{\partial d^2(u_s, v_i)}{\partial v_i} + \beta[ii] \sum_{s=1}^{n} u_{is}^2 \frac{\partial T[ii, i]}{\partial v_i} \\ &= 0. \end{aligned} \quad (15)$$

Because $d^2(u_k, v_i)$ represents the Euclidean distance between $u_k \in U$ and $v_i \in V$, we have that

$$\frac{\partial d^2(u_s, v_i)}{\partial v_i} = -2(u_s - v_i). \quad (16)$$

In terms of Eq. (9), we have that

$$\frac{\partial T[ii,i]}{\partial v_i} = 2\begin{pmatrix} \sum_{j=1}^{c_1}(v_i - \breve{v}_j[ii]) \\ \sum_{j=1}^{c_2}(v_i - \dot{v}_j[ii]) \\ \sum_{j=1}^{c_3}(v_i - \ddot{v}_j[ii]) \end{pmatrix}. \tag{17}$$

Taking Eqs. (16) and (17) into Eq. (15), we have that

$$\sum_{s=1}^{n} u_{is}^2(v_i - u_s) + \beta[ii]\sum_{s=1}^{n} u_{is}^2 \begin{pmatrix} \sum_{j=1}^{c_1}(v_i - \breve{v}_j[ii]) \\ \sum_{j=1}^{c_2}(v_i - \dot{v}_j[ii]) \\ \sum_{j=1}^{c_3}(v_i - \ddot{v}_j[ii]) \end{pmatrix} = 0. \tag{18}$$

After a series of transformation,

$$v_i = \frac{\sum_{s=1}^{n} u_{is}^2 u_s + \Phi(ii,i)}{\sum_{s=1}^{n} u_{is}^2 + \Psi(ii,i)} \tag{19}$$

comes naturally, where the parameters $\Phi(ii,i)$ and $\Psi(ii,i)$, respectively, are

$$\Phi(ii,i) = \sum_{s=1}^{n} u_{is}^2 \left(\beta[ii,c_1]\sum_{j=1}^{c_1}\breve{v}_j[ii] + \beta[ii,c_2]\sum_{j=1}^{c_2}\dot{v}_j[ii] + \beta[ii,c_3]\sum_{j=1}^{c_3}\ddot{v}_j[ii]\right),$$
$$\Psi(ii,i) = \sum_{s=1}^{n} u_{is}^2 \left(\beta[ii,c_1]c_1 + \beta[ii,c_2]c_2 + \beta[ii,c_3]c_3\right). \tag{20}$$

So far, Eqs. (14) and (19) can be applied to compute the collaborated partition matrix and cluster centers of the referential factor granule $G^{(*)}$, respectively.[2]

[2]Throughout this paper, the notation "$U[ii]$" denotes the collaborative information of the factor granule $G^{(ii)}$. To avoid unnecessary ambiguity, in the rest of this subsection we apply $U_G[ii]$ to describe the computing results of Eq. (14), Similarly, we apply $V_G[ii]$ to describe the computing results of Eq. (19).

As described at the beginning of this subsection, a linear sequence with respect to the factor granules to be ranked is what we want. Now, Eqs. (14) and (19) just provide the new clustering results of the referential factor granule $G^{(*)}$, by the collaboration of $G^{(ii)}$ for $ii = 1, 2, \ldots, p$. Here, we apply the following equation

$$S(G^{(*)}, G^{(ii)}) = \frac{S_V^2(V, V_G[ii]) + S_U^2(U, U_G[ii])}{S_V(V, V_G[ii]) + S_U(U, U_G[ii])} \tag{21}$$

to measure the similarity between $G^{(*)}$ and $G^{(ii)}$. And on this bases, the factor granules can be ranked. Up to now, the complete ranking procedure can be summarized as following algorithm.

Algorithm 1: Collaborative clustering based ranking method for factor granules

input : referential factor granule $G^{(*)}$, collaborative information $V[ii]$ for $ii = 1, 2, \ldots, p$.
output: A possible linear sequence $G^{(i1)} \succcurlyeq G^{(i2)} \succcurlyeq \ldots \succcurlyeq G^{(ip)}$

begin
On the problem of how to cluster the factor granule, we have discussed it carefully in our other research working papers.
for $ii = 1$ **to** n **do**
Determine the collaborative degree $\beta[ii]$; //*Here, equation (7) can be applied. But beyond that, other method can also be applied, such as in Refs. [24, 34].*//
Computer $U_G[ii]$ by Eq. (14), and $V_G[ii]$ by Eq. (19);
Compute the similarity of $G^{(*)}$ and $G^{(ii)}$, i.e., $S(G^{(*)}, G^{(ii)})$, by Eq. (21);
end
if $S(G, G^{(ii)}) \geqslant S(G, G^{(jj)})$ **then**
$G^{(ii)} \succcurlyeq G^{(jj)}$
else
$G^{(jj)} \succ G^{(ii)}$
end
end

4 Case Study

In this section, we use a synthetic example to show the validity of the proposed ranking approach for factor granules.

For the referential factor granule $G^{(*)}$ with 50 patterns, Fig. 1 shows the corresponding non-relationship information, Fig. 2 is its original partition matrix; and its original clustering centers are

$$V = \{v_1, v_2, v_3\} = \{(6.5, 2.9), (5.5, 2.8), (5.3, 2.4)\}.$$

For the to-be-ranked factor granules $G^{(1)}, G^{(2)}, \ldots, G^{(5)}$, Table 1 provides the corresponding collaborative information.

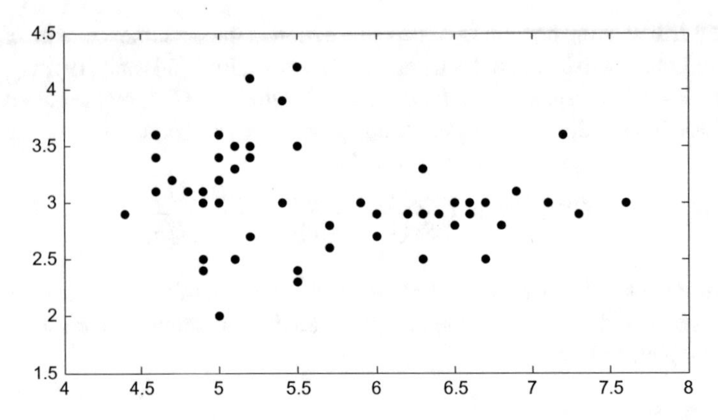

Fig. 1 The non-relationship information of the referential factor granule $G^{(*)}$

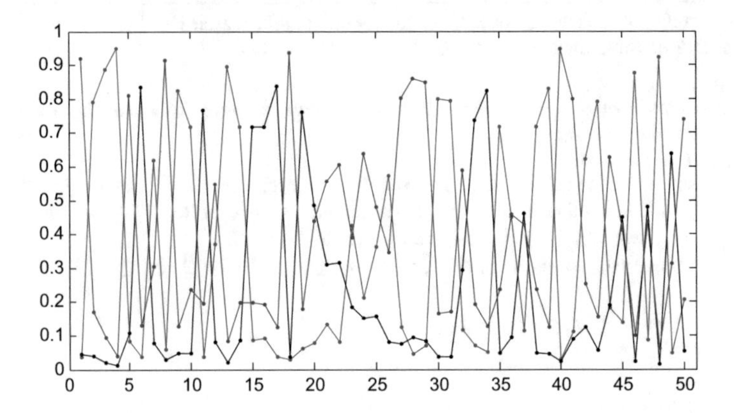

Fig. 2 The partition matrix of the referential factor granule $G^{(*)}$

Table 1 The collaborative information with respect to $G^{(ii)}$ for $ii = 1, 2, \ldots, 5$

	$\dot{V}[1]$		$\dot{V}[2]$		$\dot{V}[3]$		$\dot{V}[4]$		$\dot{V}[5]$	
v_1	5.0	2.6	4.0	3.0	7.5	4.0	5.4	4.3	4.5	3.0
v_2	5.8	3.4	3.1	4.5	5.3	2.5	5.6	1.8	6.5	2.9
v_3	6.0	4.0	5.2	2.8			5.3	2.5		

Step 1. Determine the collaborative degree $\beta[ii]$ for $ii = 1, 2, \ldots, 5$.

By comparing the number of clustering centers between V and $V[1]$ in terms of Eq. (7), one can find that $\beta[1] = (0, 1, 0)$. Similarly, we have that

$$\beta[2] = (0, 1, 0), \beta[3] = (0, \frac{2}{3}, 0), \beta[4] = (0, 1, 0), \beta[5] = (0, \frac{2}{3}, 0).$$

Step 2. Computer $U_G[ii]$ by Eq. (14), and $V_G[ii]$ by Eq. (19).

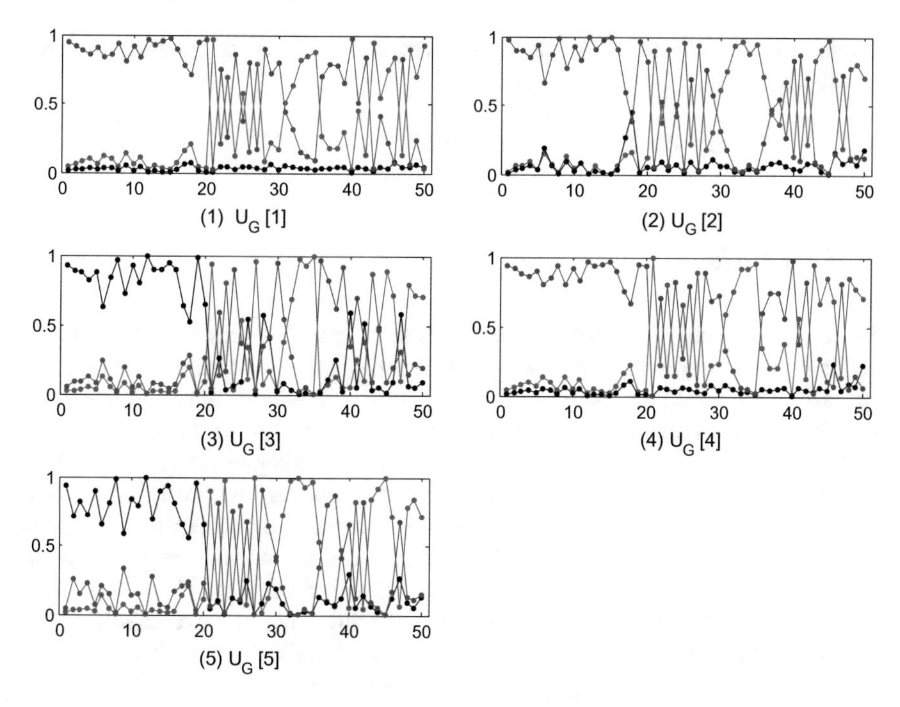

Fig. 3 The new partition matrix of the referential factor granule $G^{(*)}$

Table 2 The new clustering centers of the referential factor granule $G^{(*)}$

	$V_G[1]$		$V_G[2]$		$V_G[3]$		$V_G[4]$		$V_G[5]$	
v_1	7.0074	3.3009	6.4686	3.1775	6.7016	2.9964	6.8795	3.1434	6.5228	2.9414
v_2	8.3137	4.9306	6.0532	4.8403	5.0930	3.2486	7.9959	4.1877	5.0447	3.3159
v_3	5.4240	3.2427	5.1465	3.3466	5.8959	2.8566	5.3783	3.2068	5.3860	2.6806

With the help of Eqs. (14) and (19), one can obtain the new partition matrices of $G^{(*)}$ under the collaboration of $G^{(ii)}$ for $ii = 1, 2, \ldots, 5$ (See Fig. 3 and Table 2).

Step 3. Compute the similarity of $G^{(*)}$ and $G^{(ii)}$, i.e., $S(G^{(*)}, G^{(ii)})$, by Eq. (21).

According to Eq. (21), we have that[3]

$$
\begin{aligned}
S(G^{(*)}, G^{(1)}) &= 0.8413 \\
S(G^{(*)}, G^{(2)}) &= 0.8310 \\
S(G^{(*)}, G^{(3)}) &= 0.8278 \\
S(G^{(*)}, G^{(4)}) &= 0.8393 \\
S(G^{(*)}, G^{(5)}) &= 0.8299
\end{aligned}
$$

[3]Here, the value of $S_V(V, V_G[ii])$ and $S_U(U, U_G[ii])$, for $ii = 1, 2, \ldots, 5$, are calculated by the methods proposed in Ref. [20].

Withal, the possible linear sequence with respect to the factor granules $G^{(ii)}$ for $ii = 1, 2, \ldots, 5$ is

$$S(G, G^{(3)}) \geqslant S(G, G^{(5)}) \geqslant S(G, G^{(2)}) \geqslant S(G, G^{(4)})S(G, G^{(1)})$$

5 Conclusions

By considering the difficult of ranking highly structured data, this paper presented a collaborative clustering based ranking approach for factor granules. Before ranking, we assumed a referential factor granule which has the same patterns and factors as that of the to-be-ranked factor granules, and its information can be applied freely and openly during the whole ranking process. During the process of constructing the ranking algorithm, the privacy of each factor granule is well-preserved because only the available information, i.e., collaborative information, participated. Whatever the collaborative information are and wherever it is derived from, we regarded the ranking problem as an objective-function-based optimization problem. After that, the factor granules can be sorted by comparing the difference of the referential factor granule before and after collaboration of factor granules. Finally, a synthetic example examined the validity of the proposed ranking approach.

Acknowledgements This work has been supported by the National Natural Science Foundation of China(No. 31460297), Natural Science Foundation of Shandong Province(ZR2016AP12), Yunnan Applied Basic Research Youth Projects(No. 2015FD032), the Scientific Research Funds of Yunnan Provincial Department of Education(No. 2015Y224) and the Talent Introduction Research Project of Yunnan Minzu University.

References

1. Adler, N., Friedman, L., Sinuany-Stern, Z.: Review of ranking methods in the data envelopment analysis context. Eur. J. Oper. Res. **140**(2), 249–265 (2002)
2. Agarwal, S.: Ranking on graph data. In: The 23rd ACM International Conference on Machine learning, Pittsburgh, Pennsylvania, USA, pp. 25-32 (2006)
3. Agarwal, S.: Learning to rank on graphs. Mach. Learn. **81**(3), 333–357 (2010)
4. Agarwal, S., Dugar, D., Sengupta, S.: Ranking chemical structures for drug discovery: a new machine learning approach. J. Chem. Inf. Model. **50**(5), 716–731 (2010)
5. Bekkerman, R., Bilenko, M., Langford, J.: Scaling Up Machine Learning: Parallel and Distributed Approaches, Cambridge University Press (2012)
6. Bilsel, R.U., Büyüközkan, G., Ruan, D.: A fuzzy preference-ranking model for a quality evaluation of hospital web sites. Int. J. Intell. Syst. **21**(11), 1181–1197 (2006)
7. Chechik, G., Sharma, V., Shalit, U., et al.: Large scale online learning of image similarity through ranking. J. Mach. Learn. Res. **11**, 1109–1135 (2009)
8. Chen, H., Peng, J.T., Zhou, Y.C., et al.: Extreme learning machine for ranking: generalization analysis and applications. Neural Netw. **53**, 119–126 (2014)
9. Choudhary, L., Burdak, B.S.: Role of ranking algorithms for information retrieval. Int. J. Artif. Intell. Appl. **3**(4), 203–220 (2012)

10. Chun, Y.H., Sumichrast, R.T.: A rank-based approach to the sequential selection and assignment problem. Eur. J. Oper. Res. **174**(2), 1338–1344 (2006)
11. Coletta, L.F.S., Vendramin, L., Hruschka, E.R., et al.: Collaborative fuzzy clustering algorithms: some refinements and design guidelines. IEEE Trans. Fuzzy Syst. **20**(3), 444–462 (2012)
12. Dubois, D., Prade, H.: Operations on fuzzy numbers. Int. J. Syst. Sci. **9**(6), 613–626 (2007)
13. Džeroski, S.: Relational Data Mining, Springer (2010)
14. Figueira, J., Greco, S., Ehrgott, M.: Multiple Criteria Decision Analysis: State of the Art Surveys, Springer (2005)
15. Hsu, W.C., Liu, C.C., Chang, F., et al.: Selecting genes for cancer classification using SVM: an adaptive multiple features scheme. Int. J. Intell. Syst. **18**(12), 1196–1213 (2013)
16. Hwang, C.L., Yoon, K.: Multiple Attribute Decision Making: Methods and Applications. Springer, New York (1981)
17. Jiang, Y., Chung, F., Wang, S., et al.: Collaborative fuzzy clustering from multiple weighted views. IEEE Trans. Syst. Man Cybern. **45**(4), 688–701 (2015)
18. Lee, S., Song, S., Kahng, M. et al.: Random walk based entity ranking on graph for multidimensional recommendation. In: The 5th ACM International Conference on Recommender System, pp. 93–100 (2011)
19. Li, G., Wang, L.Y., Ou, W.H.: Robust personalized ranking from implicit feedback. Int. J. Pattern Recognit. Artif. Intell. **30**(01) (2016)
20. Liu, S.H.: Clustering analysis for data with relational information. Ph.D. Thesis, Beijing Normal University (2014)
21. Liu, T.Y.: Learning to rank for information retrieval. Found. Trends Inf. Retr. **3**(3), 225–331 (2009)
22. Mansoori, E.G.: Using statistical measures for feature ranking. Int. J. Pattern Recognit. Artif. Intell. **27**(01), 14 (2013)
23. Mihalcea, R.: Graph-based ranking algorithms for sentence extraction, applied to text summarization. In: The 2004 ACL on Interactive Poster and Demonstration Sessions, p. 4 (2004)
24. Omladic, M., Semrl, P.: On the distance between normal matrices. Proc. Am. Math. Soc. **110**(3), 591–596 (1990)
25. Pathak, A., Pal, N.R.: Clustering of mixed data by integrating fuzzy probabilistic, and collaborative clustering framework. Int. J. Fuzzy Syst. **18**(3), 339–348 (2016)
26. Pawlak, Z.: Rough sets. J. Comput. Inf. Sci. **11**(5), 341–356 (1982)
27. Pedersen, C.R., Nielsen, l.R., Andersen, K.A., et al.: An algorithm for ranking assignments using reoptimization. Comput. Oper. Res. **35**(11), 3714–3726 (2008)
28. Pedrycz, W.: Collaborative fuzzy clustering. Pattern Recognit. Lett. **23**(14), 1675–1686 (2002)
29. Przybylski, A., Gandibleux, X., Ehrgott, M.: A two phase method for multi-objective integer programming and its application to the assignment problem with three objectives. Discret. Optim. **7**(3), 149–165 (2010)
30. Qian, Y.H., Liang, J.Y., Dang, C.Y.: Interval ordered information systems. Comput. Math. Appl. **56**(8), 1994–2009 (2008)
31. Qian, Y.H., Liang, J.Y., Song, P., et al.: On dominance relations in disjunctive setvalued ordered information systems. Int. J. Inf. Technol. Decis. Mak. **9**(1), 9–33 (2010)
32. Qin, T., Zhang, X.D., Tsai, M.F., et al.: Query-level loss functions for information retrieval. Inf. Process. Manage. **44**(2), 838–855 (2008)
33. Song, F.X., You, J., Zhang, D., et al.: Impact of full rank principal component analysis on classification algorithms for face recognition. Int. J. Pattern Recognit. Artif. Intell. **26**(03), 1256005(23 pages) (2012)
34. Tiskin, A.: Fast distance multiplication of unit-monge matrices. Algorithmica **71**(4), 859–888 (2015)
35. Wang, P.Z.: Factorial analysis and data science. J. Liaoning Tech. Univ. **34**(2), 273–280 (2015)
36. Wang, H.D., Wang, P.Z., Shi, Y., et al.: Improved factorial analysis algorithm in factorspaces. In: Proceedings of 2014 International Conference on Informatics, Networking and Intelligent Computing, Shengzhen (2014)

37. Wang, P.Z., Li, H.X.: A Mathematical Theory on Knowledge Representation. Tianjin Scientific and Technical Press, Tianjing (1994)
38. Wang, P.Z., Liu, Z.L., Shi, Y., et al.: Factor space, the theoretical base of data science. Ann. Data Sci. **1**(2), 233–251 (2014)
39. Wang, P.Z., Sugeno, M.: The factors field and background structure for fuzzy subsets. Fuzzy Math. **2**(2), 45–54 (1982)
40. Webber, W., Moffat, A., Zobel, J.: A similarity measure for indefinite rankings. ACM Trans. Inf. Syst. **28**(4), 20 (2010)
41. Yu, F.S., Luo, C.Z.: Granule factors space and intelligent diagnostic expert systems. In: Proceedings of the 7th National Conference on Electric Mathematics, Advances of Electric Mathematics, China Science & Technology Press, Beijing (1999)

Learning Through Constraint Applications

Vladimir Jotsov, Pepa Petrova and Evtim Iliev

1 Introduction

The filtration of incoming data or knowledge or, in other words, focusing attention on most important details is one of the most frequently used problems in intelligent systems [1–3]. Many researchers consider that well-known statistical methods provide key solutions to the problem. In our opinion, it can't be resolved without the usage of evolutionary applications aiming at fusion of logical and statistical methods in one general system.

Anomaly detection is frequently used in intrusion detection/prevention (IDS/IPS) and other systems. Its principle is based on a conflict resolution strategy: when some of the parameters are cardinally different from the expected ones, the system must focus attention in this direction. For pity, the intruders may apply decoys, exploit some *negligible* parameters not included in the observations or gradually introduce slight changes to decept such security systems. Hence the focusing problem still stays in focus.

As a whole, statistical methods can cover the full picture of the subject area, but cannot fully represent the data/knowledge in their dynamics. Hence they are weak in terms of acquiring and post processing new information [4, 5]. In other words, they are as good as the person who uses them and one of their important weaknesses is the need for human (subjective) intervention. For example, incorrectly defined regions (strata) or selected unevenly distributed variables for cluster analysis, the results of statistical surveys can be misleading. To eliminate the human factor in the process, which most often is called the heuristic factor, it is necessary to use the logical methods combined with statistical ones. An example of a logical approach that might complement and support statistical research is the proposed Puzzle method.

V. Jotsov (✉) · P. Petrova · E. Iliev
University of Library Studies and Information Technologies (ULSIT),
P.O. Box 161, 1113 Sofia, Bulgaria
e-mail: v.jotsov@unibit.bg

V. Sgurev et al. (eds.), *Learning Systems: From Theory to Practice*, Studies in Computational Intelligence 756, https://doi.org/10.1007/978-3-319-75181-8_8

Recently, many statistical applications are called intelligent, primarily because of the increasing prevalence and popularity of the latter. Unfortunately, the use of any statistical means does not make systems intelligent [4, 6, 7]. The merging of these two groups disparate results cannot be done mechanically. To do so, we must use different special evolutionary processes, including some elements of primitive machine thinking, which is most often defined as evolutionary data and knowledge processing. Let's assume that we will use available statistical methods in situations where we have not accumulated enough knowledge on the subject in knowledge-poor domains. In this case, the main reason for the use of probabilistic methods is following: weak and unreliable information is better than the complete lack of information. Therefore, in the absence of experience in the system (in the autonomous agents) it must forcibly use probabilistic or similar assessments (fuzzy, etc.). However, in accumulating enough experience in the system (it has developed methods for such evaluations) probabilistic estimates are gradually canceled, defeated (replaced by more universal and more flexible logical methods). Regardless of the manner of realization, described processes are evolutionary by nature. Thus closer to the creation of future thinking machines, in which to improve the efficiency of the applications, a variety of synthetic methods will be used, like the considered set of Puzzle methods [8]. The processing of the information is gradually improved at each stage of the above mentioned evolutionary processes.

At the confluence of the results of the proposed two types of methods using the "most adaptable wins" (survival of the fittest). Thus, decisions that are initially assessed as weak or conflicting, in time can be developed and be revalued and become priority. For desired targets is developed an original evolutionary method which combines both statistical and logical information that is described in detail in the chapter [9] where Puzzle principles possess many features of the standards. This is not a coincidence: the Puzzle standards are good for elaborating data-driven applications [10], while most of the presented at the conference methods are based on a certain set of algorithms. There is no discrepancy in the presented descriptions: the most of contemporary data-driven applications also use algorithms.

The purpose of this chapter is to present some advantages of the Puzzle methods in building predictive models. The presented approach is compared with other well-known methods and data-driven approaches.

Next section is dedicated to the resolution of semantic conflicts as an instrument of autonomous information processing. Then the exploration is considered of newly developed types of constraints and of main data-driven principles of the proposed set of Puzzle methods. In Sect. 3 the Puzzle principles have been used for the elaboration of curious systems aiming at prediction of attacks. Logical modeling of data for security purposes is considered in Sect. 4. Sections 5 and 6 include few algorithmic examples of Puzzle-based applications. The last Sect. 7 discusses an application of Puzzle standards in SAS Enterprise Miner 14.1^{TM}.

2 Semantic Conflict Identification and Resolution

In theory of constraints there is one that is named a **focus constraint** because the goal achievement depends just on it. Analogically, we introduce a set of events considered significant, and sometimes of critical importance to the operating machine and its goals. Conditions may exist when it should stop the execution of its goals to trace and process the set of critical events. Of course, it is impossible to allocate the whole resource in this important direction because, for example, in security systems the intruders may exploit a critical event to attract the system attention far from their attack.

The main idea of the proposed conflict resolution is that the identification of any semantic conflict could be used as a tool to focus attention in this direction. Hence, the software agent or other IT application could concentrate its resources and efforts on knowledge concerning the revealed conflict and pay less attention to the other signals/information sources. In security systems any lack of collaboration in a group of agents or the intrusion could be found as an information/semantic conflict with existing models. Many methods exist where a model is constructed and every non-matching it knowledge is assumed as contradictory. Let's say, in an anomaly intrusion detection system, if the traffic has been increased, it is a conflict to the existing statistical data and an intrusion alert has been issued. The considered approach is to discover and trace different logical connections to reveal and resolve conflict information. The constant inconsistency resolution process gradually improves the system DB and KB, and leads to better intrusion detection and prevention. Models for conflicts are introduced and used, and they represent different forms of ontologies.

Let the strong (classical) negation be denoted by '¬' and the weak (conditional, paraconsistent negation be '~'. In the case of an evident conflict (inconsistency) between the knowledge and its ultimate form—the contradiction—the conflict situation is determined by the direct comparison of the two statements (the *conflicting sides*) that differ one from another by just a definite number of symbols '¬' or '~'. For example: A and ¬A; B and not B (using ¬ equivalent to 'not'), etc.

In the case of implicit (or hidden) negation between two statements, A and B can be recognized only by an analysis of preset models of the type of (1).

$$\{U\}\,[\eta: A, B] \tag{1}$$

where η is a type of negation, U is a statement with a validity including the validities of the concepts A and B, and it is possible that more than two conflicting sides may be present. It is accepted below that the contents in the figure in brackets U is called *an unifying feature*. In this way, it is possible to formalize not only the features that separate the conflicting sides but also the unifying concepts joining the sides. For example, the intelligent detection may be either automated or of a human-machine type but the conflict cannot be recognized without the investigation of the following model.

$$\{\text{detection procedures}\}\ [\neg\text{: automatic, interactive}]\ .$$

The formula (2) formalizes a model of the conflict the sides of which implicitly negate each another. In the majority of the situations, the sides participate in the conflict only under definite conditions: $\chi_1, \chi_2, \ldots \chi_z$.

$$\{U\}\left[\eta\text{: } A_1, A_2 \ldots A_p\right] < \tilde{\chi}_1 * \tilde{\chi}_2 * \cdots * \tilde{\chi}_z > . \tag{2}$$

where $\tilde{\chi}$ is a literal of χ, i.e. $\tilde{\chi} \equiv \chi$ or $\tilde{\chi} \equiv \eta\chi$, * is the logical operation of conjunction, disjunction or implication.

The present research allows a transition from models of conflicts to ontologies in order to develop new methods for revealing and resolving contradictions, and also to expand the basis for cooperation with the Semantic Web community and with other research groups. This is the way to consider the suggested ontology-type models from (2) as one of the forms of static ontologies.

The following factors have been investigated:

- T time factor: non-simultaneous events do not bear a conflict.
- M place factor: events that have taken place not at the same place, do not bear a conflict. In this case, the concept of place may be expanded up to a coincidence or to differences in possible worlds.
- N a disproportion of concepts emits a conflict. For example, if one of its parts is a small object and the investigated object is very large, then and only then it is the conflict case.
- O identical object. If the parts of the contradiction are referred to different objects, then there is no conflict.
- P the feature should be the same. If the conflicting parts are referred to different features, then there is no conflict.
- S simplification factor. If the logic of user actions is executed in a sophisticated manner, then the conflict is detected.
- W mode factor. For example, if the algorithms are applied in different modes, then the conflict is detected.
- MO conflict to the model. The conflict appears if and only if (*iff*) at least one of the measured parameters does not correspond to the meaning from the model. For example, the traffic is bigger than the maximal value from the model.

Example. We must isolate errors that are done due to lack of attention from tendentious faults. In this case we introduce the following model (3):

$$\{\text{user: faults}\}\ [\sim\text{: accidental, tendentious}]\ <T, \neg M, O; \neg S> \tag{3}$$

It is possible that the same person does sometimes accidental errors and in other cases tendentious faults; these failures must not be simultaneous on different places and must not be done by same person. On the other hand, if there are multiple errors (e.g. more than three) in short intervals of time (e.g. 10 min), for example, during authentications or in various subprograms of the security software, then we have a case of a violation, nor a series of accidental errors. In this way, it is possible to apply comparisons, juxtapositions and other logical operations to form security policies thereof.

2.1 New Ways for Conflict Resolution

Recently we shifted conflict models with ontologies that give us the possibility to apply new resolution methods. For pity, the common game theoretic form of conflict detection and resolution is usually heuristic-driven and too complex. We concentrate on the ultimate conflict resolution forms using contradictions. For the sake of brevity, the resolution groups of methods are described schematically.

The conflict recognition is followed by its resolution. The schemes of different groups of resolution methods have been presented in Figs. 1, 2, 3 and 4.

In situations from Fig. 1, one of the conflicting sides does not belong to the considered research space. Hence, the conflict may be not be immediately resolved, only a conflict warning is to be issued in the future. Let's say, if we are looking for an intrusion attack, and side 2 matches printing problems, then the system could avoid the resolution of this problem. This conflict is not necessary to be resolved automatically, experts may resolve it later using the saved information. In Fig. 2, a situation is depicted where the conflict is resolvable by stepping out from the conflict area. This type of resolution is frequently used in multi-agent systems where conflicting sides step back to the pre-conflict positions and one or both try to avoid the conflict area. In this case a warning on the conflict situation has been issued.

The situation from Fig. 3 is automatically resolvable without issuing a warning message. Both sides have different priorities, let's say side 1 is introduced by a security expert, and side 2 is introduced by a non-specialist. In this case, side 2

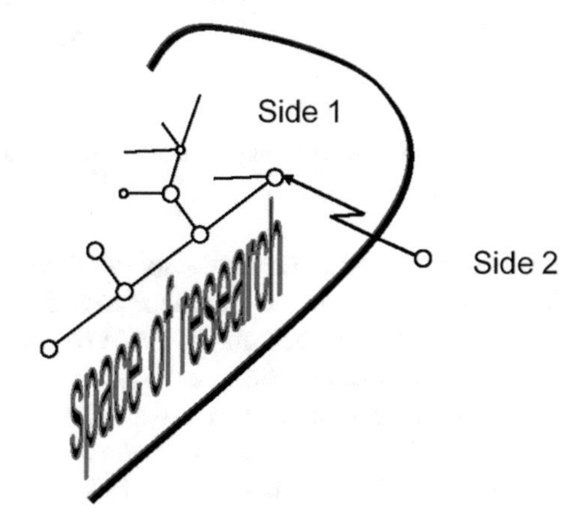

Fig. 1 Avoidable (postponed) conflicts when side 2 is outside of the research space

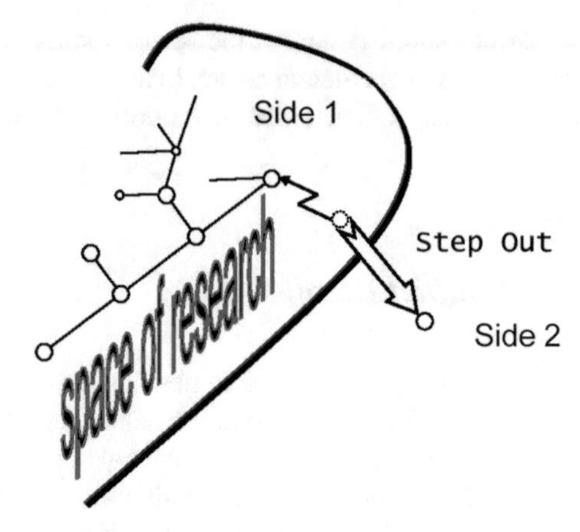

Fig. 2 Conflict resolution by stepping out of the research space (postponed or resolved conflicts)

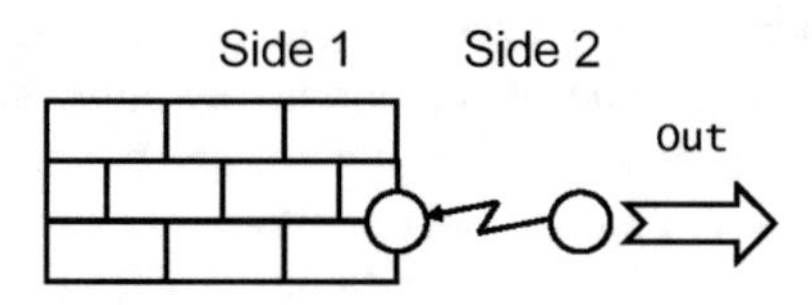

Fig. 3 Automatically resolvable conflicts

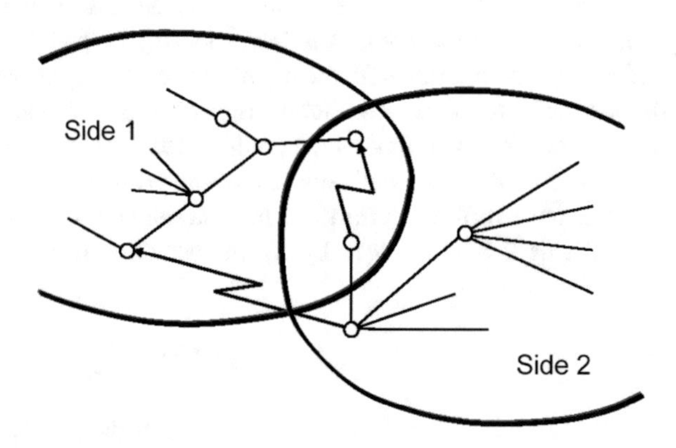

Fig. 4 Conflicts resolvable using human-machine interactions

has been removed immediately. A situation is depicted on Fig. 4 where both sides have been derived by an inference machine, let's say by using deduction. In this case, the origin for the conflict could be traced, and the process is using different human-machine interaction methods.

At last, the conflict may be caused by incorrect or incomplete conflict models (1) or (2), in this case they have been improved after same steps as shown in Figs. 1, 2, 3, 4 and 5. The resolution leads to knowledge improvement as shown in Fig. 5,

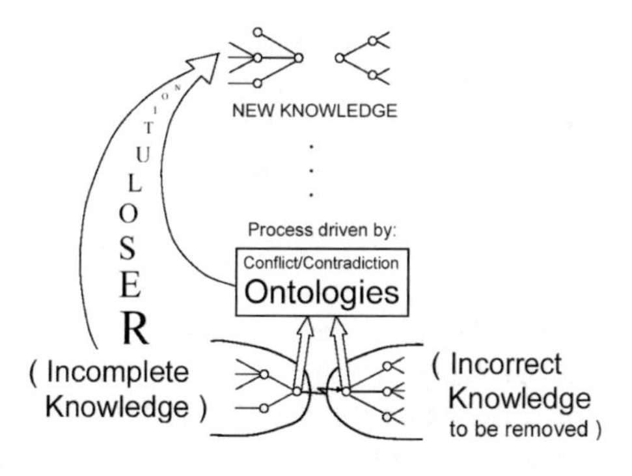

Fig. 5 Learning via resolving conflicts or with contradictions

and this improvement gradually builds a machine self-learning. Of course in many situations from Figs. 1, 2, 3, 4 and 5 the machine prompts the experts just in the moment or later, but this form of interaction is not so boring, because it appears in time and it is a consequence of a logical reasoning based on knowledge (1) or (2). Our research shows that automatic resolution processes may constantly stay resident using free computer resources OR, in other situations, they may be activated by user's command. In the first case the knowledge and data bases will be constantly improved by continuous elimination of incorrect information or by improving the existing knowledge as a result of revealing and resolving contradictions. As a result, our contradiction resolution methods have been upgraded to a machine self-learning method, i.e. learning without a teacher which is very effective in the case of ISS. This type of machine learning is novel and original in both the theoretical and applied aspects.

2.2 *Self-organization of Evolving Systems via Resolution of Conflicts/Contradictions*

Modern tools that are applied for information security possess the common drawback that they cannot process data in an evolutionary mode without the direct human-expert interference. We are speaking about the evolution without known fitness function but with a 'data-driven' one. On the other hand, subjects that cannot autonomously process data or knowledge evolutionarily, are out of perspectives to learn how to think.

The presented evolutionary processing of data or knowledge is based on two procedures. Objects including agents and their surroundings are modeled. For this purpose the usage of the introduced ontology relations answering following questions **HOW**, **WHY**, what is **IMPORTANT**, etc. becomes greater and greater. The richer the arsenal of the means used, the more effective the system operation is. The second information process registers changes in the incoming information with respect to the models. This is the place to use various methods to compare and to calculate distances via the usage of the different metrics; to reduce the complexity of the applications of the agents, it is recommended to apply more elementary methods as juxtapositions, etc. Let us assume, that, for example, the change of time for attribute A_1 of ontology O_i is $\Delta A_1(\Delta t)$:

$$\Delta A_1 (\Delta t) = \|A_1 (t + \Delta t) - A_1 (t)\| \quad (4)$$

If $\Delta A_1(\Delta t) \leq T$ where T is an a priori set threshold value then it is accepted that the change is insignificant and that there is no need to expect reactions from the agent. On the other hand when $\Delta A_1(\Delta t)$ is compatible with T then it is necessary to inspect a greater time slice Δt to explore the dynamics of the change, asking whether there is a gradual accumulation of cardinal changes. Attributes may change without using the temporal factor from (4). For example, in the case $\Delta A_1 = \llcorner A'_1 - A_1 \llcorner$, the critical mass of attacking computers from botnets will block the provider but there is no data **when** this will happen, and there is no need to include the temporal factor in the goal description.

Another example is a world-wide chart of cyber criminality. Based on it, the agents can draw useful information about the regions with lowest and highest criminality and they can search the reasons for it. With the same success the domain may be shifted: if an example from Number Theory is considered, then areas with most composite or prime numbers are established and the reason for such distribution is analyzed. The presented evolutionary information processing is based on the detection, analysis and isolation of inconsistencies or contradictions thus passing through conflicts or through contradictions, makes ISS models more improved. In spite of the theoretical nature of the application (cryptography) or the heuristic problems (information input about the reasons for cyber criminality in various regions world-wide), the attention is concentrated over extremities, e.g. the regions with highest or lowest criminality. Analysis of the data with extreme values most often lead to contradictions which in knowledge-poor environments must be resolved in an evolutionary way. Evolving methods from previous sections give new ways to resolve contradictions which are not discussed above. Their application is successful not only in separate agents but also in MAS that can be a type of collective evolutionary systems. With that, as compared to popular swarm and other strategies, the coordination applied in the present research is not controlled heuristically but via constraint satisfaction for the constraints imposed by conflicts or (better) contradictions. For more details see descriptions of the Puzzle method, as well as of the introduced evolving methods.

Let's assume that the model M* will be used where attributes of one or more objects are from the set $A = \{ A_1, \ldots, A_N \}$. The very model M* can be presented as one of the parameters of the individual from the population, so it must be replaced by M** which is chosen due to its similarity estimate with respect to M* or according to other properties in the accepted classical methods of genetic algorithms. It is possible to change the examined set in M**. With that, after one or more iterations the set A will contain just elements that are invariant to model changes. The greater the distance between M* and the current model, the more important the role of the invariant attributes will be. On the other hand, there is a difference from classical algorithms in the way that there is a possibility of the evolution, mutations and crossovers to be independent of probabilistic estimates or from different forms of heuristics, they can depend on the constraints instead.

In the interpretation proposed above, thinking is a reaction to a change and the greater the change is, the greater attention will be paid to explore the reasons leading to its appearance. According to this logic, training is not only a memorizing teacher's activities as it is in the classical case but also uses resolutions of inconsistencies, contradictions via evolutionary information processing; in other words, the strategy 'doubt in everything' is used. The transition from M* to M** inspected above, is a single case of this strategy. In this way, elements of unsupervised training are added and even of self-organization in case of a supervised training. Various practical applications of the strategy are investigated because the trainable agent often cannot change the training scenario or it cannot ask the supervisor contra questions. As a result of the proposed evolutionary way of training, acquired knowledge is constantly refined or corrected.

Following definition had been applied in the proposed research.

Definition. *Potentially interesting object* as a basis for potentially interesting event is an object $o_l \in O$ satisfying at least one of the following conditions

a. It should be investigated, it is unknown;
b. It contains an unexplored repeating part;
c. Its appearance isn't predicted;
d. It bears an inconsistent information of any kind compared to the existing model: its parameters are much lower or higher then anticipated, etc.
e. Its parameters are rapidly changing, above the a priori given threshold. In some cases too slowly change is also included in condition **e**. The acceleration rate is also important.
f. It attracts attention. For example, it is beautiful or it has an extremal [min/max] value or it isn't observable, is fuzzy, etc.

   Interdependence may exist between conditions **a** up to **f**. Every condition could be ranked according to the current goals and tasks of the system. As a result, from thousands or millions possible goals only the potentially interesting ones will be investigated altogether with the critical ones.

3 On Puzzle Methods and Standards

The essence of the Puzzle methods [9] and their various manifestations lies in mimicking the human behavior in problem-solving. For this purpose, of course, we must have a minimum required volume of background information for the task to be solved. In this situation different aid descriptive statistical methods apply, among them variable worth analysis, cluster analysis, regression analysis [5]. Through them we can extract the necessary information that would later be used as a basis for a subsequent research. Once we have processed data related in one way or another to our goal, the Puzzle methods reveal links between these pieces of knowledge that are known to us and bind them through special constraints with this knowledge. For this reason, we use Binding or Pointing (guiding) constraints, and sometimes in they are applied different combinations. For example, let the goal is: incoming data obtained in real time from sensors to be bound by previously introduced knowledge of a given subject area and thus be able to confirm or refute a hypothesis. This process is rather complicated and laborious, but in most cases allows us to optimize our search for a solution and give more precise answers to the problem. The essence of the considered method lies in the use of some basic tools which significantly increase the rate of search in a given area, and even it is possible to find a direct solution to the problem. The used constraints are:

- Spatial/classical constraints (can be both linear/nonlinear).
- Crossword constraints: a piece of knowledge that has an intersection with the solution and the more they are, the easier it is to achieve the goal.
- Binding constraint: a piece of knowledge that is not a solution but is closely around it. It signals how close the solution is. Causal relations that cannot be described by implication are widely used here.
- Pointing constraints: this kind of constraints indicates the direction in which the necessary solution should be seek. Such constraints are quite obscure, so it is convenient to combine them with other types of constraints such as Binding constraints.

It is very important for applications to take into account the interactions between different types of constraints. For example, the Pointing constraints make particularly effective use with combination of Binding or Crossword constraints. Thereby, knowledge is formed for efficiently processing large volumes of data (Big Data) by narrowing the search area and the use of combinations of classical and non-classical logic means. Currently, we are developing new modifications of the methods in which depending on the situation the meanings of Binding and Pointing constraints will be calculated and changed at any point in the study area, as well as links between subjective probabilities and Binding constraints. Overall, introduced original, non-classical logic-based systems of constraints help to significantly reduce the algorithmic complexity of logical-probabilistic applications and not just to search but to improve the precision of desired goals.

All the described types of constraints contain numerous subsets inside each type. For example, as shown on Fig. 6, when the Binding application interprets accuracy features, we should use the restriction area like the one depicted in dark grey (blue in color variant). On the other hand, when the Binding process concerns the area of influence of one or another factor, this is an absolutely different type of Binding, and it is depicted in a form of a larger, outer circle. During the evaluation process, the smaller, colorful area should diminish while the outer circle may enlarge.

The class of Binding constraints is rather extensive one. The concentric curves may represent different possibility "isolines". In other cases the probability or possibility of the considered event is equal inside the whole closed area from Fig. 6: it is shown that the solution is inside the closed area, and no more data.

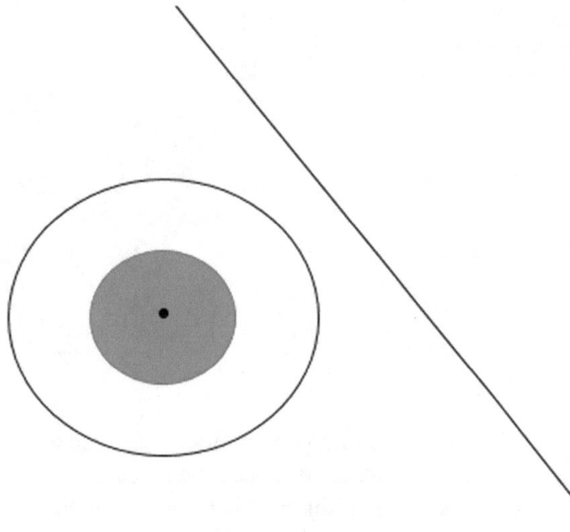

Fig. 6 Two types of Binding constraints

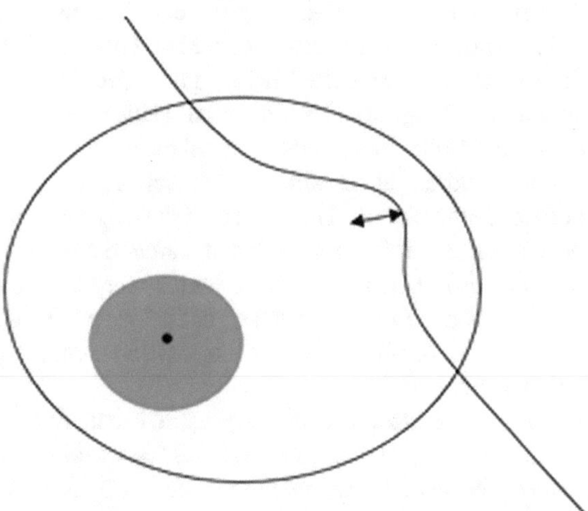

Fig. 7 The influence of Binding to classical (linear) constraints

One of the cases of influence of Binding to classical (linear) constraints is depicted in Fig. 7. When enlarged, the Binding area swallows a large part of the classical constraint, as a result the latter becomes partially non-linear. It is interesting to observe, that this process of influence can be effectively described in terms of fuzzy systems. Same influence had been detected in pointing constraint applications.

It seems that Puzzle methods can reveal the routes of the successful fuzzy applications. They effectively bind the unknown knowledge to accumulated, known ones. Actually, what is unknown is a notion that is rather fuzzy by nature. The next experiment aims to shift it up with other fuzzy notions. On Fig. 8 the case of Certain/Uncertain descriptions has been depicted.

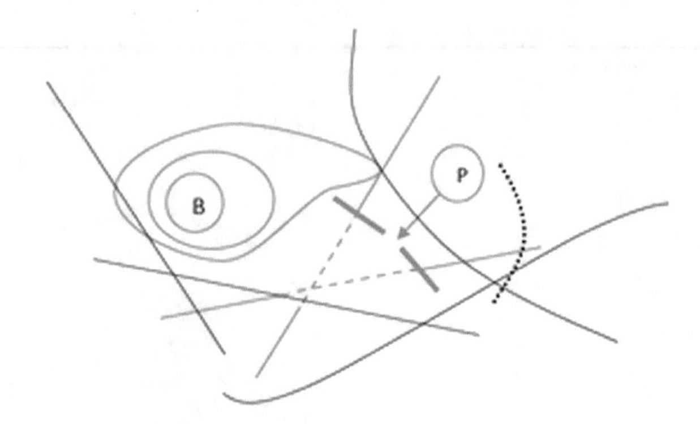

Fig. 8 The Certain/Uncertain interpretations

On Fig. 8 the Pointing constraint is depicted as P; the binding constraints, three types, have been depicted as Bs; the crossword constraints with the associated to them resolved parts of the goal have been depicted as two interrupted lines, and the other types of the constraints are the classical ones.

The whole set of the logic-oriented and the classic constraints fixes one particular piece of knowledge, called here a goal. The depicted area confines the set of the certain knowledge set about the goal. Hence all the information outside this area is an uncertain one. In a similar way other fuzzy notions could be modeled.

For example, let the situation concerning the goal 'Is he rather dangerous?' is modeled in the picture. Then the set of the depicted constraints should be analyzed for the calculation of the fuzzy value of 'rather dangerous'. The proposed rich modeling tools give us much more possibilities than the classical methods. The confined region is divided into subsets where the goal is located. This is the 'area in a focus' and the area for more detailed analysis, exploration area, the base for applications of curios systems, etc.

When the corresponding fuzzy value is calculated, the question appears: in what conditions, when, where, in what situation could be applied this value? Even if an ontology is created based on the modeling from Fig. 8, the situation/conditions could be defeated or cardinally change its semantics. The Puzzle data-driven standards not only can influence on the quoted changes but also provide possibilities to find and correct errors or knowledge incompleteness. Usually the data-driven research is much more complicated than the algorithmic one. The data-driven applications in this direction will be investigated in the nearest future.

Knowledge used in the Puzzle methods can be presented as a part of information (atoms), connected to certain relations. Usually, these non-classical (invisible) relations are obtained by logical processing like structuring, extracting meaning from text or other processing. In this terminology, a rule not necessarily containing classical implicative connections between the premises and the conclusion can be represented as follows. Denote the conjunction of the premises as the following elements: A_1, A_2 … A_z.

The conclusion is labeled as B.

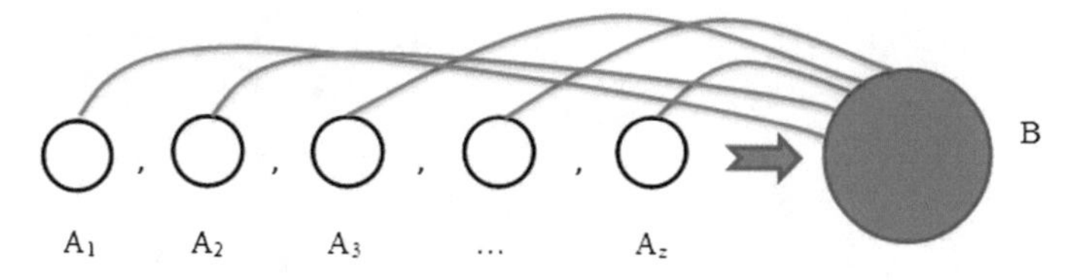

Fig. 9 The influence of Binding to classical (linear) constraints

If all conjunctions have been confirmed, the goal B is true. Hence, the task to verify that B is true is solved (Fig. 9).

When the significance of each of the atoms of the prerequisites A_i is not specified, as shown in Fig. 9, its conjunctions are deemed equivalent (1/z).

4 Logical Modeling of Data for Prediction

In this part of the paper are presented some steps and formulas used to predict the response to the question "Is he rather dangerous?" The proposed approach is data-driven in meaning of definition in [11]. Data-driven methods give answers coming from the data [1, 4]. There are some techniques for data-driven modeling in overlapping areas such as artificial intelligence, computational intelligence, soft computing, machine learning, data mining and others, described in [11]. In this paper the data-driven technology represents a symbiosis of statistics and logical data processing to reduce demand and decrease the error in the forecast. A curiosity function is defined to help identify abnormal behavior for prediction [12].

The presented example is about ways to detect people who are most likely to carry out a bombing. Everyone who comes into the departure area of the airport, is scanned by a special system. In this case, each person is identified by then characteristics from the system. All features form a set A.

$$\left.\begin{array}{l} ch_1 \\ ch_2 \\ ch_3 \\ \dots \\ ch_n \end{array}\right\} a_p \in A, \quad where\ p = \{1, m\} \tag{5}$$

Set A is presented as follows (Fig. 10).

Each characteristic is indexed and has a polarity. Every person possesses a set of positive and negative characteristics. Positive characteristics that identify a person as safe, are presented by ch_i. Negative characteristics that identify a person as dangerous are presented by ch_j.

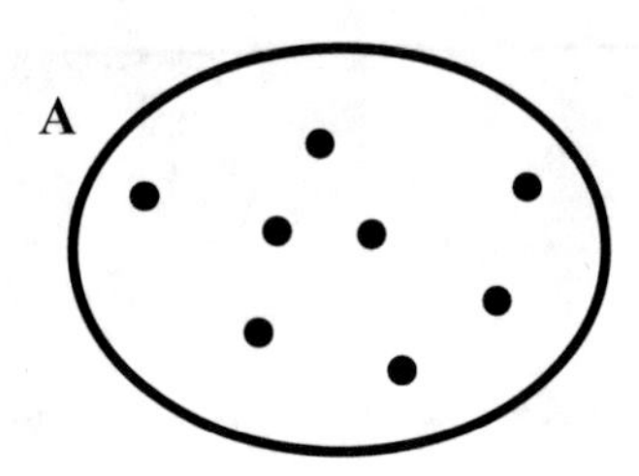

Fig. 10 The general set

$$ch_1^n \in \begin{cases} U_p - positive\, ch_i, where\, i = \{1, n\} \\ U_q - negative\, ch_j, where\, j = \{1, n\} \end{cases} \tag{6}$$

The goal is to identify whether a person is potentially dangerous or not. In this case the true value is not binary but continuous between 0 and 1. A curiosity function is used to reach the goal and find the truth about the visitors. Curiosity is calculated for each person as follows:

$$Q\left(a_p\right) = \frac{count\,(ch_i) - count\left(ch_j\right)}{count\,(ch_i) + count\left(ch_j\right)} \tag{7}$$

This function is calculated as a difference between the positive and negative characteristics divided to their sum. If positive characteristics prevail, then the value of the function is positive and tends to number one. If negative characteristics outweigh, the function has a negative value and tends to minus one.

The result for Q is calculated and represented in Table 1.

According to the polarity of the function, three cases may be considered.

First case: "The person is safe"

$$\text{If Q}\left(a_p\right) = 1, \Rightarrow the\, person\, is\, safe \tag{8}$$

Table 1 Precalculated Q for each possibility of positive and negative characteristics

Positive	Negative	Q
10	0	1
9	1	0.8
8	2	0.6
7	3	0.4
6	4	0.2
5	5	0
4	6	−0.2
3	7	−0.4
2	8	−0.6
1	9	−0.8
0	10	−1

Fig. 11 Identifying safe elements

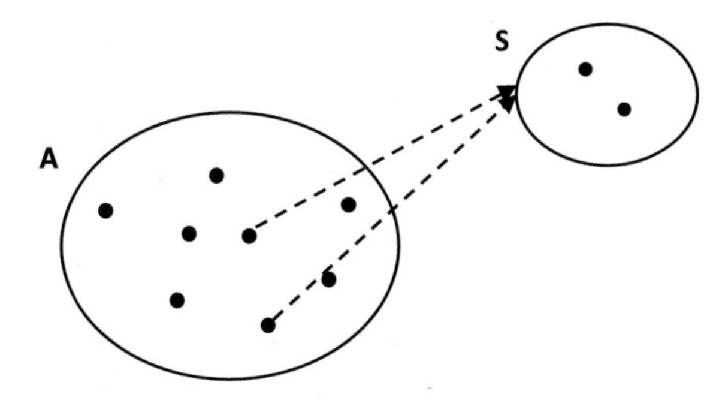

As can be seen from Table 1, the value of curiosity is 1 when a person is characterized with 10 positive and 0 negative characteristics. In this case, the only option is that the person is safe and the corresponding element is moved into subset S.

$$a_p \xrightarrow{is\, moved\, to\, subset} S, \text{ where } p = \{1, m\} \tag{9}$$

This alternative defines as an element of the safe subset S:

$$a_p \in S \tag{10}$$

In this case, there exist two safe elements with 10 positive and 0 negative characteristics. All visitors, who have 10 positive points are determined as safe and become elements of a subset S, as can be seen in Fig. 11.

Second case: "The person is not safe"

$$\text{If Q}\left(a_p\right) = -1, \Rightarrow the\ person\ is\ not\ safe \tag{11}$$

When a_p has 10 negative and 0 positive characteristics, then it is not safe and becomes an element of the subset G.

$$a_p \xrightarrow{is\, moved\, to\, subset} G, \text{ where } p = \{1, m\} \tag{12}$$

This case defines a_p as an element of the target subset G:

$$a_p \in G \tag{13}$$

In the example one dangerous person appears with 10 negative characteristics and it is moved into the target subset G (Fig. 12).

Third case: "Curiosity whether the person is dangerous or not"

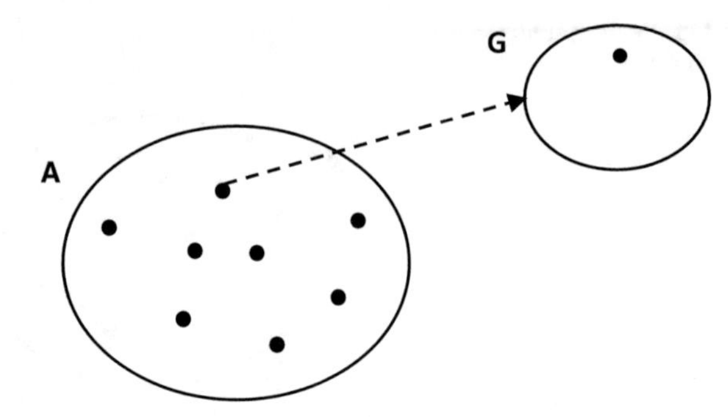

Fig. 12 Identifying dangerous elements

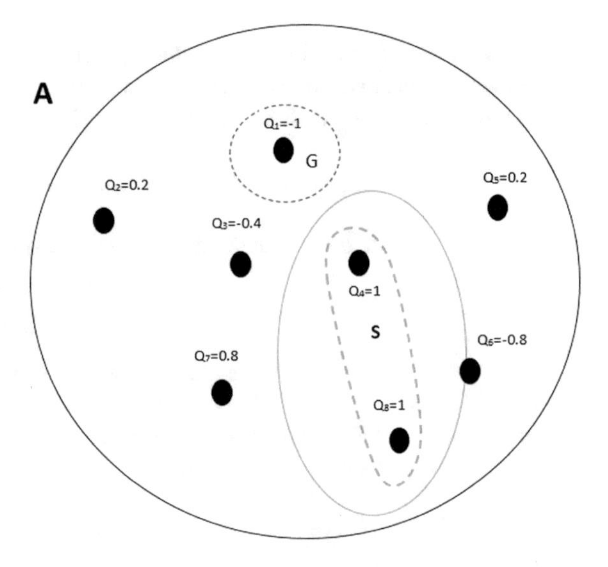

Fig. 13 Elements in set A with their status of the curiosity function

$$\text{If Q}\left(a_p\right) = 0 \Rightarrow \textit{the person has} \left[\left[\begin{array}{c} 5\,positive \\ and \\ 5\,negative \end{array}\right]\right] \textit{characteristics.} \quad (14)$$

What does it mean that $\text{Q}\left(a_p\right) = 0$?

The person may be dangerous or not and this can't be determined by statistical means. In this case, no elements exist with 5 positive and 5 negative characteristics.

The elements and their curiosity status are presented in Fig. 13.

First step: Explore the elements which have curiosity status equal or close to $\text{Q} = 0$. They are Q2 and Q5 with equal curiosity status $\text{Q} = 0.2$ (Fig. 14).

Second step: Explore the elements which have curiosity status close to minus 1. It can be seen that Q6, which was connected with the safe subset, has a curiosity status not far from minus 1 ($\text{Q6} = -0.8$). In this step the Puzzle methods are used

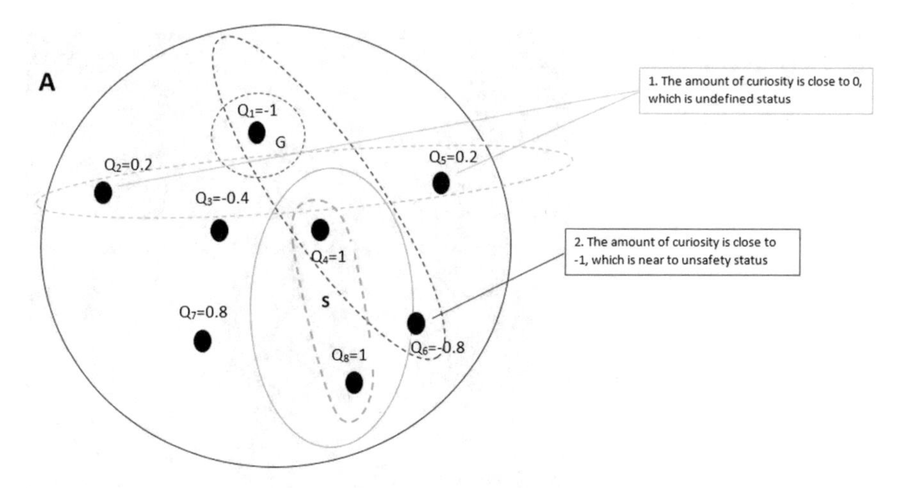

Fig. 14 The Puzzle methods for identification of not safe persons

to found out whether Q6 make connections to the target subset G. During the search process, the constraints of the Puzzle methods unlock additional information about the element.

Based on the curiosity function and constraints of the Puzzle methods, the element Q4 falls within the scope of element Q6 with the target subset G. This creates a curiosity around element Q4. In the beginning the element was classified as rather safe because of its status of curiosity equal to 1 ($Q4 = 1$). Now, thanks to a combination of approaches for data processing, the search is narrowed to a different direction. As a result, it was discovered that 100% safe visitor has a relation with dangerous visitors at the airport. Using this technique, the search can be shortened significantly and even can be applied to detect dangerous elements which the standard methods of search could miss. Figure 10a, b visualize the result of both techniques (NNS—Neural Network and Statistics; NNP—Neural Network and Puzzle).

Figure 15a presents the percentage of accurate and wrong predictions, based on 150 cases. It can be seen that using the Puzzle methods the result is 89% accurate and 11% wrong predictions against respectively 81% and 19% with statistics.

Figure 15b presents wrong versus accurate predictions. In gray is percent correlation of wrong results in relation to accurate results for each technique, in percent. In yellow is the correlation of wrong results in relation to all, 150 cases for each technique, in percent. A second dimension is added to the figure showing up accurate (blue line) and wrong (orange line) results. It can be seen that using NNP technique, accurate predictions are increased and wrong are reduced in comparison with NNS. The values of the charts can be seen in Table 2.

The application of Puzzle methods in this paper shows how statistical and logical applications may work together to unlock unseen information and adjust the search in a different way. This approach is less time consuming and gives less mistakes in prediction, compared with statistical ones.

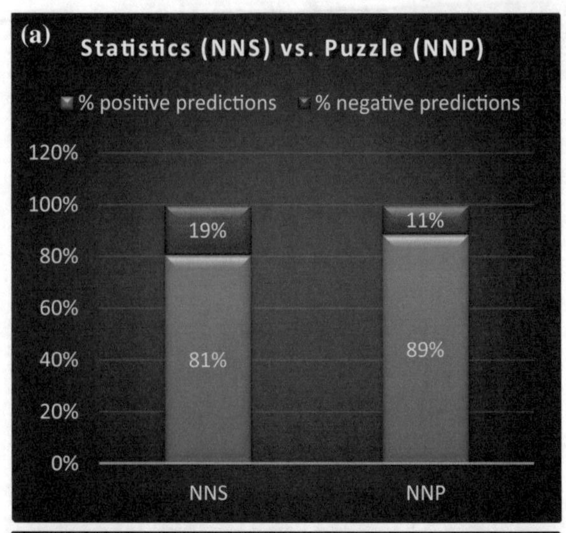

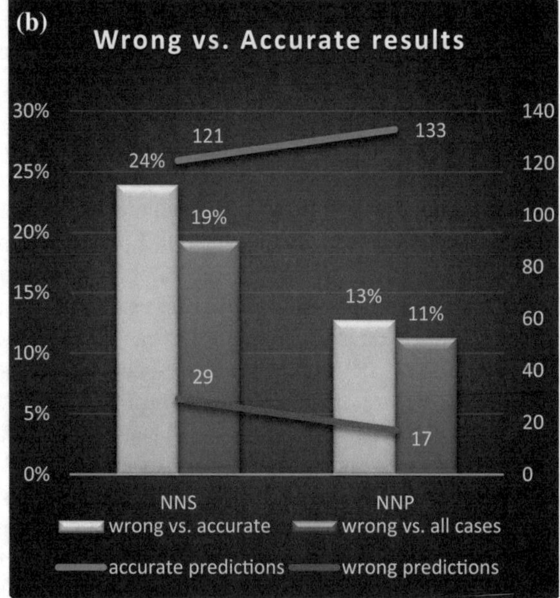

Fig. 15 a, b Comparing the results of both techniques

The comparison between Puzzle methods and well-known statistical methods for prediction of attacks is presented in the experimental part of the paper. The percentage of the mistakes as a result of Puzzle applications is lower than the results by the standard statistical methods. The described approach is implemented in SAS Enterprise Miner TM and the result is presented in the experimental part of this paper.

Table 2 Calculations based on 150 cases

	NNS	NNP
Accurate predictions	121	133
Wrong predictions	29	17
Wrong versus accurate (%)	24	13
Wrong versus all cases (%)	19	11

5 Algorithmic Introduction to the Puzzle Methods

Let the following goal have been set as: to find and identify potentially the most dangerous people in the larger group, located in a certain public place. The description of the algorithm includes the following sets:

A General Set
B Set of interesting decisions (the set after statistical processing)
C Set of potentially dangerous elements (detected by classical constraints)
D Set of the proven facts (who can be dangerous)
E Set of expert knowledge
G Set of possible solutions (which will be examined in detail)
P Set of the elements formed after applying Pointing constraints
X Set of variables with specific values related to the target detected by statistical methods
S Set of the safety elements

We assume all people within the research object form a general set (A). After the statistical analysis of this set, we form a subset of its possible solutions to the problem (B). After applying the standard constraints and different statistical methods a set (C) is formed, which is a subset of set (B). After that, we apply Pointing and Binding constraints on this set (C) and the elements that remain construct the set (G): the target set, in this case—the people who are most likely to carry out a bombing.

The algorithm is divided into several main sections as follows:

Algorithm 1:

Step 1.1: Preliminary statistical processing which form $B \subset A$.
As a result of this step, links are found between the target variable and the input variables X1, X2, …, Xn using classical statistical methods.
If the element possesses $X1 - n$ then move it to set B.
Else continue

Step 1.2: get random elements from B (Bi)
and check for another $X1 - n$ or for $Bi \cap D$
If check = true: then move the selected element to set C
Else continue.

Example

Step 1.1:

Let established links between the goal G (who is dangerous) and the following variables X1, X2 …., Xn are revealed by Chi-squared test and the variable worth analysis.

X1 nervous,
X2 wears loose clothing,
X3 aged visibly around 20–40 years,
X4 concealing his face.

Anyone who owns or possesses a weapon or explosive has either a concealed face or the corresponding feature goes into the set B. All people who are interesting solutions are included in the set B.

Step 1.2:

Let a random element of the set B is selected—(Bi). If Bi owns property X3, the check is executed:

If Bi gets at least two elements of the set X, then Bi is moved to the set C.

or

If Bi has contacts with criminals ($Bi \cap D$), then Bi goes to the set C.

Set D is presented as follows:

$$D = \begin{cases} \text{1 Activity in social networks, related to} \\ \text{terrorist acts,} \\ \text{2 Activity in the mobile network} \\ \text{3 Contacts of criminals,} \\ \text{4 Relations with arm dealers} \end{cases}$$

Algorithm 2:

Step 2.1:

Let a random element of the set C is selected—(Ci)

Then apply corresponding Puzzle constraints and move selected elements to set P

Check if ($Pi \cap D$)

Create Binding.

Else continue.

Example:

Step 2.1

A man is spotted with a partially disguised face, loose clothing, and bulky backpack. When applying the Pointing constraints, an emblem on a jacket with the initials RG of red and green background is reported and investigated. After an analysis, the emblem is identified: it belongs to the tennis tournament "Roland Garros". In the following analysis of the backpack—tennis equipment the brand is identified and labeled. This fact is enough to remove this person's feature from the set of potentially dangerous people.

Let a Binding constraint is found such as "athletes often wear clothing and equipment of specialized sports brands." This constraint can be useful for future analysis.

Algorithm 3:

Step 3.1:
Move all elements without X1 – n to set S.
In 5% of the cases
Get random element from S: (Sn) and apply Binding.
If check = true move (Sn) to set G
Else continue.

Example 1:
Step 3.1
All the people who do not meet the criteria specified above are categorized as safe. In a small percentage of the tested cases, we take one element of the set S (safety set) and the Binding constraints are applied for further examination. Upon detection of the matching element, it is directly transferred to a set G, and thus skips the need for standardized controls.
The innovative logical approach using Pointing and Binding constraints combined with the statistical basis allows the described algorithm to improve the results of classical statistical methods by an average of 10%.

Example 2:
Let such a Binding constraint is found, that bombers in many cases are nervous and often their face twitches. In this case, even in the absence of key features mentioned above, a person with specific twitches would be selected for further analysis.

Compared to other optimization algorithms, for example Gradient methods, Swarm intelligence and Evolutionary algorithms, there is no need of complex actions. For example, multi-variable functions, gene expression, adding additional calculation factors like weights, indexes, updates (for example pheromone update in Ant colony optimization algorithm) or fitness function are not necessary. The goal is to optimize the work process, giving solutions in the proximity of the goal, analogically to fuzzy approaches.

Taking advantages over other optimization algorithms we can highlight the ease of its use and its lightness, which saves time and computer resources.

6 An Example of an Application of the Puzzle Methods

Let assume that by using the methods of descriptive statistics a certain general set is examined. The experiments were realized through the intelligent data processing instrument SAS Enterprise Miner TM. For this purpose, the tool "StatExplore" is used.

The test set is examined for missing values via Chi-squared test to prove the null hypothesis: how to forecast results that are obtained from the study and analysis of

Table 3 Overall information about the data base

Data role	Variable name	Role	Number of levels	Missing values
Train	Education	Input	6	0
Train	Clothing	Input	3	0
Train	Criminal file	Input	2	0
Train	Masked	Input	4	0
Train	Motherland relations	Input	2	0
Train	Good/Bad	Target	2	0

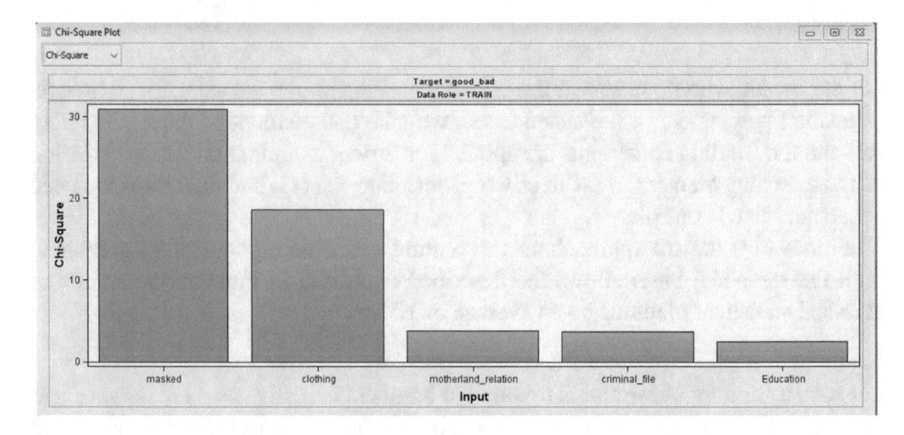

Fig. 16 Chi-square test results of variables of the general set

the relationships between the target and the input variables. Based on the experiments the following results are obtained.

Missing values in processed data, as depicted on Table 3.

The figure shows the variables in the data to analyze, separated roles are both input and target. Also of great importance is the number of different values they can take. They are shown in the column "number of Levels". The next column "Missing" indicates whether the database has missing values and how many they are for a specific variable.

The input variables were arranged in the values obtained from their chi-square test result, as depicted on Fig. 16.

The purpose of the test is to verify that the null hypothesis is confirmed or exactly how the forecast results for each variable are close to actually received.

The input variables are sorted by their degree of influence on the target variable, as depicted on Fig. 17.

The graph shows the quantitative impact of input variables on the target variable. With the greatest impact are the variables related to concealing the face, clothing and age. This means that people who have high levels of these variables (using the scale shown in Fig. 17) will be a priority of the research.

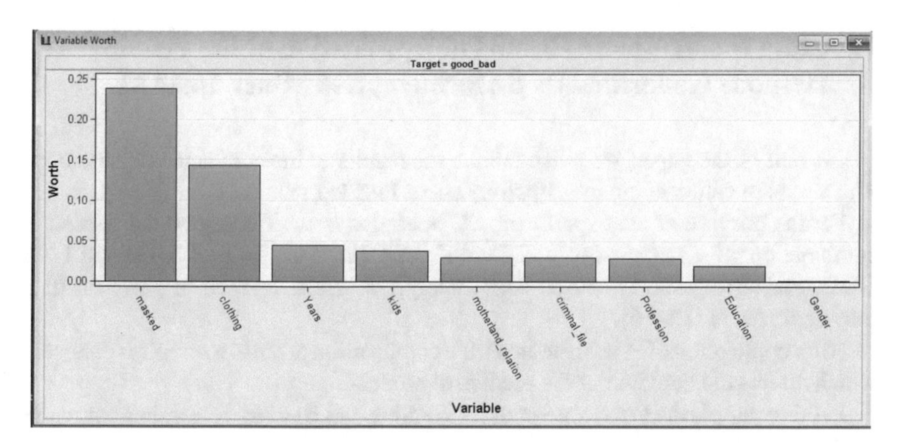

Fig. 17 Influence of input variables on the target variable

The following conclusions can be made from the conducted experiments:

- Further processing to replenish the incoming data is not necessary;
- Statistically, the most important variables for further analysis are found, namely: the most important parameters for the identification of suspicious people; how many have concealed their face, how wide their clothes are and how suitable they are to conceal weapons and explosives, the suspects' age.

Let's take a number of people in a public place (airport). Our goal is to find out which ones are potentially dangerous. We assume that we cannot explore all people, as it will take a lot of time and resources and would not be economically justified. We need a few people to examine and determine which ones are potentially dangerous to the security of the site.

Of all the people at the airport at a certain point in time, few people are selected, who possess the established parameters of survey variables—with veiled faces, loose clothes and of a certain age. Let's conventionally call the selected people Visitor 1, Visitor 10, Visitor 15 and Visitor 20, who possess high levels of corresponding parameters, according to the appropriate scale in one of the selected variables. For example, a large part of the face is hidden. We check up whether they have a high value of one of the other parameters—whether wearing loose clothing or they apparently are between 20 and 40 years. If they do not have high values of the second variable too—we discard them of further studying. If there is a presence of high amount of the second parameter, we continue to investigate the identified people by Pointing constraints that focus the research on specific subjects or to reject them as an opportunity. For example, if despite the disguise a person's face has visible scars, it is very likely he had surgery or an accident, the scars of which he is trying to hide. Thus, this person will be designated as safe and will not be analyzed further. In another case, however, if the clothes being worn do not match the season or specific weather conditions on that day, it is very suspicious and this person will be thoroughly investigated for other clues or even recommended for physical verification by employees in place.

7 Using the Template Results of Application of the Puzzle Methods Generated by SAS Enterprise Miner 14.1TM

In this part of the paper, we will examine the results of the conducted experiments. The aim is to demonstrate the effectiveness of Puzzle methods.

For the purpose of the experiment, PC operating system Windows 8.1 is used in combination with a processor Intel Core i3 3.7 GHz, 8 GB DDR 4 RAM and 1 TB HDD. For testing site is chosen SAS Enterprise Miner 14.1TM, a powerful data mining software [12–16].

After application of Puzzle methods, the composition of the test sample is changed, which affects the outcome of the predictive model.

In this example the Puzzle methods by Pointing and Binding constraints can cause the following:

- Subtract elements of the test subset
- Add elements to the test subset

Figures 18 and 19 show the samples on which we perform our research. In the first case in Fig. 18 a sample of the entire population is created by standard methods and using constraints (simple random sample).

Figure 19 shows the sample, wherein the statistical methods are applied and then the constraints of Puzzle methods are applied. As can be seen, there are some differences between the two samples—after application of the Puzzle constraints in the new sample a new element (B20) is added and two elements are removed (B2 and B9).

Experiments were conducted in SAS Enterprise Miner 14.1TM on the same set with two different samples and respectively providing two models (used node is "neural network" with default settings). The results of the two models were compared by average square error (ASE) or how often mistakes occur in prediction results.

After applying the constraints of the Puzzle methods we can compare the results obtained by standard statistical tests and those after applying logical methods.

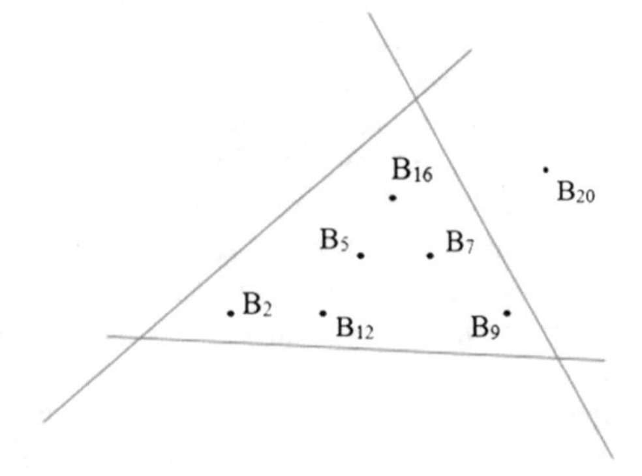

Fig. 18 Standard statistical sample

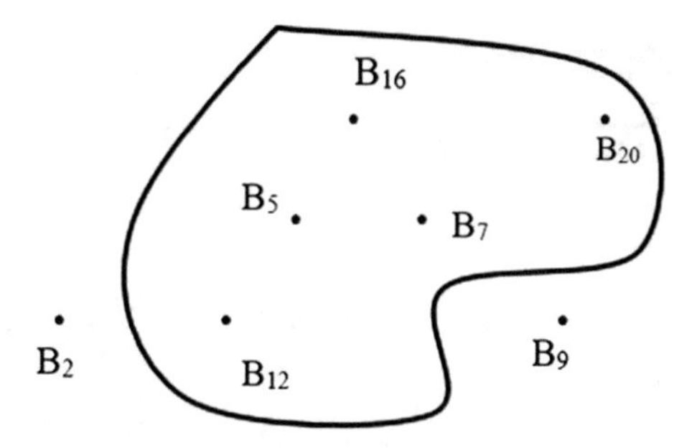

Fig. 19 The situation after the constraint application

Fit Statistics

Selected Model	Predecessor Node	Model Node	Model Description	Target Variable	Target Label	Selection Criterion: Train: Average Squared Error
Y	Neural	Neural	Neural + Puzzle	good_bad	good/bad	0.008949
	Neural5	Neural5	Neural standart	good_bad	good/bad	0.047064

Fig. 20 Comparative analysis of average squared error between predictive models using statistical methods and Puzzle methods

Figure 20 shows the results of both predictive models (artificial neural networks with the same settings) categorized in the table. We can see the target variable that we want to predict—in our case this is "good_bad", which is binary and reflects research that one passenger at airport is potentially dangerous or not. Accordingly, if—is recommended for further investigation/interrogation by the authorities. Also in Fig. 20 the first column shows the evaluation of the software SAS Enterprise Miner 14.1TM about which of the two models are chosen by the following criteria:

- Accuracy.
- Suitability for further research.

More detailed information about the difference in predicting accuracy of the two models can be seen in Table 4. Where Model node Neural represent ANN, enhanced with the Puzzle method and Neural 5 represent standard ANN.

Let us assume that the airport passenger traffic has been evaluated. Then:

- True positive represents a harmful passenger and he/she is rightly evaluated as such;
- True negative is a harmless passenger and he/she is rightly evaluated as such;
- False positive is a harmless passenger and he/she is wrongly evaluated as such;
- False negative is a harmful passenger and he/she is wrongly evaluated as harmless;

Table 4 Table based event classification on average squared error

Event classification table								
Model selection based on train: average squared error (_ASE_)								
Model node	Model description	Data role	Target	Target label	False negative	True negative	False positive	True positive
Neural 5	Neural network (5)	TRAIN	good_bad	good/bad	14	731	43	212
Neural	Neural network	TRAIN	good_bad	good/bad	11	774	0	215

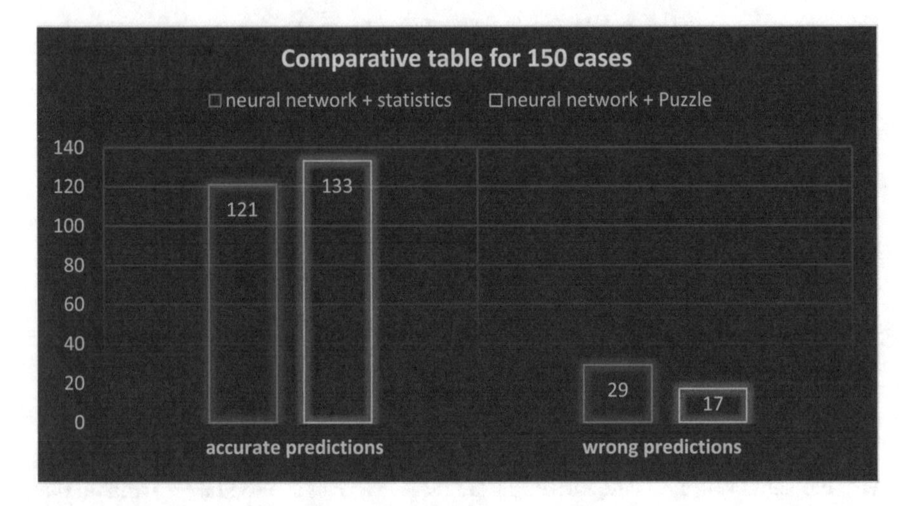

Fig. 21 Comparative analysis of forecasts between the application of the constraints of Puzzle methods and standard statistical methods for evaluation in 150 cases

Therefore, the sensitivity of the artificial neural network (ANN) is also better when applying Binding constraints, as shown in Table 4.

The figure shows a 10% improvement in accuracy of the results of the predictive model after applying the constraints of the Puzzle methods at the expense of processing time required calculations and using system resources as shown in Figs. 21 and 22.

Of course, mistakes can never be completely excluded from the estimates.

As shown by the experiments conducted by applying the constraints of the Puzzle methods in combination with recognized statistical methods, the accuracy of predictive models in SAS Enterprise Miner 14.1TM can be improved by up to 10% compared to results of the classical statistical methods.

A comparison is made between the average error function for up to 50 iterations between predictive models, using artificial neural networks and standard statistical methods (Fig. 24) and the same model using Puzzle methods (Fig. 23).

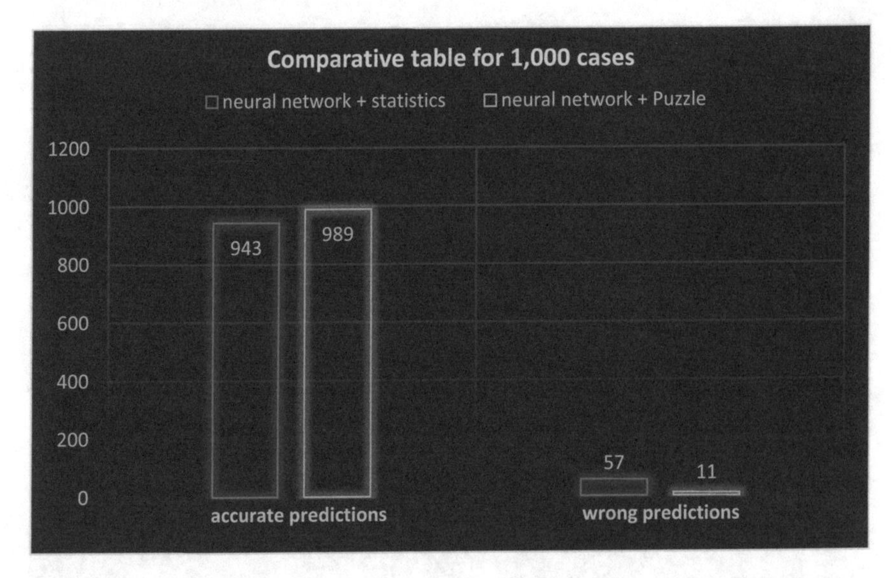

Fig. 22 Comparative analysis of forecasts between the application of the constraints of Puzzle methods and standard statistical methods for evaluation in 1,000 cases

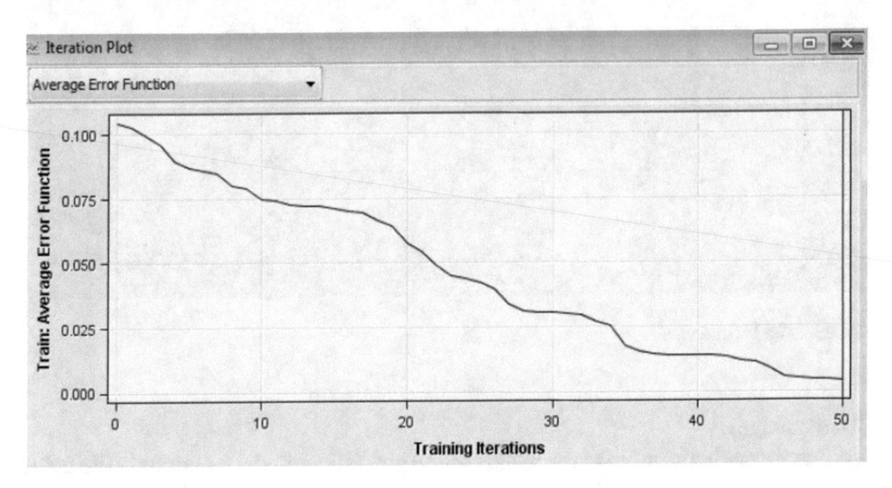

Fig. 23 The Puzzle case

Let's investigate a different comparison of the same models, but this time let the number of wrong classifications be restricted to 50. In this case, artificial neural network node is used and it is combined with standard statistical methods (Fig. 26) and with the usage of Puzzle methods (Fig. 25).

It is observable that the predictive results of the model by using Puzzle methods have lower rate of wrong classifications and error function compared to the model which uses only standard statistical methods.

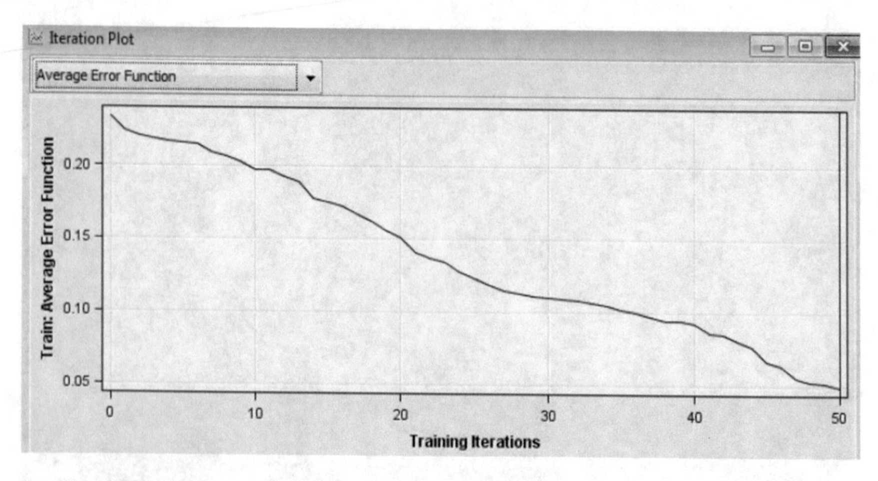

Fig. 24 The standard case

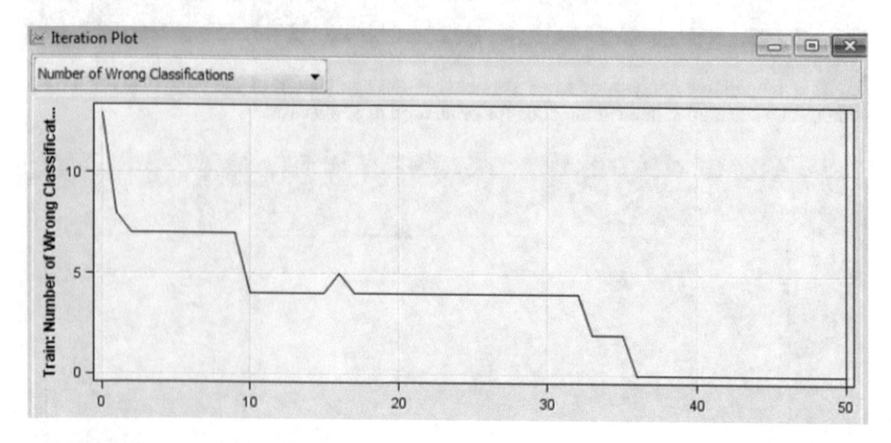

Fig. 25 The Puzzle case

Disadvantages:
The Puzzle methods improve the results of the predictive model by around 10% at the expense of the time required for analysis and system resource. Also as a shortcoming, we can quote the need for a specially prepared set with the expert knowledge from the scope aiming to allow the algorithm the full functionality.

8 Conclusions

The statistical methods do not fully cover and investigate a number of complex real-world problems. Sometimes they can cover the subject area, but cannot completely process the data in their dynamics, as they are weak in terms of acquiring new

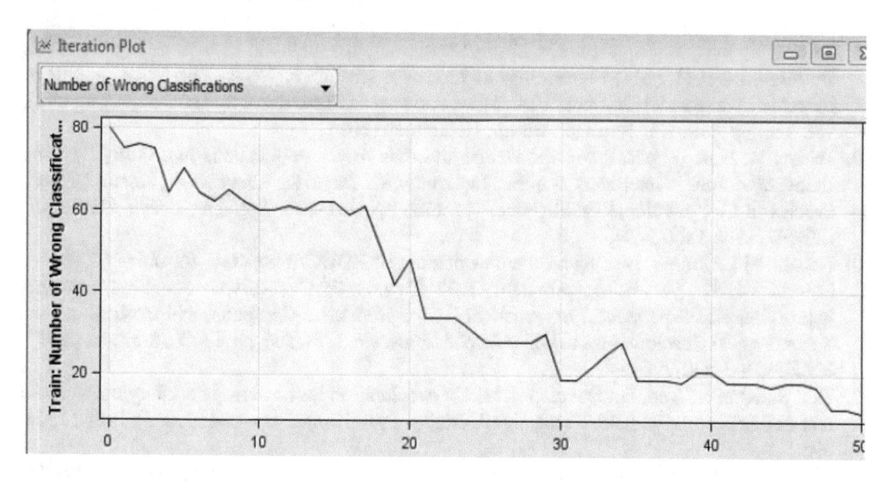

Fig. 26 The standard case

information [17]. Logical methods themselves also have some disadvantages to handle the task. The solution is a combination of statistical and logical methods to achieve the maximum efficiency. This research examines the application of Puzzle methods as a standard for addition and improvement of the results obtained by statistical methods. Through the experiments were shown the advantage of its implementation, leading to improved forecast results by an average of 10% compared to using only statistical methods. The results are tested in SAS Enterprise Miner 14.1TM framework.

References

1. Misselhorn, C.: Collective Agency and Cooperation in Natural and Artificial Systems: Explanation, Implementation and Simulation. Springer, Berlin, Heidelberg (2015)
2. Danny, W.: Architecture-Based Design of Multi-Agent Systems. Springer, Berlin, Heidelberg (2010)
3. Yokoo, M., Hirayama, K.: Algorithms for distributed constraint satisfaction: a review. Auton. Agent. Multi-agent Syst. **3**, 185–207 (2000)
4. Christopher, B., Harith, A., Srinandan, D., Yorick, W.: Data driven ontology evaluation. In: International Conference on Language Resources and Evaluation (LREC 2004), Lisbon, Portugal, 24–30 May 2004
5. Descriptive and Inferential Statistics. https://statistics.laerd.com/statistical-guides/descriptive-inferential-statistics.php. Accessed 30 March 2011
6. Dong, G., Kuang, G.: Target recognition via information aggregation through Dempster–Shafer's evidence theory. IEEE J. Geosci. Remote Sens. Lett. **12**(6), 1247–1251 (2015)
7. Yong, W., et al.: Research on evaluation method of power quality for wind plant based on probability theory and evidence theory. In: 2011 International Conference on Transportation, Mechanical, and Electrical Engineering (TMEE), Changchun, China, pp. 1003–1006, 16–18 Dec 2011

8. Jotsov, V., Iliev, E.: Applications of advanced analytics methods in SAS enterprise miner. In: Filev, D., et al. (eds.) Proceedings of the IEEE 7th International Conference 'Intelligent Systems', Warsaw, 24–26 Sept 2014. Series Advances in Intelligent Systems and Computing, vol. 323, pp. 413–430. Springer, Berlin, Heidelberg (2014)
9. Jotsov, V.: New proposals for knowledge and data driven applications in security systems. Innovative issues in intelligent systems. In: Sgurev, V., Yager, R., Kacprzyk, J., Jotsov, V. (eds.) Studies in Computational Intelligence, vol. 623, pp. 231–294. Springer, Berlin, Heidelberg (2016). ISSN: 1860-949X
10. Nelson, M.L.: Data-driven science: a new paradigm? EDUCAUSE Rev. **44**(4):6–7 (2009)
11. Solomatine, D., See, L.M., Abrahart, R.J.: Chapter 2: Data-driven modeling: concepts, approaches and experiences. In: Practical Hydroinformatics Computational Intelligence and Technological Developments in Water Applications, vol. XVI, 506, pp. 17–30. Springer (2008). ISBN: 978-3-540-79880-4
12. Tak, S., Kim, S., Yeo, H.: Development of a deceleration-based surrogate safety measure for rear-end collision risk. IEEE Trans. Intell. Transp. Syst. **16**(5), 2435–2445 (2015). ISSN: 1524-9050
13. Sinco, B.R., Arbor, A., Chapman, P.L., Collins, F.: Adventures in path analysis and preparatory analysis, Paper AA 05-2013. In: SAS Conference Proceedings: Midwest SAS Users Group 2013, Columbus, Ohio, 122 papers, 22–24 Sept 2013
14. Wetzel, C., O'Keefe, A., Frederick, J.: Improving data collection efficiency and programming process flow using SAS® Enterprise Guide® with Epi Info™, Paper SA-10-2015. In: SAS Conference Proceedings: Midwest SAS Users Group 2015, Omaha, Nebraska, 96 papers, 18–20 Oct 2015
15. Martinez, E.Z., Achcar, J.A., Aragon, D.C.: Parameter estimation of the beta-binomial distribution: an application using the SAS software. Ciencia e Natura, Santa Maria, vol. 37 no. 4, pp. 12–19. Revista do Centro de Ciencias Naturais e Exatas - UFSMISSN impressa: 0100-8307 (2015). ISSN on-line: 2179-460X
16. Larsen, E., O'Brien, T.E.: SAS® Software as an Essential Tool in Statistical Consulting and Research, Paper CD-02—2015, Conference: Midwest SAS Users Conference, At Omaha, NE, October 2015
17. Derby, N.: Managing and Monitoring Statistical Models, Paper AA-06-2013. In: SAS Conference Proceedings: Midwest SAS Users Group 2013, Columbus, Ohio, 122 papers, 22–24 Sept 2013

Autonomous Flight Control and Precise Gestural Positioning of a Small Quadrotor

Nikola G. Shakev, Sevil A. Ahmed, Andon V. Topalov, Vasil L. Popov and Kostadin B. Shiev

1 Introduction

Unmanned aerial vehicles (UAVs) are becoming increasingly popular for different commercial and noncommercial applications in the areas of industry, entertainment, research, and military. Recent technological achievements have led to their further miniaturization and increased capability for carrying on number of sensors while the duration of the flight can now last longer (several decades of minutes or even hours). On the other hand, Cyber Physical Systems (CPS) are getting more and more popular and powerful tool for implementing advanced solutions of old problems, relying on technologies such as embedded computing, Internet of Things (IoT), Internet of Service (IoS), cloud computing, big data, etc.

The number of applications of CPS is growing in last few years. These systems help industry to become "smart", precise, competitive, globally controllable and visible. CPS are the basis of "smart cities" with smart building, intelligent traffic control and security systems. CPSs are successfully integrated with smart multi agent robotized systems for environmental surveillance, rescue robots, etc. The role of these systems in such missions is extremely important because they can provide

N. G. Shakev (✉) · S. A. Ahmed · A. V. Topalov · V. L. Popov · K. B. Shiev
Control Systems Department, TU – Sofia, Branch Plovdiv, Plovdiv, Bulgaria
e-mail: shakev@tu-plovdiv.bg

S. A. Ahmed
e-mail: sevil.ahmed@tu-plovdiv.bg

A. V. Topalov
e-mail: topalov@tu-plovdiv.bg

V. L. Popov
e-mail: vasil_liubenov_popov@abv.bg

K. B. Shiev
e-mail: k.shiev@gmail.com

V. Sgurev et al. (eds.), *Learning Systems: From Theory to Practice*, Studies in Computational Intelligence 756, https://doi.org/10.1007/978-3-319-75181-8_9

timely intervention without endangering the health and lives of specialized team members that they replaces.

The algorithms proposed in this work are inspired by the concept of CPSs. Using gestures for precise positioning of small UAVs is an innovative way of making the human part of real-time working CPSs. Most of the UAVs, currently available on the market, have been designed to work in outdoor environments, relying on GPS data to determine their current position. Although well-developed GPS systems and services, momentary loose of coverage and missing data could became a serious problem. Moreover, there are many practical tasks where UAVs will be required to operate in indoor environments or along narrow city streets where GPS data cannot be received or are unreliable. The inspection of buildings and equipment with damaged structures and the decommissioning of nuclear power plants are among them. Since such monitoring tasks can be difficult or dangerous for the engaged personnel, a distant inspection is frequently considered. Most of the currently existing robots capable to perform such activities are ground based vehicles and can inspect only the lower zones in these environments [1]. Furthermore, inspections could be difficult and time consuming process that may include several stages like characterization, decontamination, demolition and waste management. It requires collection of significant amount of information about the ruggedness and structural integrity of the building or the monitored equipment [2]. There might be a necessity to determine also the quantity, disposition and intensity of the emissions of radioactive materials or dangerous chemicals.

An UAV capable to perform such inspection tasks must maintain stable flight, avoid obstacles, localize position and orientation and plan its way without a GPS signal. Human operator must have a possibility to send instructions for additional maneuvers in order to obtain better information about the inspected items. Several basic problems have to be solved at the design stage of such aerial robot: (i) construction of suitable vehicle platform, capable of carrying the required payload; (ii) UAV stabilization and control of the movements during the flight; (iii) calculation of the current position and orientation; (iv) planning the path in order to reach a specified target position.

The first problem could be resolved through the purchase of an available on the market aerial platform. A number of well-designed quadrotor rotorcrafts, ranging from the professional ones [3], to those intended for hobbyists [4] are currently available. Many of them are stabilized and have good control on the dynamics of the flight. In spite of that, for environments where the GPS signal is absent or unreliable, the position and orientation determination issues as well as the path planning are currently an active area of research. Solutions, ranging from applying external sensors [5, 6] to the usage of simple sets of sensors [7] to perform simultaneous localization and mapping (SLAM) based on LiDAR or video camera have been proposed recently In order to boost the progress in this research area, the International Association for Unmanned Vehicle Systems (AUVSI) has expanded the organized International Aerial Robotics Competition (IARC) with inclusion of autonomous aerial vehicles for indoors [8].

Gestural control is getting more and more popular with introduction of collaborative features of modern robots. In last 5 years many researchers devote big attention to gestures defining and recognition with idea to use them in different robot control strategies—beginning from basic movements to fine effector control [9–13]. Including gestures in addition to autonomous flight, allows precise post-positioning, which could be used for sensor reading, data gathering, video streaming, emerge intervention (dependent of the UAV's equipment), etc. This type of remote control applied by an operator is stated as a variety of natural user interface (NUI) used in man-machine communication [14, 15]. The idea is not a new one. In particular, a group of researchers from the Institute for Dynamics and Control at ETH Zurich has proposed a way to dynamically interact with quadrotors based on the hands position of the operator [16]. The local 3D coordinates of the helicopter are placed in accordance with the 3D coordinates of the operator's hands and thus a direct correspondence between the coordinates of the hands and the position and orientation of the quadrotor can be established. The approach adopted here differs from [16] in this that the recognized gestures are used to launch discrete control commands to the quadrotor and absence of external tracking system to determine its position and orientation.

2 Quadrotor Autonomous Flight

The proposed flight control system enables the rotorcraft to implement navigation tasks autonomously (by performing missions where the path following is defined as a sequence of predefined points through which to pass) and permits to control remotely the aerial vehicle (the AR. Drone quadrotor) after switching it in semi-autonomous mode. The human-rotorcraft interaction is done through gestures and body postures recognized by Microsoft Kinect.

2.1 The Parrot AR. Drone Quadrotor

The Parrot's AR. Drone quadrotor is a commercially available electrically powered four-rotor mini-rotorcraft with dimensions of 525 mm by 515 mm and weight of 420 g, initially designed for playing augmented reality games [17]. The quadrotor consists of a carbon-fiber support structure, four high-efficiency brushless motors, sensor and control board, two cameras, and is equipped with indoor and outdoor removable hulls. The power supply is provided by a 3-cell, 2000 mAh lithium-polymer battery that suffices for approximately 15 min long flights. The rotorcraft has a set of onboard sensors including ultrasonic altimeters measuring the distance to the land surface, and thus helping to maintain stability when hovering at up to 6 m, a 3-axis accelerometer together with a 2-axis gyro used to provide accurate pitch and roll angles, and 1-axis high precision gyro to measure the yaw angle.

The AR. Drone's stabilization system is implemented on its on-board embedded computer system based on ARM9 32-bit RISC CPU running at 468 MHz with 128 MB of DDR RAM that uses a BusyBox based GNU/Linux distribution with 2.6.27 kernel. The communication with the drone can be done via an ad hoc WiFi network using the provided software interface. An external computer can connect to AR. Drone using the IP address fetched from the DHCP server of the quadrotor. The communication can be done via three channels, each with a different UDP port. Over the command channel the drone, can be requested to take off and land, to change configuration of its controllers, to calibrate sensors etc. The computer can send via this channel commands to the internal controller of the aerial vehicle to set required pitch and roll angles, drone's vertical speed and yaw rate. The navdata channel sends information about the status of the quadrotor and pre-processed sensory data containing current yaw, pitch and roll angles, the altitude, battery state and 3D speed estimates. Both channels update the transmitted data at 30 Hz rate. The stream channel provides images from the frontal and bottom cameras. The control board of the AR. Drone is accessible by telnet and changes can be made in the settings of the on-board operating system and in the configuration files of the internal controllers.

2.2 *Quadcopter Dynamics, Simulation, and Control*

A quadrotor is a rotorcraft vehicle that is capable of hovering, vertical take-off and landing, as well as four-way (forward/backward/sideward) flight.

For the description of the quad-rotor rotorcraft motion two coordinate frames have to be used: an inertial earth frame $\{R_E\}$ (I, X, Y, Z) and a body frame fixed to the center of mass of the aerial vehicle $\{R_B\}$ (G, x, y, z). The relation between the frame $\{R_B\}$ and the frame $\{R_E\}$ can be obtained by the position vector $\xi = (X, Y, Z)^T \in \Re^3$, which describes the displacement of the center of mass G in the inertial frame and the orientation vector $\eta = (\psi, \theta, \phi)^T \in \Re^3$ where (ψ, θ, ϕ) are the Z-Y-X Euler angles (yaw, pitch, and roll) representing the orientation (Fig. 1). Yaw, pitch, and roll angles describe the orientation of the frame $\{R_B\}$ with respect to the frame $\{R_E\}$ by defining the angles of consecutive rotations around the $\{R_B\}$ axes. The corresponding vector of time derivatives is $\dot{\eta} = (\dot{\psi}, \dot{\theta}, \dot{\phi})^T$. However, note that the angular velocity vector $\omega \neq \dot{\eta}$. The angular velocity ω is a vector pointing along the axis of rotation, while $\dot{\eta}$ is just the time derivative of yaw, pitch, and roll angles. In order to convert the time derivatives of Z-Y-X Euler angles (yaw, pitch, and roll) into the angular velocity vector, we can use the following relation:

$$\omega = \begin{bmatrix} 1 & 0 & -S_\theta \\ 0 & C_\phi & C_\theta S_\phi \\ 0 & -S_\phi & C_\theta C_\phi \end{bmatrix} \dot{\eta} \tag{1}$$

where ω is the angular velocity vector in the body frame and $S_{(\cdot)}$ and $C_{(\cdot)}$ represent $\sin(\cdot)$ and $\cos(\cdot)$ respectively.

For *Z-Y-X* Euler angles the transition between $\{R_B\}$ and $\{R_E\}$ can be done by the rotational matrix R [18].

$$R = \begin{bmatrix} C_\theta C_\psi & C_\psi S_\theta S_\phi - C_\phi S_\psi & C_\phi C_\psi S_\theta + S_\phi S_\psi \\ C_\theta S_\psi & S_\theta S_\phi S_\psi + C_\phi C_\psi & C_\phi S_\theta S_\psi - C_\psi S_\phi \\ -S_\theta & C_\theta S_\phi & C_\theta C_\phi \end{bmatrix} \quad (2)$$

The rotation matrix R is orthogonal thus $R^{-1} = R^T$ which is the rotation matrix from the inertial frame to the body frame.

The generalized coordinates for the rotorcraft are $q = (X, Y, Z, \psi, \theta, \phi)^T \in \Re^6$

We will consider the following assumptions:

(1) The four rotors are in one plane, which is orthogonal to the rotors' axes;
(2) The structure of the quad-rotor is symmetric and the center of gravity is placed in the geometric center of the helicopter.

There are two popular types of quad-rotor flying configuration: "Plus-type" and "X-type".

- The first configuration is considering one frontal motor, one rear and two side motors (Fig. 1a). The positive direction of the $\vec{x}$ axis in the body frame $\{R_B\}$ is aligned with the front motor.
- In "X-type" configuration considers two frontal motors and two rear motors (Fig. 1b). The positive direction of the $\vec{x}$ axis in the body frame $\{R_B\}$ is aligned with the angle bisector between two frontal motors.

The lifting force is generated by four propellers. Unlike most rotorcrafts, quadrotors use fixed pitched propellers. A couple of the propellers rotate clockwise (CW) as the other couple rotates counter clockwise (CCW). The angular velocity variations between the propellers cause lift and torque thus motion.

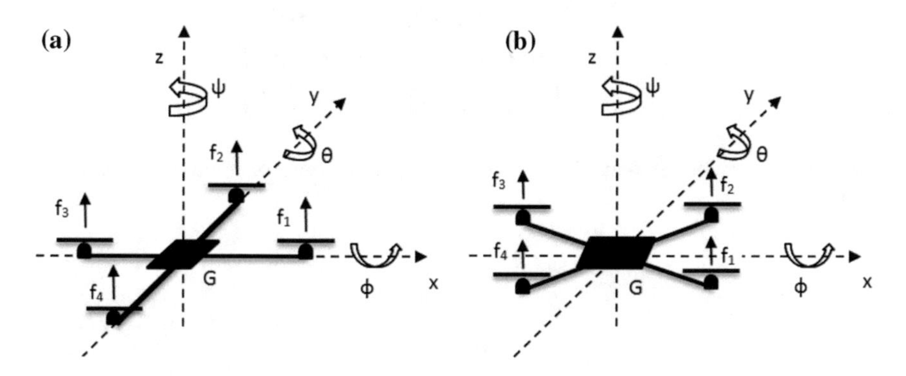

Fig. 1 Coordinate frames: **a** "Plus" type and **b** "X" type quadrotor constructions

2.2.1 Forces and Torques

The angular velocity of rotor i, denoted with ω_i, creates force f_i in the direction of the rotor axis. The angular velocity and acceleration of the rotor also create torque τ_{Mi} around the rotor axis

$$f_i = k\omega_i^2, \tag{3}$$

$$\tau_{Mi} = b\omega_i^2 + I_M\dot{\omega}_i \tag{4}$$

in which the lift constant is k, the drag constant is b and the inertia moment of the rotor is I_M. Usually the effect of $\dot{\omega}_i$ is considered small and thus it is neglected.

The overall lifting force will act along the axes Gz in $\{R_B\}$ frame, and its magnitude will be:

$$u = f_1 + f_2 + f_3 + f_4 \tag{5}$$

Then the force applied to the mini-rotorcraft relative to frame $\{R_B\}$ is defined as

$$F_B = [0 \quad 0 \quad u]^T \tag{6}$$

Torque vector τ_B consists of the torques τ_ϕ, τ_θ, τ_ψ in the direction of the corresponding body frame angles.

For the "Plus-type" configuration the generalized moments on the η variables are

$$\tau = \begin{bmatrix} \tau_\phi \\ \tau_\theta \\ \tau_\psi \end{bmatrix} = \begin{bmatrix} (f_4 - f_2)l \\ (f_3 - f_1)l \\ (\omega_2^2 + \omega_4^2 - \omega_1^2 - \omega_3^2)d \end{bmatrix} \tag{7}$$

Respectively for the "X-type" configuration the generalized moments on the η variables are

$$\tau = \begin{bmatrix} \tau_\phi \\ \tau_\theta \\ \tau_\psi \end{bmatrix} = \begin{bmatrix} (f_1 + f_4 - f_2 - f_3)l \\ (f_3 + f_4 - f_1 - f_2)l \\ (\omega_2^2 + \omega_4^2 - \omega_1^2 - \omega_3^2)d \end{bmatrix} \tag{8}$$

where l denotes the distance from any motor to the center of mass and d is a drag constant.

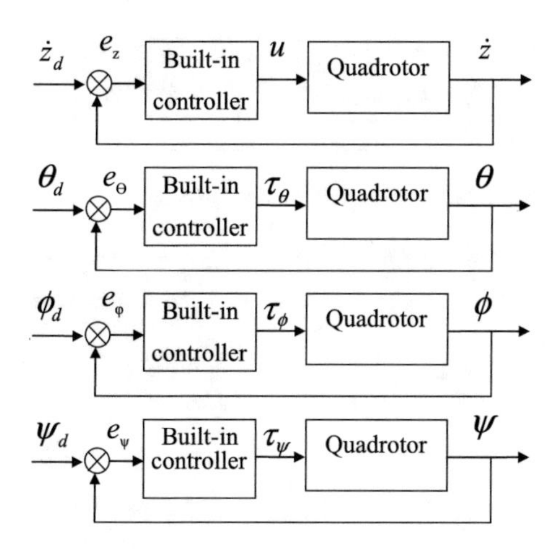

Fig. 2 The built-in control loops

2.3 *Close-Loop Dynamic Estimation of the Parrot AR. Drone*

The position of the rotorcraft can be described by its generalized coordinates $q = (X, Y, Z, \psi, \theta, \phi)^T \in \Re^6$ and typically the control signals are the overall lifting force u that is normal to the plane of propellers and the moments $\left(\tau_\psi, \tau_\theta, \tau_\phi\right)$. Considering Parrot's AR. Drone quadrotor there are built-in onboard control loops for the attitude and vertical speed stabilization. Thus the primary control signals u, τ_ψ, τ_θ, τ_ϕ which are inner signals for these loops are inaccessible for external control. The existence of the build-in controllers imposes a requirement to modify the control problem since it becomes necessary to model the dynamics of the entire aerial vehicle together with the built-in control loops system. In this case the trajectory tracking control problem can be defined as to find the appropriate reference values of the vertical speed, yaw, pitch, and roll angles that will allow the rotorcraft to follow desired positions in the space (see Fig. 2).

2.3.1 Dynamic Modeling and Controller Design for Vertical Displacements

The vertical position can be controlled by changing the overall lifting force u. Since u is not directly accessible it cannot be used as a control signal. The desired vertical speed $\dot{z}_d$ can be considered instead as a control input. An external proportional plus derivative (PD) controller is implemented to calculate the value of $\dot{z}_d$ based on the error $e_z = z_d - z$. The block diagram is shown on Fig. 3.

To achieve a stable closed loop dynamic system with acceptable dynamic properties the gains of the controller have to be tuned appropriately. The dynamics of the

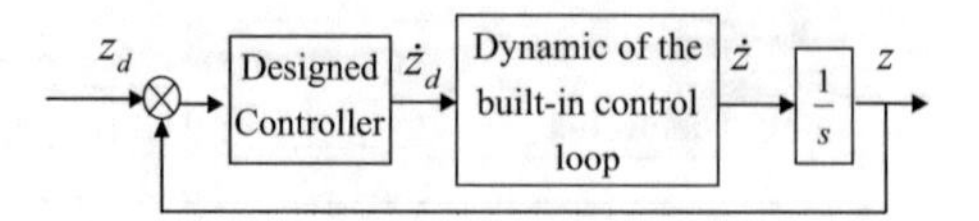

Fig. 3 The designed control loop for the vertical displacement

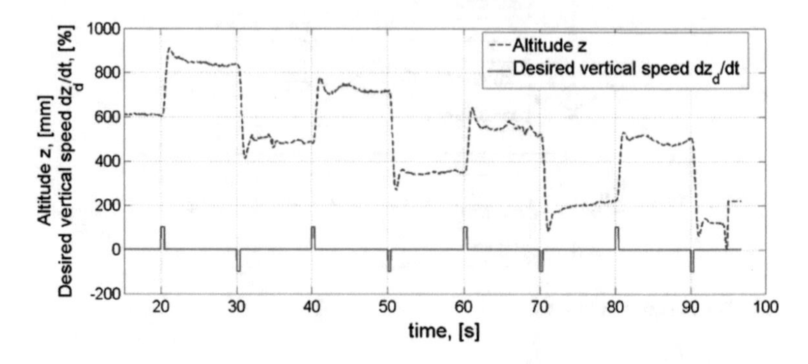

Fig. 4 Experimental results for vertical movement dynamics modeling

quadrotor along the altitude channel together with the built-in control loop can be estimated as a first order transfer function:

$$G_z^{\text{in}}(s) = \frac{\dot{z}(s)}{\dot{z}_d(s)} = \frac{k_z}{T_z s + 1} \tag{9}$$

To determine the values of parameters k_z and T_z in (9), the following experiment is done: The control signal $\dot{z}_d$ is generated as series of rectangular pulses, with amplitude $\pm\dot{z}_{\max}$. The selected pulse width was 40 ms, and the pulse period was 10 s. The graphical results are represented on Fig. 4.

According to these results the values of k_z and T_z are estimated to: $k_z = 700$; $T_z = 0.7$ s.

We assume the transfer function of the PD controller in the following form:

$$G_z^{ctrl} = k_{pz}\left(1 + T_{dz}s\right); \tag{10}$$

Thus the transfer function of the overall open loop system, depicted on Fig. 3, can be expressed as:

$$G_z^{open}(s) = \frac{z(s)}{z_d(s)} = G_z^{ctrl}(s)G_z^{in}(s)\frac{1}{s} = \frac{k_{pz}k_z\left(1 + T_{dz}s\right)}{(1 + T_z s)s}; \tag{11}$$

And respectively the transfer function of the overall closed loop system is:

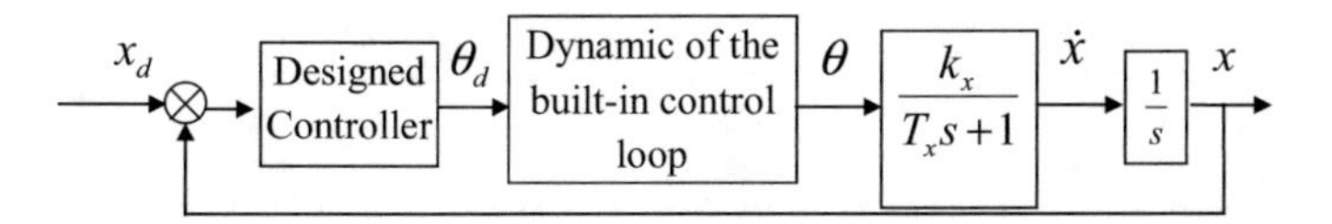

Fig. 5 The designed control loop for forward movements

$$G_z^{closed}(s) = \frac{G_z^{ctrl} G_z^{in} \frac{1}{s}}{1 + G_z^{ctrl} G_z^{in} \frac{1}{s}} = \frac{1 + T_{dz} s}{\frac{T_z}{k_{pz} k_z} s^2 + \left(\frac{1}{k_{pz} k_z} + T_{dz} \right) s + 1} \tag{12}$$

The parameters of the proposed PD controller k_{pz} and T_{dz} are chosen in such a way that the closed loop system is stable and the value of damping ratio is close to 1: $k_{pz} = 0.2; \quad T_{dz} = 0.7$ s.

2.3.2 Dynamic Modeling and Controller Design for the Forward Movements

The forward movement is caused by the rotational moment τ_θ which allows to change the quadrotor pitch angle θ. If θ is different than zero the overall lifting force u, which is normal to the plane of the propellers, can be decomposed to a vertical component and a horizontal component. The horizontal component results in forward/backward movement. As it was mentioned above the signal τ_θ is not accessible. Thus the desired value for the pitch angle θ_d might be considered as a control signal. In order to obtain precise trajectory tracking it is important to estimate the dynamics of the quadrotor forward/backward movement channel together with the built-in control loop (Fig. 2). The transfer function between θ and θ_d is also considered to be of first order.

$$G_\theta^{in}(s) = \frac{\theta(s)}{\theta_d(s)} = \frac{k_\theta}{T_\theta s + 1} \tag{13}$$

The dynamic relation between the pitch angle θ and the position x for small values of θ (when $\theta \approx \sin(\theta)$) [19] can be estimated as $\ddot{x} = g\theta$. By considering the influence of the drag, the appropriate model is:

$$G_x(s) = \frac{\dot{x}(s)}{\theta(s)} = \frac{k_x}{T_x s + 1} \tag{14}$$

The block diagram of the designed control loop is shown on Fig. 5.

To provide precise trajectory tracking it is necessary to generate appropriate values of θ_d. This is done by an external PD controller that calculates the value of θ_d based on the error $e_x = x_d - x$.

The following experiment has been carried: the control signal has been set to $-3°$ and the values of the pitch angle θ and velocity $\dot{x}$ have been measured.

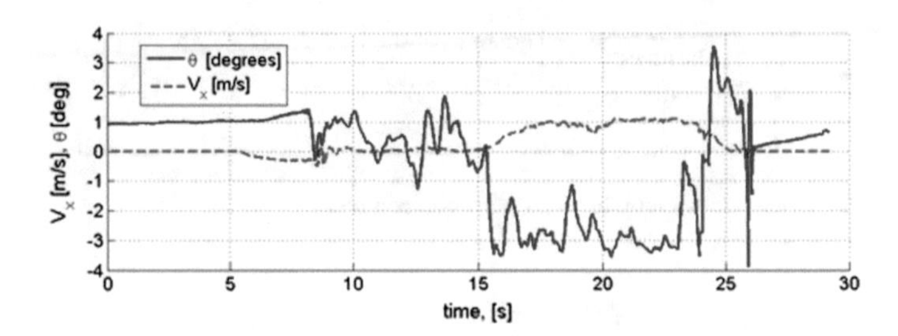

Fig. 6 Experimental results about the forward movement dynamics

The experimental results are shown on Fig. 6 and the following values for model parameters have been obtained: $T_\theta = 0{,}3$ s; $T_x = 2{,}45$ s; $k_\theta = 1$; $k_x = -0{,}37$.

The transfer function of the PD controller is:

$$G_x^{ctrl} = k_{px}\left(1 + T_{dx}s\right); \tag{15}$$

The transfer function of the overall open loop system, depicted on Fig. 5 can be expressed as:

$$\begin{aligned} G_x^{open}(s) &= \frac{x(s)}{x_d(s)} = G_x^{ctrl}(s)G_\theta^{in}(s)G_x(s)\frac{1}{s} \\ &= k_{px}\left(1 + T_{dz}s\right)\frac{k_\theta}{\left(T_\theta s + 1\right)}\frac{k_x}{\left(T_x s + 1\right)}\frac{1}{s}; \end{aligned} \tag{16}$$

And the transfer function of the closed loop system is:

$$\begin{aligned} G_x^{closed}(s) &= \frac{x(s)}{x_d(s)} = \frac{G_x^{ctrl}(s)G_\theta^{in}(s)G_x(s)\frac{1}{s}}{1 + G_x^{ctrl}(s)G_\theta^{in}(s)G_x(s)\frac{1}{s}} \\ &= \frac{k_{px}k_\theta k_x\left(1 + T_{dz}s\right)}{\left(T_\theta s + 1\right)\left(T_x s + 1\right)s + k_{px}k_\theta k_x\left(1 + T_{dz}s\right)} \end{aligned} \tag{17}$$

Based on simulations with the above model the appropriate parameters of the external PD controller have been selected as shown on Table 1.

Table 1 The obtained parameters for the controllers

	k	T_d (s)
Forward movements PD controller	0.3	0.38
Lateral movements PD controller	0.3	0.38
Altitude movements PD controller	0.2	0.7

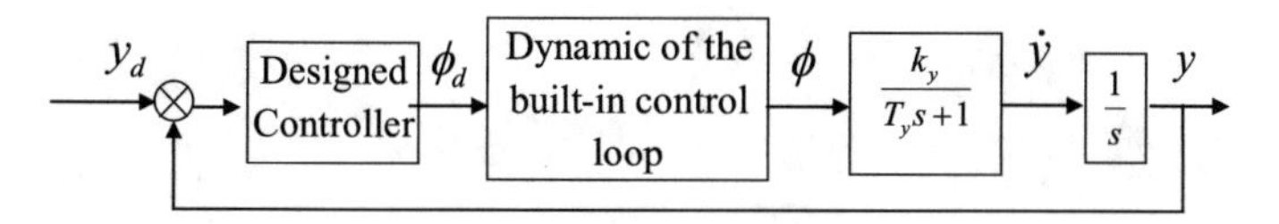

Fig. 7 Designed control loop for side movements control

2.3.3 Dynamic Modeling and Controller Design for Quadrotor Lateral Movements

The lateral movements of the rotorcraft are similar to forward/backward movements. They can be done by changing the roll angle ϕ. The block diagram showing the control loop for the lateral movements is presented on Fig. 7. The obtained values for the parameters are: $T_\phi = 0{,}3$ s; $T_y = 2{,}45$ s; $k_\phi = 1$; $k_y = 0{,}37$.

The parameters of the external PD controller for lateral movements are similar to those obtained for forward movements controller as shown on Table 1.

3 Gestures Controlled Semi-autonomous Flight

The semi-autonomous mode of quadrotor flight is designed for precise positioning of the rotorcraft by operator using gestural interface. Since the control task is very complex (the operator has to control six degrees of freedom using four control signals), it was found that the system achieves better performance if the control signals are with fixed values. This means that if the rotorcraft has to move forward, the operator sends command "move forward" and the control system generates a control signal which causes a forward transition with a predefined linear velocity. This approach obtains at least two important advantages:

- The operator easy manages only the direction of movement instead of direction and velocity.
- Obtaining only the desired direction from the natural user interface reduces significantly the noise and disturbances due to adopted approximations and imprecisions in image recognition process.

Thus the semi-autonomous control system has to receive from the gestural interface system only the movement directions, to execute the desired movements with predefined constant velocities while keeping the rotorcraft attitude.

The necessary requisites for implementation of a system capable to control the flight of a rotorcraft using gestures of the operator are: (i) retrieval of spatial information from certain parts of the operator's body; (ii) identification of the positions (body postures, gestures) based on the obtained spatial information; (iii) association of the recognized postures and gestures with specific commands that can be sent to the quadrotor. The structure of the developed gestural control strategy is presented on Fig. 8. It could be seen that the communication between the C# application and

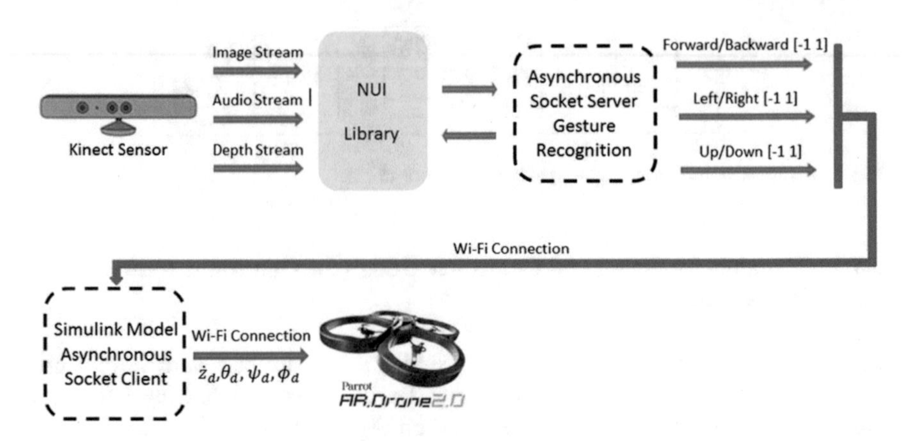

Fig. 8 Basic structure of the developed gesture control system

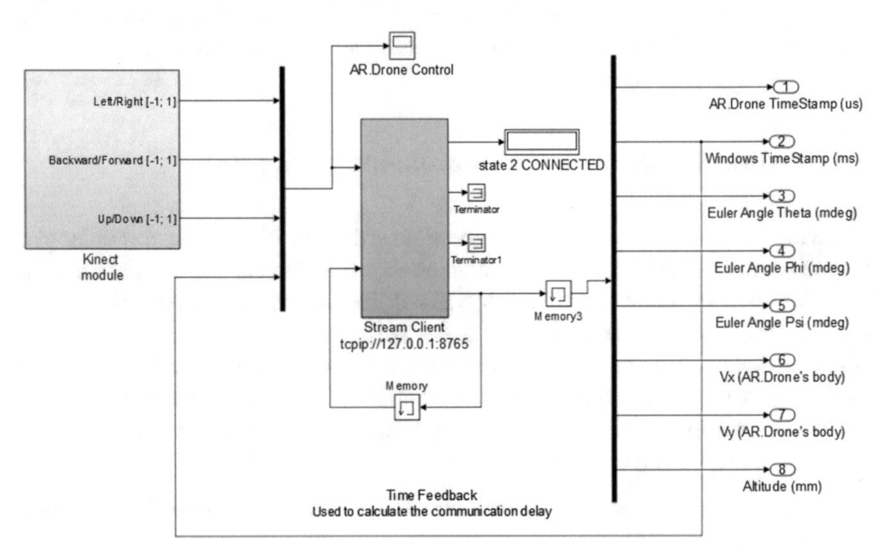

Fig. 9 *Kinect* subsystem receives gesture data from the C# application, *AR Drone* subsystem receives the current position from the quadcopter

the quadcopter consists of three subsystems: (i) C# application exchanging data with Microsoft Kinect device, (ii) a Simulink model created to control the rotorcraft and (iii) the quadcopter itself. The C# program is configured as an asynchronous socket server and sends gesture data to the Simulink model, which controls the drone by passing the recognized gestures. The environment of Matlab/Simulink is augmented with the ad-on software *QuarC* developed by *Quanser Consulting Inc.*

Implementation of the proposed concept is made in the environment of Matlab/Simulink. Thus, a control system consisting of two main modules—*Kinect* and *AR Drone* (see Fig. 9) is developed.

Table 2 Conditions for lateral movement commands

Angles	Command
$\varepsilon_r < 65°$	Right movement
$\varepsilon_r > 115°$	Left movement
$65° < \varepsilon_r < 115°$	No movement

Table 3 Conditions for forward/backward movement commands

Angles	Commands
$\varepsilon_l > 115°$	Forward movement
$\varepsilon_l < 65°$	Backward movement
$65^o < \varepsilon_l < 115°$	No movement

Table 4 Conditions for vertical movement commands

Angles	Command
$\varepsilon_r < 40°$ and $\delta_r > 70°$	Up movement
$\varepsilon_r > 40°$ and $\delta_r > 70°$	Down movement
$\varepsilon_r < 40°$ and $\delta_r < 70°$	No movement

All of the created C# functions are united in the *Kinect module*. They read the skeleton information over the described above data streams and recognize the gestures according to the active joints of the skeleton and the rules from Tables 2, 3 and 4.

3.1 Retrieval of Spatial Information from Certain Parts of the Operator's Body

Microsoft Kinect [20] is used as a gestures tracking device in this investigation. The recognized gestures and body postures are then used for the remote control of the AR. Drone quadrotor.

The information from Kinect is structured in three basic data streams: depth, color, and audio. Each frame of the depth data stream is made up of pixels that contain the distance (in millimeters) from the camera plane to the nearest object. The color stream consists of image data which is available at different resolutions and formats. The audio stream is provided by a microphone array captures audio data at a 24-bit resolution. In addition to the hardware capabilities, the Kinect software runtime implements a software pipeline that can recognize and track a human body. The runtime converts depth information into the skeleton joints in the human body. This makes it possible to track up to two people in front of the camera. The image of the skeleton consists of 20 points (placed in the body joints) that are connected with lines (links), Fig. 10.

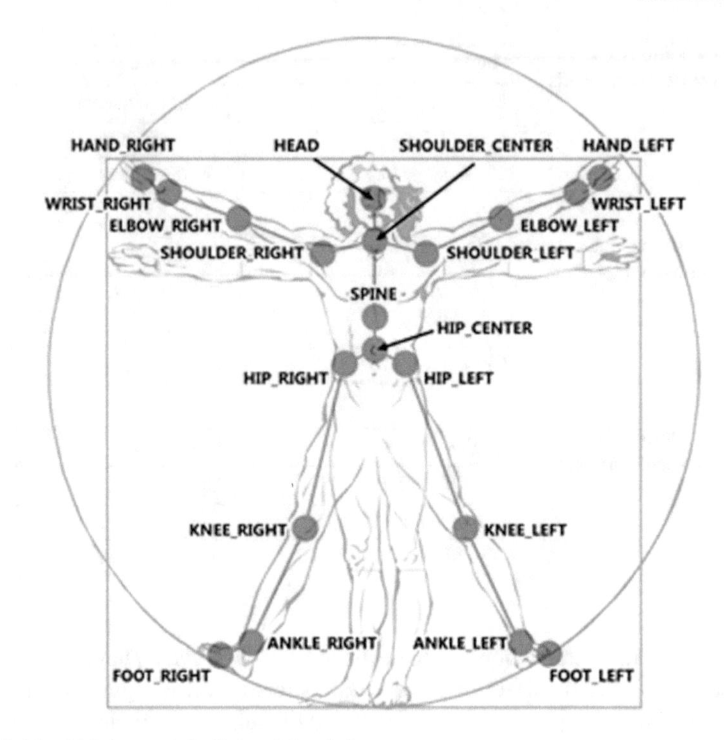

Fig. 10 The 20 joints and the links of the skeleton

The coordinates of each joint are available along the three axes X_j, Y_j, Z_j. The origin of the coordinate system is located in the image sensor of the color camera (Kinect CMOS color sensor).

Given the coordinates of the joints, the distance between them can be determined and the angles between the bones (links) of the skeleton can be calculated. On the basis of the obtained coordinates and angles a number of selected gestures can be analyzed and by imposing certain conditions they are accepted uniquely.

3.1.1 Identification of the Positions (Body Postures, Gestures) Based on the Obtained Spatial Information

The image of the skeleton is constructed as a result of the information about the remoteness of the object. To be recognized, the object should be frontally located in front of the Kinect. A WPF application is created using C# and the necessary libraries are used after the installation of the Kinect for Windows (SDK). The C# application informs about the Kinect's status and starts streaming data out of the sensor. It is called every time when a new frame with information appears. This allows tracking in real time the changes of the coordinates for the selected points. By finding the coordinates of several points a logical tracing of their position can be done and appropriate gestures can be defined. The angle between two skeleton links connected with a joint can be determined too.

3.2 Association of the Recognized Gestures with Specific Commands to the Quadrotor

According to the goals of the semi-autonomous gesture control mode three different discrete signals have to be send to the quadrotor rotorcraft. These signals specify the direction of movement along the three axes in the body reference frame $\{R_B\}$ (G, x, y, z). Each signal accepts one of the values -1, 0, or 1. The zero value means no movement along this axis has to be executed. The value of -1 or 1 means that the rotorcraft has to move with a fixed predefined speed in negative or positive direction of the current axis.

If ε_r and ε_l denote the angle of the right arm elbow and the left arm elbow respectively; δ_r and δ_l denote the angle between the shoulders and the right elbow and the left elbow respectively then the conditions for the different commands used are presented on Tables 2, 3 and 4.

4 Trajectory Tracking and Gesture Control Experimental Tests

The proposed concept for autonomous flight control and precise gestural positioning of a small quadrotor is validated via several indoor experiments with A.R. Drone Parrot quadrotor. Performed experiments consists of two stages: a trajectory tracking task based on the autonomous flight control system described in Sect. 2 and precise prepositioning by gestural control introduced in Sect. 3.

4.1 Trajectory Tracking Tests During Autonomous Flights

The trajectory tracking results obtained from two real-time experiments are presented in this section. In the first experiment the tracked trajectory is a rectangle in the *X-Z* plane. The movement of the rotorcraft is counter-clockwise and two simultaneous rounds are done. The results are shown on Fig. 11. The second experiment consists of following a circular trajectory with radius 300 mm. The results are shown on Fig. 12. Both experiments demonstrate that the rotorcraft have been able to follow quite satisfactorily the desired trajectories. The first experiment allows an independent analysis of forward and vertical movement. An overshoot of approximately 20 cm

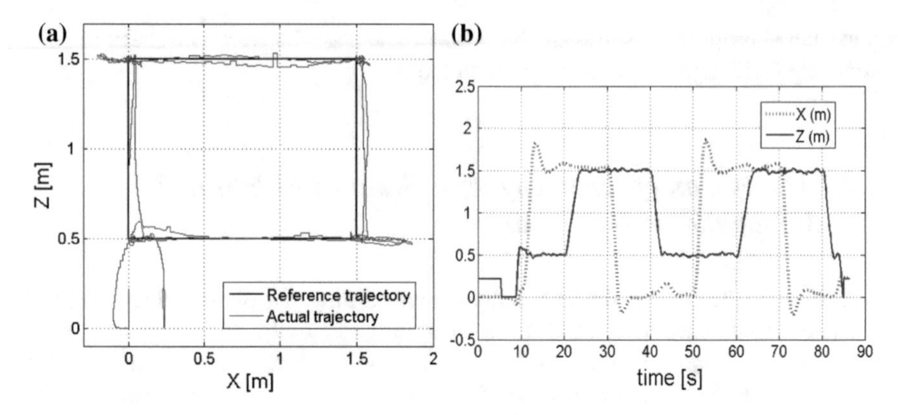

Fig. 11 Rectangular trajectory tracking presented: **a** in X-Z plane; **b** as a time process

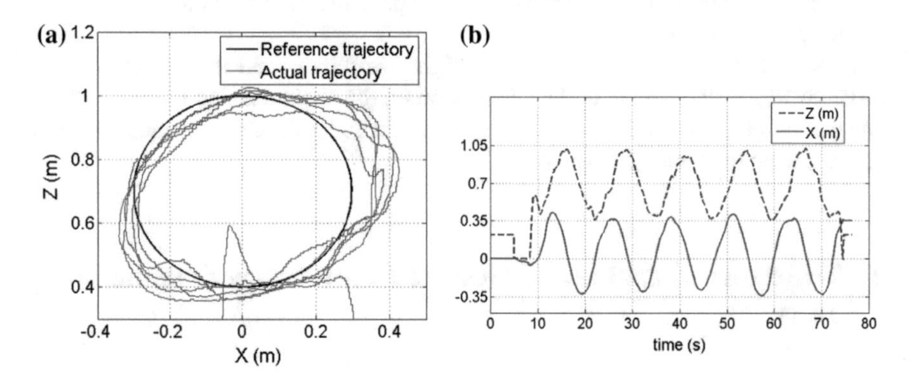

Fig. 12 Circular trajectory tracking presented: **a** in X-Z plane; **b** as a time process

exists during the horizontal movement, while the precision in the vertical movement is high.

4.2 Flight Control Tests Based on the Recognition of Gestures

Several experiments have been carried with the designed gesture-oriented interface to control the flight of the small quadrotor (Fig. 13).

The results from the control of the lateral movement are shown on Fig. 14. The velocity and the roll angle of the vehicle have been recorded. Until the 254th sec. there is no change of the vehicle position. At the 254th sec. a signal to change the position is generated. When a "zero" command is received it take some time until

Fig. 13 Experimental studies on the remote control of the flight of the A.R. Drone v.1.0 quadrotor through a gesture-oriented interface

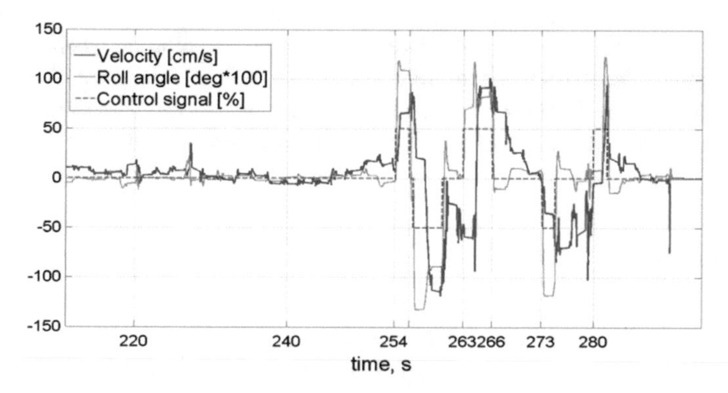

Fig. 14 Time diagram of the measured signals during the experiment: gesture control signal, roll angle of the rotorcraft, and lateral velocity of the rotorcraft

the roll angle and the velocity become constant. A command to change the direction is given next.

It can be seen that after the end of a command for movement the roll angle is changing for a while in opposite direction before it stabilizes to a constant value (some overshoots are observed).

The data shown on Fig. 15 are recorded from the same experiment but are related to the altitude control channel. It can be seen that the gesture-based control begins at 214,4 s. Before this an initial altitude is reached due to the command generated by the computer. At the 214,4 s there is a very short command to increase the altitude but it has a little effect to the quadrotor. It is followed by a command for descend and the aerial vehicle reaches the ground. From the 230 s there is a signal for elevation lasting until 244 s. During these 14 s the quadrotor altitude reaches 2 m.

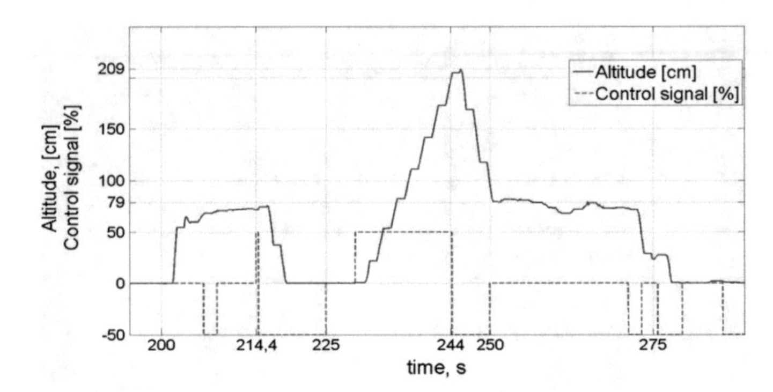

Fig. 15 Time diagram of the measured signals during the experiment: altitude of the rotorcraft and gesture control signal

From the duration of the control signal and the reached altitude the average vertical speed can be calculated, that in this time is 14.28 cm/s. From the 244th sec. there is a command signal to descend with duration of 6 s and the altitude reaches 79 cm. A "zero" signal follows after that in order to stop the vertical movement. At the end of the experiment there is a sequence of descends and zero control signals until the zero altitude is reached. This is done in order to obtain a smooth landing of the drone.

5 Conclusions

A new approach to control a quadrotor rotorcraft is designed. It is suitable for inspection tasks in hard to reach areas of large industrial facilities at risk and for buildings, located in heavily built-up urban environments regarding missing of GPS signal. The proposed control scheme allows two different flight modes: an automatically controlled (autonomous) flight, which allows precise following of the predefined trajectories and a semi-autonomous mode for precise positioning via gesture-based instructions given by a human operator. The performed experimental tests have shown the efficient control in both modes.

Major advantage of the autonomous robots controlled by gestures is the fact that the human/operator is not responsible and does not care about the main control algorithms. The proposed gesture control is a promising approach allowing closer interaction between the rotorcraft and the operator and can be implemented to series of specific applications. It could be also easily implemented into gestural control of other kind of autonomous mobile robots (ground-based for example) in order to ensure human-machine collaboration features. An example could be an autonomous robotized mobile platform serving in an industrial environment supplying details according operator's gestures or personal robot supporting physical activity of elder people accompanying them during daily tasks.

Acknowledgements The authors gratefully acknowledge the financial support provided within the Ministry of Education and Science of Bulgaria Research Fund Projects: FNI I 02/6/2014 and FNI M 07/03/2016.

References

1. Kawatsuma, S., Fukushima, M., Okada, T.: Emergency response by robots to Fukushima Daiichi accident: summary and lessons learned. Ind. Robot **39**(5), 428–435 (2012)
2. Cragg, L., Hu, H.: Application of mobile agents to robust tele-operation of internet robots in nuclear decommissioning. In: Proceedings of the IEEE International Conference on Industrial Technology, Maribor, Slovenia, 10–12 Dec 2003, pp. 1214–1219 (2003)
3. Microdrones GmbH (2016). www.microdrones.com
4. Ascending Technologies GmbH (2016). www.asctec.de
5. Martinez, J.L., Pequeno-Boter, A., Mandow, A., Garcia-Cerezo, A., Morales, J.: Progress in mini-helicopter tracking with a 3D laser range finder. In: Proceedings of the 16th IFAC Triennial World Congress, Prague, Check Republic, vol. 38, no. 1, pp. 648–653 (2005)
6. Roberts, J., Stirling, T., Zufferey, J.-C., Floreano, D.: 3-D relative positioning sensor for indoor flying robots. Auton. Robots **33**(1/2), 1–16 (2012)
7. Roberts, J., Stirling, T., Zufferey J.-C.: Quadrotor using minimal sensing for autonomous indoor flight. In: Proceedings of the European Micro Air Vehicle Conference and Flight Competition (EMAV2007), Toulouse, France, 17–21 Sept 2007, pp. 1–8 (2007)
8. Winkvist, S., Rushforth, E., Young, K.: Towards an autonomous indoor aerial inspection vehicle. Ind. Robot: Int. J. **40**(3), 196–207 (2013)
9. Wongphati, M., Osawa, H., Imai, M.: User-defined gestures for controlling primitive motions of an end effector. J. Adv. Robot. **29**(4), 225–238 (2015). https://doi.org/10.1080/01691864.2014.978371
10. Kılıboz, N., Güdükbay, U., Hand A.: Gesture recognition technique for human-computer interaction. J. Vis. Commun. Image R (2015). http://dx.doi.org/10.1016/j.jvcir.2015.01.015
11. Bhuyan, M.K., MacDorman, K.F., Kar, M.K., Neog, D.R., Lovell, B.C., Gadde, P.: Hand pose recognition from monocular images by geometrical and texture analysis. J. Vis. Lang. Comput. (2015). http://dx.doi.org/10.1016/j.jvlc.2014.12.001
12. Xu, D., Wu, X., Chen, YL., Xu, Y.: Online dynamic gesture recognition for human robot interaction. J. Intell. Robot Syst. **77**, 583 (2015). https://doi.org/10.1007/s10846-014-0039-4
13. Cao, X.Q., Liu, Z.Q.: Type-2 fuzzy topic models for human action recognition. IEEE Trans. Fuzzy Syst. **23**(5), 1581–1593 (2015). https://doi.org/10.1109/tfuzz.2014.2370678
14. Aly, A., Tapus, A.: Towards an intelligent system for generating an adapted verbal and nonverbal combined behavior in human–robot interaction. Auton. Robot **40**, 193 (2016). https://doi.org/10.1007/s10514-015-9444-1
15. Pavlovic, V.I., Sharma, R.T., Huang, S.: Visual interpretation of hand gestures for human—computer interaction: a review. IEEE Trans. Pattern Anal. Mach. Intell. **19**, 677–695 (1997)
16. Interaction with a Quadrotor via the Kinect, ETH Zurich. http://www.youtube.com/watch?v=A52FqfOi0Ek
17. A.R. Drone web site. http://www.parrot.com/
18. Craig, J.J.: Introduction to Robotics: Mechanics and Control, 3rd edn. Pearson Education Inc., Prentice Hall (2005)
19. Shakev, N.G., Topalov, A.V., Kaynak, O.K., Shiev, K.B.: Comparative results on stabilization of the quad-rotor rotorcraft using bounded feedback controllers. J. Intell. Robot. Syst.: Theory Appl. **65**(1–4), 389–408 (2012)
20. Microsoft Kinect web site. http://www.xbox.com/en-US/xbox-360/accessories/kinect/

Introduction to the Theory of Randomized Machine Learning

Yuri S. Popkov, Yuri A. Dubnov and Alexey Y. Popkov

1 Introduction

Modern Computer Science has often faced problems connected to efficient utilizing of the information under uncertainty conditions. Uncertainty management is based on uncertainty modeling. The most advanced are stochastic models with theoretically proven probabilistic properties, and randomized models in which uncertainty is simulated by an ensemble of events. Problem of filtering the signal by its noised observations is an example of the first case. The nature of the signal and the noise is unknown and is modeled by an appropriate dynamic system excited by Wiener process. As the properties of the process are known than we are able to build a filter (for instance, of Kalman type), which will separate the signal and the noise in terms of used functional. Monte Carlo Method (MMC) can be an example of the second case. Using standard sequences of random (quasi-random) numbers, MMC allows to generate ensembles of randomized events, which are interpreted as the model of uncertainty. Thus, in these methods of uncertainty modeling the information about

This work is partially supported by Russian Fund for Basic Research (project no. 16-07-00743).

Y. S. Popkov (✉) · Y. A. Dubnov · A. Y. Popkov
Institute for Systems Analysis of Federal Research Center "Computer Science and Control" of Russian Academy of Sciences, Moscow, Russia
e-mail: popkov@isa.ru

Y. A. Dubnov
e-mail: yury.dubnov@phystech.edu

A. Y. Popkov
e-mail: apopkov@isa.ru

Y. S. Popkov · Y. A. Dubnov · A. Y. Popkov
Moscow Institute of Phisics and Technology, Moscow, Russia

Y. S. Popkov · Y. A. Dubnov
National Research University "Higher School of Economics", Moscow, Russia

V. Sgurev et al. (eds.), *Learning Systems: From Theory to Practice*, Studies in Computational Intelligence 756, https://doi.org/10.1007/978-3-319-75181-8_10

the object being studied (real data) is not used. This leads to reducing an utility of such models.

Here we propose to use the entropy to develop optimal randomized models under maximal level of uncertainty (in terms of entropy). Methods developed in the work are aimed at machine learning problems which are called Randomized Machine Learning (RML) problems.

Machine Learning (ML) has more than sixty-year history and rich experience in solution of many problems. The first publications on the subject dates back to 1957 when F. Rosenblatt created an algorithm called "the perceptron" [19], implemented as the computing machine "Perceptron MARK I". Rosenblatt's work laid the foundations of research on neural networks. The notion of empirical risk, a basic one for ML, was introduced by Ya. Tsypkin in the monograph [21]. The method of potential functions for classification and pattern recognition problems was published in 1970, see the book [27].

Modern ML concept is based on the *deterministic* parametrization of models and estimation using data arrays with hypothetical properties. Estimation quality is characterized by empirical risk (ER) functions, and minimization yields the corresponding optimal estimates in terms of the chosen ER; the details can be found in [2, 23, 24]. Machine Learning problems are devoted by a large amount of literature. Some incomplete view of this area can be fetched from monographs [3–5, 26].

Generally, the real problems solved by ML procedures are involved in some uncertain environment. If the matter concerns data, they are obtained with errors, omissions, dubious validity. Model design and parametrization represents a subjective non-formalizable process depending on the individual knowledge of a researcher. Therefore, mass application of ML procedures causes rather high uncertainty. Empirical risk minimization gives parameter estimates for an existing data array and an adopted parameterized model; in other words, these estimates are conditional. There is no clarity about behavior with other data arrays or another parametrization model.

The above circumstances call for linkage of uncertainty simulation and available real data. Methodological basis for solving the problem is the entropy functional considered as measure of uncertainty. RML procedures allow to determine entropy-optimal estimates of probability density functions (PDF) of random parameters and measurement errors. RML uses generalized informational entropy as optimality criterion, which maximization is implemented on the set, defined by a system of empirical balances. In RML procedure, the "learned" model generates an ensemble of random events (vectors, trajectories) according to the entropy-optimal PDFs of parameters and noise.

2 The Structure of RML Procedure

Randomized Machine Learning procedure aimed to maximize entropy functionals on the sets which configuration depends on available data obtained with errors. The results of RML procedure are entropy-optimal probability density functions for the

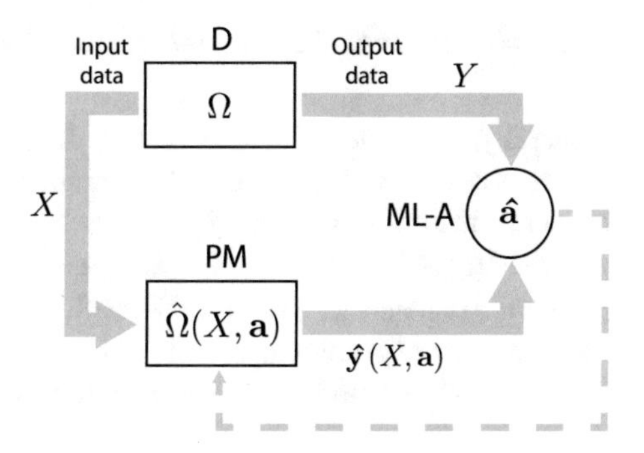

Fig. 1 General structure of the Randomized Machine Learning concept

model parameters and the noise. An important feature of RML procedure is the generation of randomized model events according to entropy-optimal PDFs. The structure of RML procedure is presented at Fig. 1.

2.1 Data (D)

There is a set of data composed of two groups, namely, the "input" data array—matrix $X = [\mathbf{x}^{(1)}, \dots, \mathbf{x}^{(s)}]$ and the "output" data array—matrix $Y = [\mathbf{y}^{(1)}, \dots, \mathbf{y}^{(s)}]$. Vectors $\mathbf{x}^{(j)} \in R^n$, $\mathbf{y}^{(j)} \in R^m$, where s is the number of observations of input vectors. It is assumed that there exists a connection between the "input" and "output" determined by the problem. For instance, in classification problem input data is the learning matrix of term weights, output is the matrix of the "learner" responses. In technological processes modeling input data is the matrix of numerical properties of raw components, output data is the matrix of properties of the products developed during technological process. Input data is considered to be exact and output is considered to have errors of interval type, i.e. with known range of values. Output errors are simulated by a matrix of random noises $\Xi = [\xi^{(1)}, \dots, \xi^{(s)}]$. Random vectors $\xi^{(j)}$, $j = \overline{1, s}$ are from vector intervals

$$\xi^{(j)} \in \mathscr{E}_j = [\xi^-(j), \xi^+(j)], \quad j = \overline{1, s}. \tag{1}$$

Probabilistic properties of the noises are characterized by PDFs $Q_j(\xi^{(j)})$, $j = \overline{1, s}$. PDFs are considered to be continuously differentiable. Random vectors $\xi^{(j)}$ are generated according to PDFs found during RML procedure, i.e. ensembles $\mathscr{E}(j \mid Q_j(\xi^{(j)}))$ of random vectors $\xi^{(j)}$ are being built.

2.2 Randomized Parameterized Model (RPM)

Generally, the input X and output Y data have a dynamic relationship. In other words, the output data observed at moment j depend on the input data observed on some historical interval $j-\rho, j-\rho+1, \ldots, j$. In mathematical terms, this relationship is described by a nonrandom vector functional $\hat{\Omega}(\tilde{X}_\rho^{(j)} \mid \mathbf{a}, P(\mathbf{a}))$ with random parameters **a** and their PDF $P(\mathbf{a})$. Input data is a matrix $\tilde{X}_\rho^{(j)}$ consists for every observation j of ρ column-vectors $\mathbf{x}(j-\rho), \mathbf{x}(j-\rho+1), \ldots, \mathbf{x}(j)$. Parameters of RPM $\mathbf{a} \in R^d$ are random and of interval type

$$\mathbf{a} \in \mathscr{A} = [\mathbf{a}^-, \mathbf{a}^+]. \tag{2}$$

Probabilistic characteristics of the parameters are characterized by PDF $P(\mathbf{a})$ which is considered to be continuously differentiable functions.

Consequently, the model output at observation j represents an ensemble $\hat{\mathscr{Y}}(j \mid P(\mathbf{a}))$ of the random vectors $\hat{\mathbf{y}}(j \mid P(\mathbf{a}))$relating to the input data and random parameters through the vector functional $\hat{\Omega}(\tilde{X}_\rho^{(j)} \mid \mathbf{a}, P(\mathbf{a}))$, i.e.,

$$\hat{\mathscr{Y}}(j \mid P(\mathbf{a})) = \hat{\Omega}(\tilde{X}_\rho^{(j)} \mid \mathbf{a}, P(\mathbf{a})), \qquad j = \overline{1, s}. \tag{3}$$

As it was mentioned above, the errors in the output data are modeled by an ensemble $\mathscr{E}(j \mid Q_j(\xi^{(j)}))$ of the random vectors $\xi^{(j)}$ PDFs $Q_j(\xi^{(j)})$, which is added to the ensemble of the RPM output:

$$\mathscr{V}(j \mid P(\mathbf{a}), Q_j(\xi^{(j)})) = \hat{\mathscr{Y}}(j \mid P(\mathbf{a})) + \mathscr{E}(j \mid Q_j(\xi^{(j)})), \qquad j = \overline{1, s}. \tag{4}$$

Ensemble $\mathscr{V}(j \mid P(\mathbf{a}), Q_j(\xi^{(j)}))$ consists of an array of the random vectors

$$\mathbf{v}(j \mid P(\mathbf{a}), Q_j(\xi^{(j)})) = \hat{\mathbf{y}}(j \mid P(\mathbf{a})) + \xi(j \mid Q_j(\xi^{(j)})), \qquad j = \overline{1, s}. \tag{5}$$

Let us declare the numerical characteristics of the ensemble. Let us define a vector $M^{(k)}(j \mid P(\mathbf{a}), Q_j(\xi^{(j)}))$, using the k-th moments of the components of the vector $\mathbf{v}(j \mid P(\mathbf{a}), Q_j(\xi^{(j)}))$[1]:

$$\begin{aligned} M_i^{(k)}(j \mid P(\mathbf{a}), Q_j(\xi^{(j)})) = \mathscr{M}_P\{\hat{y}_i^k(j \mid P(\mathbf{a}))\} + \\ \mathscr{M}_{Q_j}\{\xi_i^k(j \mid Q_j(\xi^{(j)})\}, \end{aligned} \tag{6}$$

$$i = \overline{1, m}, \ j = \overline{1, s}.$$

[1]In particular, option transactions employ the mean values of financial tools having power dependence on random parameters [1].

Similarly, let us define a vector $\mathbf{m}^{(k)}(j \mid P(\mathbf{a}), Q_j(\xi^{(j)}))$ using the k-means of the components of the vector $M_i^{(k)}(j \mid P(\mathbf{a}), Q_j(\xi^{(j)}))$

$$m_i^{(k)}(j \mid P(\mathbf{a}), Q_j(\xi^{(j)})) = \left(\mathscr{M}_P\{\hat{y}_i^k(j \mid P(\mathbf{a}))\}\right)^{1/k} + \left(\mathscr{M}_{Q_j}\{\xi_i^k(j \mid Q_j(\xi^{(j)}))\}\right)^{1/k}, \quad (7)$$

$$i = \overline{1, m}, \ j = \overline{1, s}.$$

The k-mean vector $\mathbf{m}^{(k)}(j \mid P(\mathbf{a}), Q_j(\xi^{(j)}))$ will be applied as a numerical characteristic of the ensemble $\mathscr{V}(j \mid P(\mathbf{a}), Q_j(\xi^{(j)}))$.

2.3 RML Algorithms (RML-A)

For learning quality assessment, here we introduce *the information entropy functional* [14, 15] defined on the PDF $P(\mathbf{a})$ of the random parameters of RPM and on the vector PDF $Q(\xi) = \{Q_1(\xi^{(1)})), \ldots, Q_s(xi^{(s)})\}$ of the noises as follows:

$$\mathscr{H}[P(\mathbf{a}), Q(\xi)] = -\int_{\mathscr{A}} P(\mathbf{a}) \ln \frac{P(\mathbf{a})}{P^0(\mathbf{a})} d\mathbf{a} - \\ -\sum_{j=1}^{s} \int_{\Xi_j} Q_j(\xi^{(j)}) \ln \frac{Q_j(\xi^{(j)})}{Q_j^0(\xi^{(j)})} d\xi^{(j)}, \quad (8)$$

where $P^0(\mathbf{a})$; $Q_1^0(\xi^{(1)}), \ldots, Q_s^0(\xi^{(s)})$ denote the prior PDFs of the parameters and noises, respectively.

This consideration concerns the methodological interpretations of the notion of informational entropy as a measure of uncertainty. Its maximization yields best solutions under maximum uncertainty. This logical chain was first declared in [9] (see also [10, 12, 18]). Informational entropy characterizes uncertainty connected with the random parameters of RPM and, moreover, with the measurement noises; the last property guarantees best estimates under maximum uncertain noises (in entropy units). Therefore, the PDF estimates obtained by informational entropy maximization can be interpreted as **robust** ones.[2]

Recall that the PDFs $P(\mathbf{a})$; $Q_1(\xi^{(1)})), \ldots, Q_s(\bar{xi}^{(s)})$ are continuously differentiable and normalized, i.e.,:

$$\int_{\mathscr{A}} P(\mathbf{a}) d\mathbf{a} = 1, \quad \int_{\Xi_j} Q_j(\xi(j))\, d\xi^{(j)} = 1, \quad j = \overline{1, s}. \quad (9)$$

[2]This treatment differs from the classical definition of robustness given in [7].

Let us introduce the notion of *empirical balances* stating that the k-means of the observable RPM output are balanced with the output data—the vectors $\mathbf{y}^{(j)},\ j = \overline{1, s}$. The empirical balance conditions are described by the following system of equalities:

$$\mathbf{m}^{(k)}(j \mid P(\mathbf{a}), Q_j(\xi^{(j)})) = \mathbf{y}(j), \quad j = \overline{1, s}. \tag{10}$$

Here the vector $\mathbf{m}^{(k)}$ contains the components (7). For the existing data array, equalities (10) define the set $\mathbb{B}_s$ of admissible PDFs:

$$\mathbb{B}_s = \mathscr{P} \bigcap \mathscr{Q} \bigcap \mathscr{D}_s, \tag{11}$$

where $\mathscr{P}$ is the set of continuously differentiable normal functions (the left equality in (9)); $\mathscr{Q} = \bigcap_{j=1}^{s} \mathscr{Q}_j$, where $\mathscr{Q}_j$ indicates the set of continuously differentiable normal functions (the right equality in (9)); and finally, $\mathscr{D}_s$ means the set of continuously differentiable functions $P, Q_1, \ldots, Q_s$, satisfying the empirical balances (10) for the existing output data array.

And so, **RML-algorithm** is written in the form

$$\mathscr{H}[P(\mathbf{a}), Q(\xi)] \Rightarrow \max, \tag{12}$$

subject to the condition

$$(P(\mathbf{a}), Q(\xi)) \in \mathbb{B}_s. \tag{13}$$

This is a functional entropy-linear programming problem defined on the set of continuously differentiable functions satisfying the normality conditions (9) and the empirical balances (10) and (13).

3 Randomized Dynamical Model in RML Procedure

An important component of the RML algorithm (12) that is connected with the empirical balances comes to formation of the set $\mathbb{B}_s$ (8–11). The structure of this set depends on the k-mean empirical balances and the chosen RPM.

Here we will use the description of the set $\mathbb{B}_s$ induced by the system of 1-mean empirical balances, and the RPM is described by the functional *monomial* of degree R [22, 25]:

$$\hat{\mathbf{y}}^{(j)} = \sum_{h=1}^{R} \sum_{(k_1,\ldots,k_h)=0}^{\rho} A^{(k_1,\ldots,k_h)} \mathbf{x}^{(j-k_1,\ldots,j-k_h)}. \tag{14}$$

In this expression, the components of the vector $\mathbf{x}^{(j-k_1,\ldots,j-k_h)}$ are the lexicographically ordered h-products for the components of the vectors $\mathbf{x}^{(j-k_1)}, \ldots, \mathbf{x}^{(j-k_h)}$. Denote by n_h the dimension of this vector. The matrices

$$A^{(k_1,\dots,k_h)} = \begin{pmatrix} a_{1,1}^{(k_1,\dots,k_h)} & \cdots & a_{1,n_h}^{(k_1,\dots,k_h)} \\ \cdots & \cdots & \cdots \\ a_{m,1}^{(k_1,\dots,k_h)} & \cdots & a_{m,n_h}^{(k_1,\dots,k_h)} \end{pmatrix}, \quad (k_1,\dots,k_h) = \overline{0,\rho}. \tag{15}$$

These matrices contain random independent interval-type elements of the form

$$\begin{aligned} a_{i,s_h}^{(k_1,\dots,k_h)} &\in \mathscr{A}_{i,s_h}^{(k_1,\dots,k_h)} = \left[{}^{(-)}a_{i,s_h}^{(k_1,\dots,k_h)}, {}^{(+)}a_{i,s_h}^{(k_1,\dots,k_h)}\right]; \\ A^{(k_1,\dots,k_h)} &\in \mathscr{A}^{(k_1,\dots,k_h)} = \left[{}^{(-)}A^{(k_1,\dots,k_h)}, {}^{(+)}A^{(k_1,\dots,k_h)}\right], \\ (k_1,\dots,k_h) &= \overline{0,\rho}. \end{aligned} \tag{16}$$

Designate by

$$\mathbb{A}_h = \{A^{(k_1,\dots,k_h)}\},\ (k_1,\dots,k_h) = \overline{0,\rho} \tag{17}$$

the set of the matrices (15). The elements of this set possess values from the interval

$$\mathscr{A}_h = \bigcup_{(k_1,\dots,k_h)=0}^{\rho} \mathscr{A}^{(k_1,\dots,k_h)}. \tag{18}$$

On the above intervals, there exists the PDF $P_h(\mathbb{A}_h)$ from the class of continuously differentiable functions.

According to (5) and (14), the ensemble of the observable RPM output consists of the random vectors

$$\mathbf{v}^{(j)} = \sum_{h=1}^{R} \mathbf{v}_h^{(j)}(\mathbb{A}_h) + \xi^{(j)}, \quad j = \overline{1,s}. \tag{19}$$

where

$$\mathbf{v}_h^{(j)}(\mathbb{A}_h) = \sum_{(k_1,\dots,k_h)=0}^{\rho} A^{(k_1,\dots,k_h)}\mathbf{x}^{(j-k_1,\dots,j-k_h)} \tag{20}$$

Recall that the noise vectors $\xi^{(j)},\ j = \overline{1,s}$ are random independent with the intervals (1) and the PDF $Q_j(\xi^{(j)}),\ j = \overline{1,s}$.

RL-algorithm (12) and (13) for the RPM (14) has the form:

$$\begin{aligned} \mathscr{H}[P(\mathbb{A}), Q(\xi)] = &-\sum_{h=1}^{R} \int_{\mathscr{A}_h} P_h(\mathbb{A}_h) \ln \frac{P_h(\mathbb{A}_h)}{P_h^0(\mathbb{A}_h)}\, d\mathbb{A}_h - \\ &-\sum_{j=1}^{s} \int_{\Xi_j} Q_j(\xi^{(j)}) \ln \frac{Q_j(\xi^{(j)})}{Q_j^0(\xi^{(j)})} d\xi^{(j)} \Rightarrow \max; \end{aligned} \tag{21}$$

subject to

- *the normalization conditions of the PDFs*

$$\int_{\mathscr{A}_h} P_h(\mathbb{A}_h)\, d\mathbb{A}_h = 1, \quad h = \overline{1, R}; \qquad \int_{\Xi_j} Q_j(\xi^{(j)})\, d\xi^{(j)} = 1, \quad j = \overline{1, s}; \tag{22}$$

- *and the* 1*-mean empirical balances*

$$\sum_{h=1}^{R} \int_{\mathscr{A}_h} P_h(\mathbb{A}_h)\mathbf{v}_h^{(j)}(\mathbb{A}_h) d\mathbb{A}_h + \int_{\Xi_j} Q_j(\xi^{(j)})\, \xi^{(j)}\, d\xi^{(j)} = \mathbf{y}^{(j)}, \qquad j = \overline{1, s}. \tag{23}$$

4 Optimality Conditions of RML Algorithm. Structure of Entropy-Optimal PDFs

The RML algorithm (21–23) is formulated in terms of a functional entropy-linear programming problem [17] which belongs to the Lyapunov-type optimization problems [8, 11] (all components are described by integral functionals). For such problems, optimality conditions can be obtained by using the Lagrange functional and multipliers. Since the RML algorithm involves continuously differentiable PDFs, i.e. belong to $\mathbb{C}^1$ class, the variation of the Lagrange functional can be defined via the Gateaux derivatives [13].

Using the denoted scheme we can obtain following expressions for entropy-optimal PDFs of the parameters and noises:

$$P_h^*(\mathscr{A}_h) = \frac{P_h^0(\mathscr{A}_h) \exp\left[-\sum_{j=1}^{s} \langle \theta^{(j)}, \mathbf{v}_h^{(j)}(\mathbb{A}_h)\rangle\right]}{\mathscr{P}_h(\theta)},$$

$$Q_j^*(\xi^{(j)}) = \frac{Q_j^0(\xi^{(j)}) \exp\left[-\sum_{j=1}^{s} \langle \theta^{(j)}, \xi^{(j)}\rangle\right]}{\mathscr{Q}_j(\theta)}, \tag{24}$$

$$h = \overline{1, R}, \quad j = \overline{1, s}.$$

In these equalities,

$$\mathscr{P}_h(\theta) = \int_{\mathbb{A}_h} P_h^0(\mathscr{A}_h) \exp\left[-\sum_{j=1}^{s} \langle \theta^{(j)}, \mathbf{v}_h^{(j)}(\mathbb{A}_h)\rangle\right] d\mathbb{A}_h, \tag{25}$$

$$\mathscr{Q}_j(\theta) = \int_{\Xi_j} Q_j^0(\xi^{(j)}) \exp\left[-\sum_{j=1}^{s} \langle \theta^{(j)}, \xi^{(j)}\rangle\right] d\xi^{(j)}, \tag{26}$$

$$h = \overline{1, R}, \quad j = \overline{1, s}.$$

Here $\theta = \{\theta^{(1)}, \dots, \theta^{(s)}\}$ is the vector of Lagrange multipliers for the conditions (23). By (24),the entropy-optimal PDFs belong to the exponential family parameterized by the Lagrange multipliers responsible for the empirical balance conditions. The above Lagrange multipliers are defined by the equations resulting from the empirical balances (23) by substituting the PDFs from (24):

$$\sum_{h=1}^{R} \mathscr{P}_h^{-1}(\theta) \int_{\mathscr{A}_h} P_h^0(\mathbb{A}_h) \exp\left[-\sum_{j=1}^{s} \langle \theta^{(j)}, \mathbf{v}_h^{(j)}(\mathbb{A}_h) \rangle \right] d\mathbb{A}_h +$$

$$+ \mathscr{Q}_j^{-1}(\theta) \int_{\Xi_j} Q_j^0(\xi^{(j)}) \exp\left[-\sum_{j=1}^{s} \langle \theta^{(j)}, \xi^{(j)} \rangle \right] d\xi^{(j)} = \mathbf{y}^{(j)}, \tag{27}$$

$$j = \overline{1, s}.$$

Equations (27) represent a special class of nonlinear equations with integral components parameterized by Lagrange multipliers θ and admit no analytic definition. The alternative is to perform numerical calculation, e.g., based on Monte Carlo trials [17].

5 Applications

5.1 Entopic "2-Soft" Classification of Texts

In the sequel, "n-soft" classification implies the allocation of objects (documents, texts) among n classes with probability defined by application of the RML procedure.[3]

Consider a randomized linear classifier for 2 classes. Let there exist two collections of text documents, namely, a learning collection $\mathbb{E} = \{e_1, \dots, e_m\}$ and a classification (test) collection $\mathbb{T} = \{t_1, \dots, t_s\}$. The documents in the learning collection are marked by their belonging to class 1 or to class 2. The documents in the test collection have no marking.

The documents in both collections are described by the vectors containing the weights of the words (terms) in a given document. Weight admits many interpretations and quantitative characterizations that are not discussed in the paper. The sets of vectors in corresponding collections are somehow ordered, i.e.,

$$\mathscr{E} = \{\mathbf{e}^{(1)}, \dots, \mathbf{e}^{(m)}\}, \qquad \mathscr{T} = \{\mathbf{t}^{(1)}, \dots, \mathbf{t}^{(s)}\}. \tag{28}$$

[3]Distribution of objects among n classes is reduced to C_n^2 distributions among 2 classes of n.

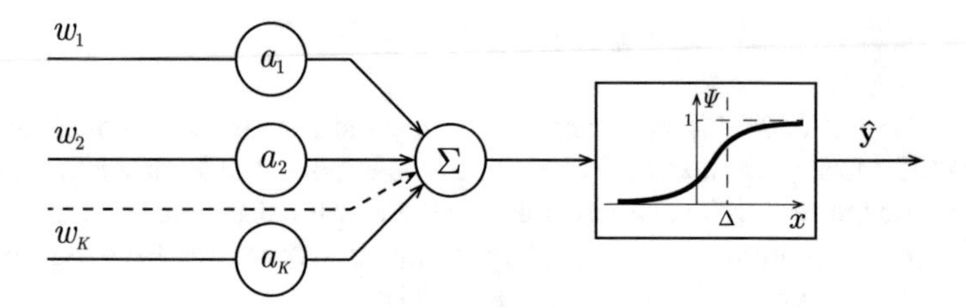

Fig. 2 Decision rule for the "2-soft" classification problem

By assumption, these vectors have a same dimension, that is, $(\mathbf{e}^{(i)}, \mathbf{t}^{(j)}) \in R^n$, where R^n denotes the attribute space (in this case, weights).

5.1.1 Learning Stage

Recall that the documents are marked at this stage. And so, there exists a learning sequence representing a sequence of pairs $\mathscr{O} = \{(1, 1), (2, 1), \ldots, (m, 1)\}$ of length m, where the first element is document number and the second element gives its class. Transform it into a sequence of numbers from the interval [0,1], where element position in the sequence stands for document number, while the numbers $[1/2, 1]$ and $[0, 1/2)$ mean document belonging to class 1 and class 2, respectively. Therefore, the "learning" vector $\mathbf{y}$ has values from the interval [0,1] as its components and dimension coinciding with the number m of documents in the collection:

$$\mathbf{y} = \{y^{(1)}, \ldots, y^{(m)}\}. \tag{29}$$

The randomized model (decision rule) is defined by a random vector $\hat{\mathbf{y}}(\mathbf{a})$ that depends on the random parameters $\mathbf{a}$ of a single-layer neural network (see Fig. 2). The independent components of the vector $\hat{\mathbf{y}}(\mathbf{a})$

$$\hat{y}^{(i)}(\mathbf{a}) = \psi(x^i), \qquad i = \overline{1, m}, \tag{30}$$

where

$$x^{(i)} = \langle \mathbf{e}^{(i)}, \mathbf{a} \rangle, \tag{31}$$

and function $\psi(x^{(i)}))$ have the form:

$$\text{sigm}(x^{(i)}) = \frac{1}{1 + \exp[-\alpha(x^{(i)} - \Delta)]}. \tag{32}$$

Parameters α, Δ are fixed. The values of sigm(x) from the interval $[\frac{1}{2}, 1]$ correspond to class 1, while the values from the open interval $[0, \frac{1}{2})$ to class 2.

The parameters $\mathbf{a} = \{a_1, \ldots, a_n\}$ in the randomized model (28) are independent interval-type:

$$a_k \in \mathscr{A}_k = [a_k^-, a_k^+], \quad \mathscr{A} = \bigcup_{k=1}^{n} \mathscr{A}_k, \quad k = \overline{1, n}. \tag{33}$$

There exists the joint probability density function $P(\mathbf{a})$ on these interval sets.

Since the parameters $\mathbf{a}$ are random with the PDF $P(\mathbf{a})$ for each document with number (i) we obtain an ensemble $\hat{\mathscr{Y}}^{(i)}$ of random values from the interval (0, 1) (30). The mean values of the components of vector $\hat{\mathbf{y}}(\mathbf{a})$ will have the form:

$$\mathscr{M}\{\hat{y}^i(\mathbf{a})\} = \int_{\mathscr{A}} P(\mathbf{a})\text{sigm}\left(\langle \mathbf{e}^{(i)}, \mathbf{a}\rangle\right) d\mathbf{a}. \tag{34}$$

Thus, the “2-soft” classification problem in terms of RML is stated as

$$\mathscr{H}[P(\mathbf{a})] = -\int_{\mathscr{A}} P(\mathbf{a}) \ln P(\mathbf{a}) d\mathbf{a} \Rightarrow \max, \tag{35}$$

subject to the conditions

$$\int_{\mathscr{A}} P(\mathbf{a}) d\mathbf{a} = 1, \tag{36}$$

$$\int_{\mathscr{A}} P(\mathbf{a})\text{sigm}\left(\langle \mathbf{e}^{(i)}, \mathbf{a}\rangle\right) d\mathbf{a} = y^{(i)}, \qquad i = \overline{1, m}. \tag{37}$$

We introduce the Lagrange multipliers $\theta = \{\theta_1, \ldots, \theta_m\}$ for the constraints (37). According to (24) the solution of this problem that is adapted to the model (30) acquires the form:

$$P^*(\mathbf{a}) = \frac{W^*(\theta, \mathbf{a})}{\mathscr{P}(\theta)}, \tag{38}$$

where

$$W^*(\theta, \mathbf{a}) = \exp\left(-\langle \theta, \hat{\mathbf{y}}(\mathbf{a})\rangle\right), \tag{39}$$

$$\mathscr{P}(\theta) = \int_{\mathscr{A}} \exp\left[-\langle \theta, \hat{\mathbf{y}}(\mathbf{a})\rangle\right] d\mathbf{a}, \tag{40}$$

and the vector $\hat{\mathbf{y}}(\mathbf{a})$ consists of the components (30–32). Lagrange multipliers θ are defined by the system of balance equations (37) of the form

$$\mathscr{P}^{-1}(\theta) \int_{\mathscr{A}} \exp\left[-\langle \theta, \hat{\mathbf{y}}(\mathbf{a})\rangle\right] \hat{y}^{(i)}(\mathbf{a}) d\mathbf{a} = y^{(i)}, \qquad i = \overline{1, m}. \tag{41}$$

Interestingly, the dimension of this system coincides with the number of documents in the training collection. Therefore, in the sense of computing effort, the randomized learning procedure for the classification problem is more efficient under limited-volume data collections. Consider two examples demonstrating the main idea of "2-soft" classification of texts without discussion of computational aspects.

5.1.2 Classification Stage

This stage employs the collection $\mathbb{T} = \{t_1, \ldots, t_s\}$, where each element is characterized by a vector $\mathbf{t}^{(j)} \in R^n$. Consider classification procedure for an arbitrary document $\mathbf{t}^{(j)}$.

Step 1-i. Generate an ensemble $\hat{\mathscr{Y}}^{(i)}$ of the randomized model output (decision rules) (30) with the function $P^*(\mathbf{a})$ (38). The ensemble contains N random values from the interval [0,1].

Step 2-i. If a random value from this ensemble exceeds $1/2$, then document $\mathbf{t}^{(i)}$ is assigned class 1; otherwise, class 2.

Step 3-i Suppose that N_1 values are assigned class 1 and N_2 values class 2. Since the number of trials N is sufficiently large, the quantities $p_1^{(i)} = N_1/N$ and $p_2^{(i)} = N_2/N$ yield the empirical probabilities of assigning appropriate classes to document $\mathbf{t}^{(i)}$.

By repeating steps $2 - i, 3 - i$ for the whole collection $\mathbb{T}$, we obtain the probability distribution of assigning class 1 or 2 to the document. These distributions are obtained by the entropy-optimal decision rule (30–32, 38) and (39).

To estimate the quality of classification we use real information of the distribution of documents among classes which is than compares to accepted quantile for empirical probabilities $p_1^{(i)}, p_2^{(i)}$.

5.1.3 Examples

Example 1 For classification we use a collection of 503 documents, 3 of which are used for learning, 500—for testing. Each document is represented by a four-dimensional vector of weights.

Learning. Learning collection is shown in Table 1.

The randomized model (32) has the parameters $\alpha = 1.0$ $\Delta = 0$. The "learner" responses are $\mathbf{y} = \{0.18;\ 0.81;\ 0.43\}$ ($y_i < 0.5$ corresponds to class 2, $y_i \geq 0.5$ to

Table 1 Learning collection for Example 1

i	$e_1^{(i)}$	$e_2^{(i)}$	$e_3^{(i)}$	$e_4^{(i)}$
1	0.11	0.75	0.08	0.21
2	0.91	0.65	0.11	0.81
3	0.57	0.17	0.31	0.91

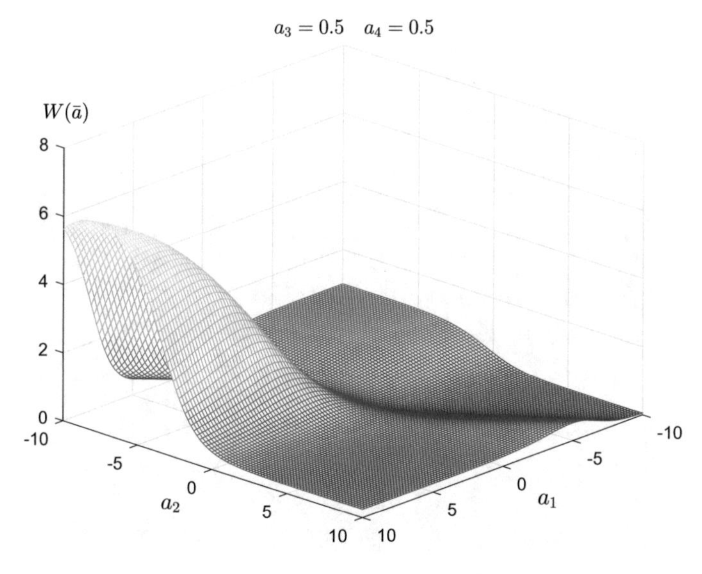

Fig. 3 Joint PDF for Example 1

class The Lagrange multipliers for the entropy-optimal PDF (38) are $\theta = \{3.2807; -3.5127; 1.6373\}$. The parameters $a_i \in [-10, 10]$, $i = \overline{1, 4}$. For this learning collection, the entropy-optimal function $W^*(\mathbf{a})$ (39) takes the form

$$W^*(\mathbf{a}) = \exp\left(-\sum_{i=1}^{3} \theta_i y_i(\mathbf{a})\right),$$

$$y_i(\mathbf{a}) = \left(1 + \exp(-\sum_{k=1}^{4} e_k^{(i)} a_k)\right)^{(-1)}. \tag{42}$$

Fig. 3 shows the two-dimensional section of the function (42) under $a_3 = 0.5$; $a_4 = 0.5$.

Classification. For classification we use a collection of 500 documents represented by an array of the four-dimensional random vectors $\mathbf{t}^{(i)}$, $i = \overline{1, 500}$ with independent components obeying the uniform distribution on the interval [0,1]. For each element of this sample, generate the random parameters of the model (30) according to the PDF (38) ($N = 1000$) using the acceptance-rejection method [20] and calculate its output. Perform steps $2 - i, 3 - i$.

Figure 4a, b demonstrates the empirical probabilities $p_1^{(i)}$, $p_2^{(i)}$ of assigning class 1 and 2 to document t_i. For different documents their assigning probabilities vary from 15 to 85%.

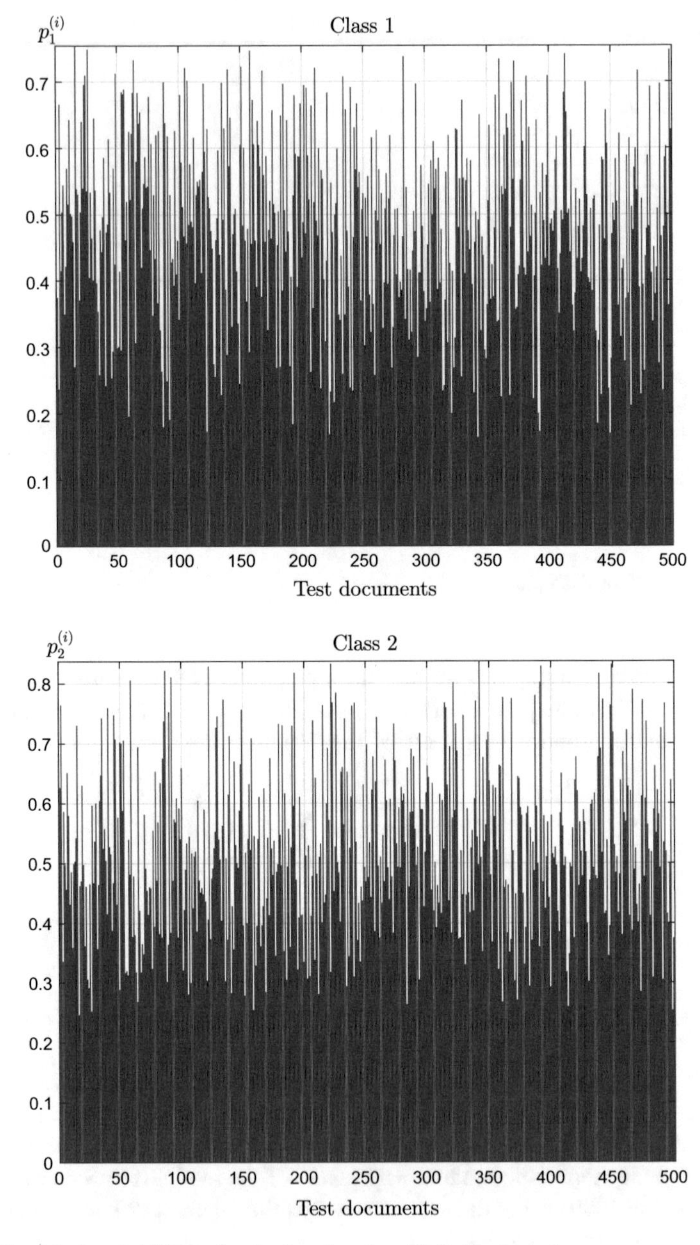

Fig. 4 Empirical probabilities of assigning class 1 and 2 for Example 1

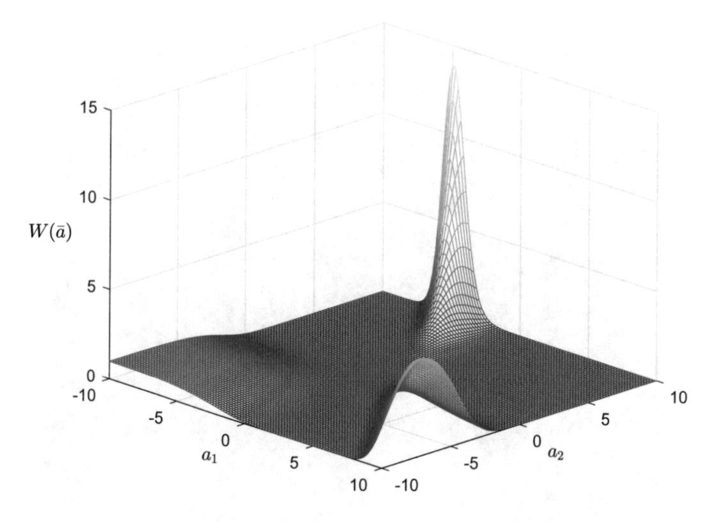

Fig. 5 Joint PDF for Example 2

Example 2 Let us consider a collection of 503 documents, 3 of which are used for learning, 500—for testing. Each document is represented by a two-dimensional vector of weights.

Learning. Learning collection is shown in the first two columns of Table 1.The values of the parameters α, Δ and the intervals for the random parameters $\mathbf{a}$ are the same as in Example 1. The Lagrange multipliers for the entropy-optimal PDF (38) are $\theta = \{9.6316;\ -18.5996;\ 16.7502\}$. For this learning collection, the entropy-optimal function $W^*(\mathbf{a})$ (39) takes the form

$$W^*(\mathbf{a}) = \exp\left(-\sum_{i=1}^{3} \theta_i y_i(\mathbf{a})\right),$$

$$y_i(\mathbf{a}) = \left(1 + \exp(-\sum_{k=1}^{2} e_k^{(i)} a_k)\right)^{(-1)}. \qquad (43)$$

Fig. 5 illustrates the function $W^*(\mathbf{a})$ (43).

Classification. All parameters are same as in *Example 2 (Learning)*. Figure 6a, b shows the empirical probabilities $p_1^{(i)}$, $p_2^{(i)}$ of assigning class 1 and 2 to document t_i. In this case, for different documents their assigning probabilities vary from approx 10–90%.

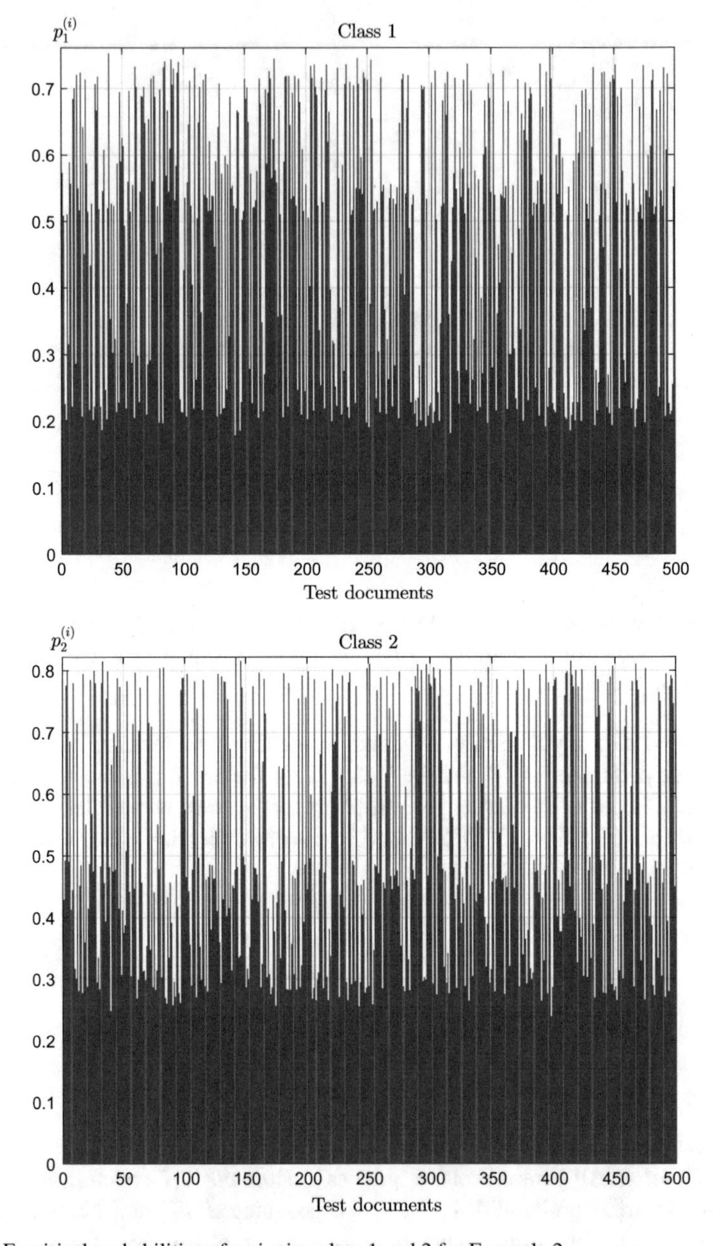

Fig. 6 Empirical probabilities of assigning class 1 and 2 for Example 2

5.2 RML Algorithms for Dynamic Regression Problems

The general form of the randomized dynamic model (RDM) for dynamic regression problems is given by (3), and it is possible to apply the general RML procedure stated in Sect. 3. Consider this procedure in detail to restore the characteristics of the randomized World population dynamics model using it as a forecasting tool [16].

Let the World population evolve according to the discrete-form exponential randomized model of population dynamics with measurement errors

$$v[ih] = E_i(r, u_r | E_0) + \xi[ih], \qquad i \in [0, I], \tag{44}$$

with the function

$$E_i(r, u_r \mid E_0) = E_0 \exp[(r + u_r i)ih], \quad i \in [0, I]. \tag{45}$$

Here $h = 5\,[years]$ is the measurement interval, E_0 specifies the World population at the beginning of the observation interval, I indicates the length of the measurement interval (an integer number), r means reproduction rate (difference between fertility and mortality flows), u_r is the velocity of its changing.

The measurement errors are modeled by a random vector $\bar{\xi} = \{\xi[0], \ldots, \xi[Ih]\}$ with independent interval-type components and a PDF $Q(\bar{\xi})$ defined on the set

$$\Xi = \bigcup_{j=0}^{I} \Xi_j, \qquad \Xi_j = [\xi_j^-, \xi_j^+], \tag{46}$$

where $I + 1$ gives the number of World population measurements on a corresponding interval. By assumption, the intervals are same for all measurements: $[-0.5; 0.5]$ *billion people*. And so,

$$Q(\bar{\xi}) = \prod_{j=0}^{I} q_j(\xi[jh]). \tag{47}$$

A common approach in World population forecasting is to consider model parameters like fertility and mortality rates as constant on definite intervals, analyzing different scenario values (see [6], http://data.un.org).

However, in a more realistic situation, one knows only the intervals covering the values of the above parameters, i.e.,

$$\mathscr{I}_r = [r^-, r^+], \quad \mathscr{I}_{u_r} = [u_r^-, u_r^+]. \tag{48}$$

Assume that reproduction rate and velocity of its changing are random, taking values from the intervals $\mathscr{I}_r$ and $\mathscr{I}_{u_r}$ with a joint PDF $P(r, u_r)$, where $(r, u_r) \in \mathscr{I} = \mathscr{I}_r \cup \mathscr{I}_{u_r}$. In the sequel, $\mathscr{I}_r = [-0.025; 0.075]$, $\mathscr{I}_{u_r} = [-0.002; 0.001]$. Selection

Table 2 World population in billions of people

i	0	1	2	3	4	5	6	7
year	1960	1965	1970	1975	1980	1985	1990	1995
$E^{ml}_{real}[i]$	3.026	3.358	3.691	4.070	4.449	4.884	5.320	5.724

of such intervals for model parameters assumes the possibility for both positive and negative trends of World population growth.

5.2.1 Learning Stage

Find the entropy-optimal PDFs of the model parameters and noises for the retrospective data corresponding to the period from 1960 to 1995 with step $h = 5$ *years* (see Table 2, http://data.un.org).

The RML algorithm (4–6) yields the following entropy-optimal PDFs:

- model parameters

$$P^*(r, u_r) = \frac{1}{\mathscr{R}(\bar{\theta})} \prod_{i=0}^{7} p_i^*(r, u_r|\theta_i),$$
$$p_i^*(r, u_r|\theta_i) = \exp\left(-\theta_i E_i(r, u_r|E^{ml}_{real}[0])\right); \tag{49}$$

- noise

$$Q^*(\xi) = \frac{1}{\mathscr{Q}(\bar{\theta})} \prod_{j=0}^{7} q_j^*(\xi[jh]|\theta_j),$$
$$q_j^*(\xi[jh]|\theta_j) = \exp\left(-\theta_j \xi[jh]\right). \tag{50}$$

where

$$\mathscr{R}(\theta \mid E^{ml}_{real}[0]) = \int_{\mathscr{I}} \prod_{i=0}^{7} \exp\left(-\theta_i E_i(r, u_r|E^{ml}_{real}[0])\right) dr\, du_r \tag{51}$$

and

$$\mathscr{Q}(\bar{\theta}) = \prod_{j=0}^{7} \int_{\xi_j^-}^{\xi_j^+} \exp(-\theta_j \xi[jh]) d\xi[jh] =$$
$$= \prod_{j=0}^{7} \frac{1}{\theta_j} \left(\exp(-\theta_j \xi_j^-) - \exp(-\theta_j \xi_j^+)\right). \tag{52}$$

Table 3 Calculated Lagrange multipliers

Measurement points	0	1	2	3	4	5	6	7
θ	0.0000	0.3833	0.3984	0.5839	0.3802	0.4679	0.1812	0.8881

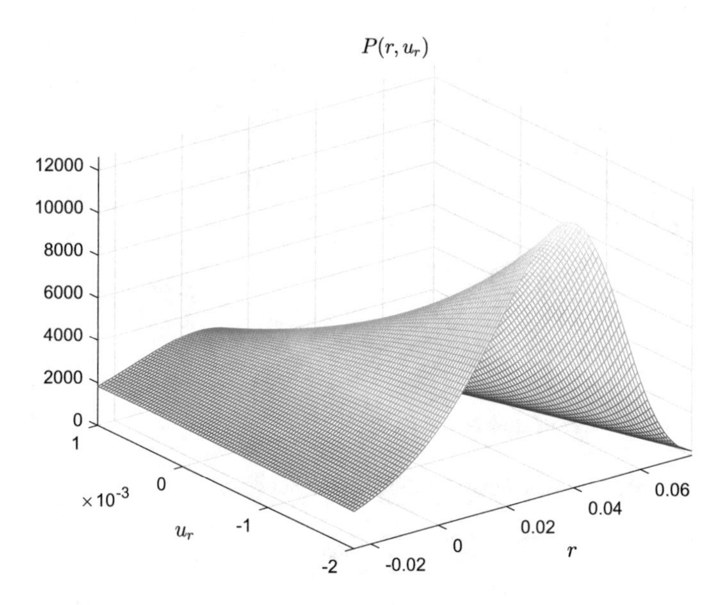

Fig. 7 Joint PDF of r u_r in interval $\mathscr{I}_r \bigcup \mathscr{I}_{u_r}$

The Lagrange multipliers θ are defined by solving the system of balance equations, see (7). Table 3 shows the Lagrange multipliers calculated for the above intervals.

The curves of the PDF $P^*(r, u_r)$ of the model parameters and the PDFs $q^*(\xi[ih])$, $i \in [0, 7]$ of the noise components are shown by Figs. 7 and 8.

5.2.2 Testing Stage

The model has been tested using the data from Table 4.

The World population is evaluated by formula (44) and (45), where r, u_r represent the random parameters with the PDF $P^*(r, u_r)$ (49); $\xi[ih]$ are the random noises with the PDFs $q_i(\xi[ih])$, $i \in [0, 4]$ (50).

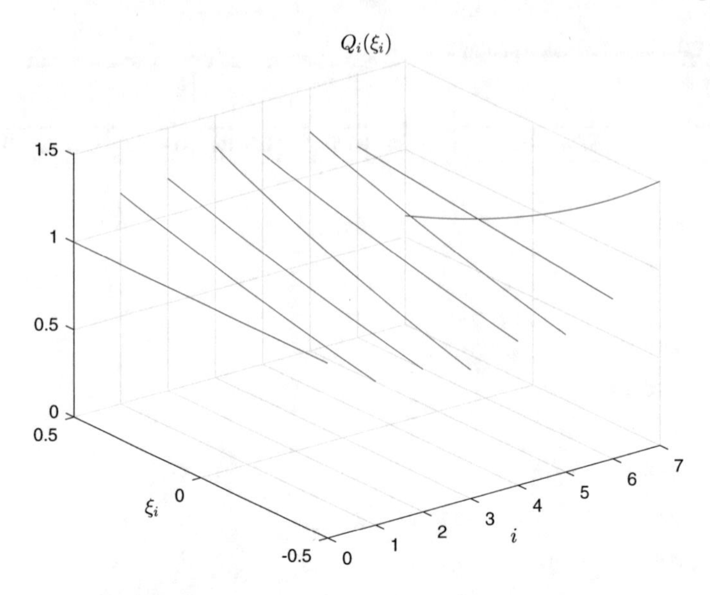

Fig. 8 Ensemble of PDFs of noise ξ_i, $i \in [0, 7]$

Table 4 World population and forecast at 1985 in billions of people

i	0	1	2	3	4
year	1995	2000	2005	2010	2015
E^{tst}_{real}	5.724	6.128	6.514	6.916	7.359
E^{prn}_{1985}	5.666	5.962	6.450	6.985	7.469

To generate an ensemble of the random variables (the model parameters r, u_r and the noises $\xi[ih]$, $i \in [0, 4]$), we have adopted the generalized two-dimensional modification of the Ulam-von Neumann method (acceptance-rejection method [20]). The size of the generated sample is $k = 10^5$. Figure 9 illustrates the resulting ensemble with the following notation: (1) are the ensemble-average trajectories of population dynamics, (2) is the real population dynamics; (3) is the population dynamics on the testing interval (1995–2015) according to the UN forecast made at 1985, (4) is the standard deviation. The relative mean-square deviation between the real trajectory and the ensemble-average one is 0.3%. The relative mean-square deviation for the UN forecast is 0.8%.

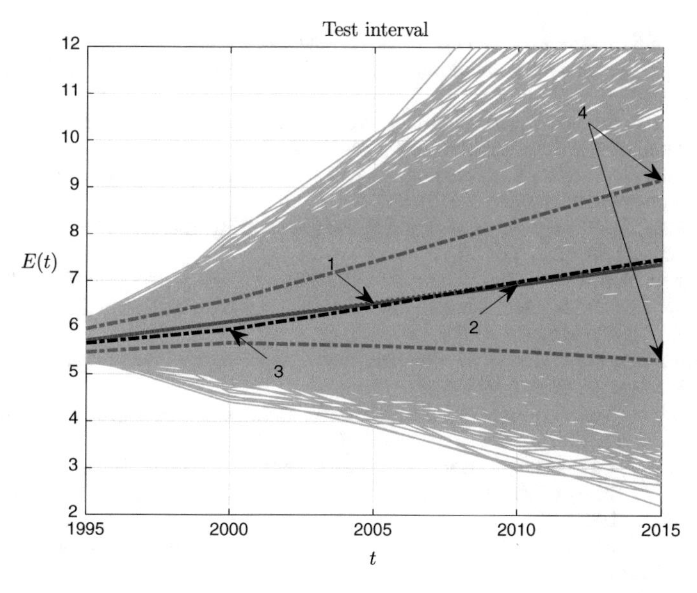

Fig. 9 Ensemble of forecast trajectories on the test interval

References

1. Avellaneda, M.: Minimum-relative-entropy calibration of asset-pricing models. Int. J. Theor. Appl. Finance **1**(04), 447–472 (1998)
2. Bishop, C.: Pattern Recognition and Machine Learning (Information Science and Statistics), 1st edn. 2006. corr. 2nd printing edn. Springer, New York (2007)
3. Boucheron, S., Bousquet, O., Lugosi, G.: Theory of classification: A survey of some recent advances. ESAIM: probab. Stat. **9**, 323–375 (2005)
4. Flach, P.: Machine Learning: The Art and Science of Algorithms that Make Sense of Data. Cambridge University Press (2012)
5. Friedman, J., Hastie, T., Tibshirani, R.: The elements of statistical learning. In: Springer Series in Statistics, vol. 1. Springer, Berlin (2001)
6. Gonzalo, J.A., Muñoz, F.F., Santos, D.J.: Using a rate equations approach to model world population trends. Simulation **89**(2), 192–198 (2013)
7. Huber, P.J.: Robust Statistics. Springer (2011)
8. Ioffe, A.D., Tikhomirov, V.M.: Teoriya ekstremal'nykh zadach (Theory of Extremal Problems). Nauka, Moscow (1974)
9. Jaynes, E.T.: Information theory and statistical mechanics. Phys. Rev. **106**(4), 620–630 (1957)
10. Jaynes, E.T.: Probability Theory: The Logic of Science. Cambridge University Press (2003)
11. Kaashoek, M.A., et al.: Recent Advances in Operator Theory and its Applications: the Israel Gohberg Anniversary Volume, vol. 160. Springer Science & Business Media (2005)
12. Kapur, J.N.: Maximum-Entropy Models in Science and Engineering. Wiley (1989)
13. Kolmogorov, F.N., Fomin, S.V.: Elements of the Theory of Functions and Functional Analysis, vol 1. Courier Corporation (1999)
14. Kullback, S., Leibler, R.A.: On information and sufficiency. Ann. Math. Stat. **22**(1), 79–86 (1951)
15. Popkov, Y.S.: Macrosystems theory and its applications (equilibrium models). In: Lecture Notes in Control and Information Sciences. Springer (1995)

16. Popkov, Y.S., Dubnov, Y.A., Popkov, A.Y.: New method of randomized forecasting using entropy-robust estimation: application to the world population prediction. Mathematics **4**, 1–16 (2016a)
17. Popkov, Y.S., Popkov, A.Y., Darkhovsky, B.S.: Parallel monte carlo for entropy robust estimation. Math. Models Comput. Simul. **8**(1), 27–39 (2016b)
18. Racine, J.S., Maasoumi, E.: A versatile and robust metric entropy test of time-reversibility, and other hypotheses. J. Econom. **138**(2), 547–567 (2007)
19. Rosenblatt, F.: The perceptron, a Perceiving and Recognizing Automaton Project Para. Cornell Aeronautical Laboratory (1957)
20. Rubinstein, R.Y., Kroese, D.P.: Simulation and the Monte Carlo Method, vol. 707. Wiley (2011)
21. Tsypkin, Y.Z.: Osnovy teorii obuchayushchikhsya sistem (Foundations of Theory of Learning Systems). Nauka, Moscow (1970)
22. Tsypkin, Y.Z., Popkov, Y.S.: Teoriya nelineinykh impul'snykh sistem (Theory of Nonlinear Impulse Systems). Nauka, Moscow (1973)
23. Vapnik, V.N.: Vosstanovlenie zavisimostei po empiricheskim dannym (Restoration of Dependencies Using Emprirical Data). Nauka, Moscow (1979)
24. Vapnik, V.N., Chervonenkis, A.Y.: Teoriya raspoznavaniya obrazov. Nauka, Moscow (1974)
25. Volterra, V.: Theory of Functionals and of Integral and Integro–Differential Equations. Dover Publications (2005)
26. Witten, I.H., Frank, E.: Data Mining: Practical Machine Learning Tools and Techniques. Morgan Kaufmann (2005)
27. Yzerman, M.A., Braverman, E.M., Rozonoer, L.I.: Metod potentsialnykh funktsii v teorii obucheniya mashin. Nauka, Moscow (1970)

Grid-Type Fuzzy Models for Performance Evaluation and Condition Monitoring of Photovoltaic Systems

Gancho Vachkov and Valentin Stoyanov

1 Introduction

The solar energy is a typical source of "clean", renewable energy that can be relatively easily converted into electrical energy by using the photovoltaic (PV) systems. The classical PV systems consist of multiple *solar panels* that convert the solar energy into a DC current. Additional electrical circuits and devices (inverters) serve to convert the DC current into AC current that is used to supply the energy either to an individual (self-standing) customer or to the general purpose smart grid system.

Unlike the most typical industrial process systems, the PV systems have their specific properties and peculiarities that create respective problems and difficulties in their everyday functioning, performance evaluation and maintenance. For example the PV systems work under wide range of changing atmospheric conditions, such as solar radiation, ambient temperature and wind speed. Such highly fluctuating conditions directly affect the final productivity (the output electric power), which also fluctuates in wide margins within the *24* hourly cycle (day and night). Usually the solar energy is available for only about *9–14* h a day and the duration of the "sunny period" can be predicted relatively easily. However the other factors, like wind speed and ambient temperature are much harder to predict. While all these fluctuations are unavoidable, because of the nature of the source energy, they lead to respective undesirable effects to the stability of the smart grid system, to which the PV systems are connected.

G. Vachkov (✉)
International Business School, Distance Learning Center, 1527 Sofia, Bulgaria
e-mail: gancho.vachkov@gmail.com

V. Stoyanov
Department of Automation and Mechatronics, University of Ruse "Angel Kanchev", 7017 Ruse, Bulgaria
e-mail: vstojanov@uni-ruse.bg

V. Sgurev et al. (eds.), *Learning Systems: From Theory to Practice*, Studies in Computational Intelligence 756, https://doi.org/10.1007/978-3-319-75181-8_11

The frequent and periodical fluctuations in the productivity of the PV systems are observed not only within the *day-night* time period, but also within much longer periods of time, such as *seasons:* summer, autumn, winter and spring. As a result the performance of the PV system varies also in wide margins during one whole calendar year. For example the output electric power from the PV system could be quite similar during the late autumn and during the early spring, while winter and summer performances are usually quite different.

The strong seasonal influence to the performance of the PV systems possesses a real challenge to establish an unbiased and plausible modeling method and criterion for evaluating the real performance (productivity) of the whole PV system. A good modeling method should be able to distinguish the normal seasonal fluctuations in the performance of the PV system from other non-typical changes (fluctuations) that are due to a kind of anomaly in the system, such as partial malfunction of some solar panels or some electrical devices in the system.

The following Fig. 1 illustrates the effect of the different seasonal atmospheric conditions (during winter and summer periods) over the performance of a real PV system, in terms of output power in [kW]. The measurements are taken with a sampling period of *15* min. The night periods, when the output power is *zero*, are omitted in the plots. The periodical daily behavior of the whole PV system, as well as the difference between the summer and winter performances are easily noticed in Fig. 1.

Here it should be noted that in both seasons (winter and summer) shown in Fig. 1, the PV system was working normally, without any anomaly or malfunction. The significant changes in the measured parameters are only due to the different atmospheric conditions in these two seasons of the year. Therefore a good modeling method and strategy for performance evaluation and monitoring of the PV system should be able to distinguish the seasonal fluctuations from those, caused by an existing anomaly in the system.

From the above illustration it becomes clear that the PV systems belong to the type of *multi-mode* (mult-regime) systems, since they operate under different conditions in the different seasons of the year. Then it is obvious that a kind of a *composite model* of the whole PV system should be created in order to estimate properly the behavior of the system under the different operating conditions. Such composite model will consist of several *sub-models*, each of them estimating in a plausible way the performance of the system in one concrete operating mode.

Once all the sub-models are created, then the whole composite model will be used in a special strategy that analyses the *similarity* between the current observed operating mode and a number of pre-defined and saved operating modes. Then, according to the similarity level to each of the pre-defined modes, the new observed mode will be classified to one of these modes. In the case of very little similarity to any of the predefined modes, the existing mode will be judged as a *new* or *abnormal* operating mode and can be added to the list of the pre-defined modes of the PV system.

From all the above considerations, it becomes clear that the way of creating each of the sub-models in the composite model of the PV system plays important role for the final accuracy (resolution) in defining the type of the current operating condition.

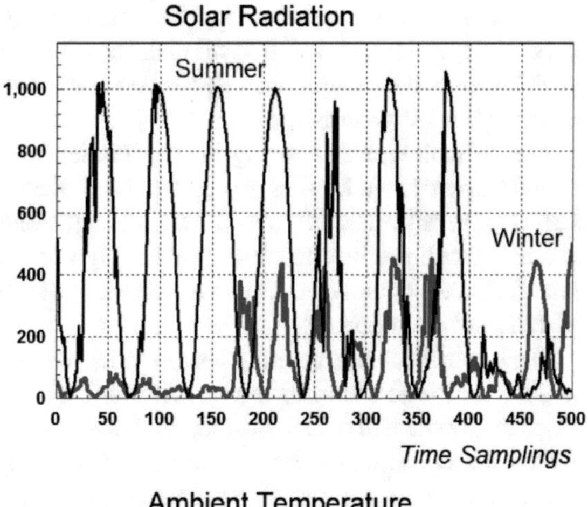

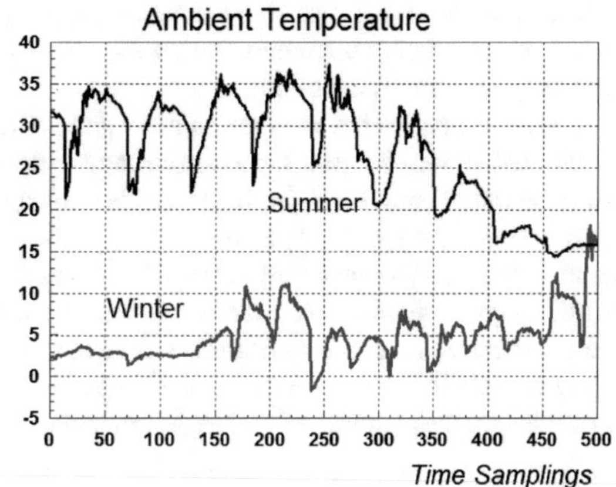

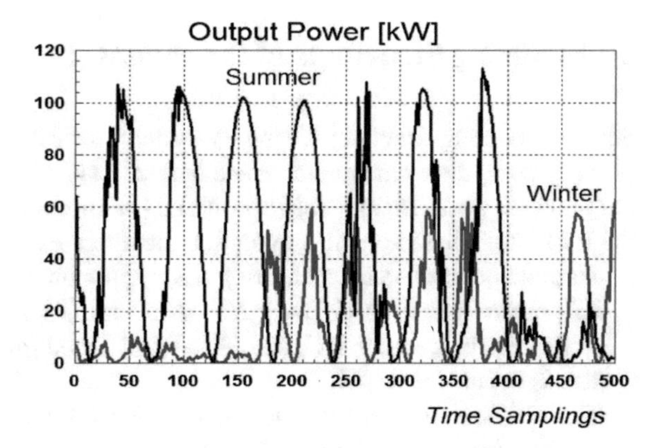

Fig. 1 Illustration of the influence of the seasonal conditions (summer-winter) over the overall performance of the PV system. **a** Solar Radiation [W/m2]; **b** Ambient (Air) temperature [°C]; **c** Output Power [kW]

Obviously, the priority in creating each sub-model will not be given to achieving the highest possible accuracy, because the sub-model will not be used for activities like real-time control. Rather than that, the sub-model has to possess a good *general-*

ization capability that makes it capable of capturing the most essential characteristics of a given operating mode. This will make possible for such model to properly distinguish between the current operating mode and other previously recorded operating modes of the PV system.

The difficult task in creating the sub-model for each operating mode is to find the most appropriate level of generalization of the model. In fact a model with *high level* of generalization will lose the accuracy (resolution) of the approximation and as a result will produce results, similar to the approximation of some other (even quite different) operations. This will produce inaccuracy in distinguishing between the different operations.

The opposite case is creating a sub-model with *low level* of generalization, i.e. a model with a *high accuracy* (high resolution). This will lead to another situation, in which the model will detect all other operations, even very similar to the current one, as different operation modes, because of the high approximation error. The result is that such model will not be able to *classify* properly the current operation as *similar* to one of the existing operating modes, but rather will point to a discovery of a *new* (probably abnormal) operation.

It is clear that we have to make a good and appropriate choice for the type of the model that should be used for representing the operating modes of the PV system. Here the fuzzy models [1–3] are a suitable choice for such models that can be used for describing the specific operations of the process, based on available experimental input-output data, because they are well known as universal approximators. Therefore in this chapter we introduce fuzzy models with a new grid-type structure along with their respective learning algorithm. After that we show the use of these types of fuzzy models for performance evaluation and condition monitoring of multi-mode processes on the example of real experimental data taken from PV systems.

2 Modeling Structures of the Photovoltaic Systems

Several modeling structures for the PV systems have been proposed and analyzed in [4, 5] by using different number of available measurable parameters of the system. In our previous works [6, 7] we have proposed two input-output structures for modeling the PV system, based on the assumption that there are four available sensors for the inputs and another one for measuring the output power, as follows:

X1—Solar Radiation [W/m2]; *X2*—Ambient (Air) Temperature [°C]; *X3*—the Solar Panels Temperature [°C] and *X4*—Wind Speed [m/s]. The output power of the system is denoted as *Y* [kW].

Then the complete model structure of the PV system with *4* inputs and one output, as described in [5, 6] is depicted in Fig. 2.

It is easy to notice that the Solar Panel temperature *X3* depends on two other inputs, namely: the Air Temperature *X2* and the Wind Speed *X4*. Therefore the following *3*-dimensional PV system model, as shown in Fig. 3 with three inputs, namely: *Solar*

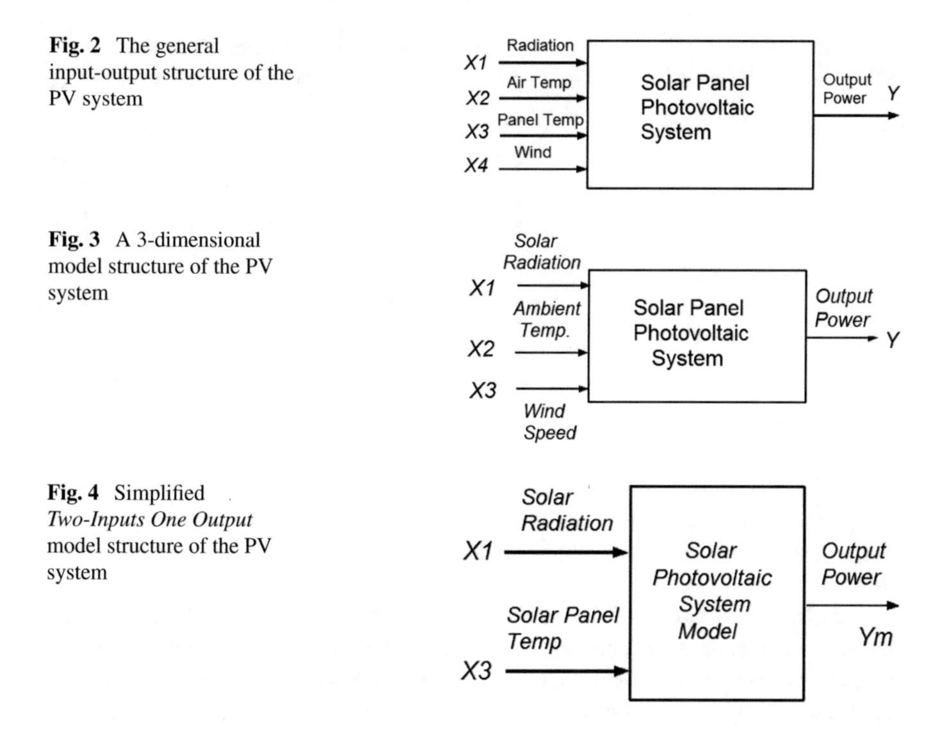

Fig. 2 The general input-output structure of the PV system

Fig. 3 A 3-dimensional model structure of the PV system

Fig. 4 Simplified *Two-Inputs One Output* model structure of the PV system

radiation, *Ambient (Air) temperature* and the *Wind speed* represents one possible model structure of the PV system.

Another possible *simplified* model structure of the PV system is shown in Fig. 4. It uses only two inputs, as follows: the Solar Radiation *X1* and the Solar Panel temperature *X3*. It has been used for modeling of the PV system in our previous work [6, 7]. It is also used for producing all modeling results in this paper.

3 The Concept of Creating the Complete Grid-Type Fuzzy Models

The classical fuzzy models are known as universal approximators of any kind of nonlinearity in the input-output relationship of the process under investigation While this is a good property of the fuzzy models, they also have a certain demerit in terms of a large number of parameters [2, 3] that have to be tuned in order to achieve the desired model accuracy. To alleviate the problem of tuning multiple parameters, different heuristics are usually used, For example, some parameters (mostly those that represent the structure of the model) can be fixed (assumed) beforehand, while the others are tuned iteratively during the identification (learning) process. This obviously leads to a near-optimal solution, but it is still widely acceptable in solving

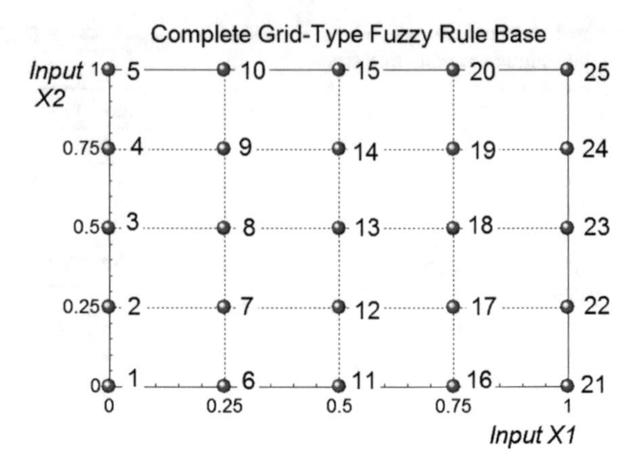

Fig. 5 Complete fuzzy rule base of *25* fuzzy rules used by the classical grid-type fuzzy model with two inputs

real practical problems. Here "the art of modeling" is left to the knowledge of the human expert, who has to properly select the Ratio (balance) between these two groups of parameters.

The classical structure of the fuzzy models consists of a Fuzzy Rule Base (FRB) that has a *complete* grid-type structure [1, 2]. Therefore we will refer to them as *complete* Grid-type Fuzzy Models (complete GFM). Each node of the complete grid represents one *if—then* fuzzy rule, which is defined by the respective *membership functions with centers* located at this node of the grid.

In the widely used *Takagi-Sugeno* (TS) fuzzy model the outputs (consequents) of the rules are assumed to be in a polynomial form with their respective parameters. The simplest case of *zero-order* TS fuzzy model uses *singletons* as consequents of the rules. These are "one-parameter-per-rule" numerical constants, which can be tuned during the iterative learning of the model, based on the available experimental (input-output) data.

An example of a Grid-type Fuzzy Model with a complete fuzzy rule base, consisting of *25* fuzzy rules that are evenly distributed in the *2*-dimensional input space *X1-X2* is presented in Fig. 5. The numbers beside the grid points refer to the assigned number (index) of the fuzzy rules.

It is well known that the classical fuzzy models, which have a Fuzzy Rule Base (FRB) with a *complete* grid-type structure, suffer from a serious computational problem, known as *combinatorial rules explosion* [8]. This means that the model complexity, in terms of number N of all fuzzy rules, grows exponentially with increasing the number K of the model inputs. In fact the total number N of the fuzzy rules is obtained as *product* of the preliminary *assumed* number of the membership functions MF_i for each of the all K inputs, as follows:

$$N = \prod_{i=1}^{K} MF_i \tag{1}$$

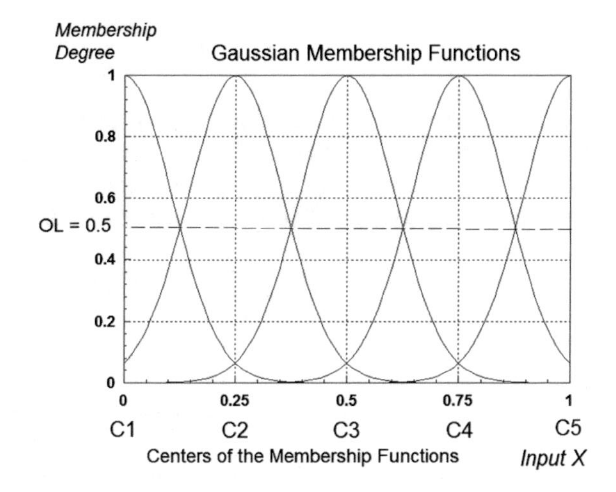

Fig. 6 Example of five evenly distributed Gaussian membership functions in the input space X, with overlapping level $OL = 0.5$

For example, if a *4*-dimensional fuzzy model ($K = 4$) should be created for the PV system with the general structure shown in Fig. 2, and if 7 membership functions are assumed for each input ($MF_i = 7, i = 1, 2, 3, 4$), then the total number of the fuzzy rules (with their respective consequents that have to be tuned) will be: $N = 7 \times 7 \times 7 \times 7 = 2491$.

To deal with the problem of the rapidly growing number of the fuzzy rules (and their respective parameters), different methods have been proposed and developed as in [2, 3]. Some of them use hierarchical tree-structure of the fuzzy rule base, other use appropriate input space partition to reduce the number of fuzzy rules.

Other approaches to reducing the complexity of the fuzzy models include genetic algorithms, evolutionary strategies and different fuzzy similarity measures in order to discard (prune) the unnecessary fuzzy rules, as in [2, 3]. In our previous work [7], we have used clustering of the input data in order to optimally locate the centers of a fixed small number of fuzzy rules, thus achieving locally optimal fuzzy models.

One general way to reduce the large number of fuzzy rules and their respective parameters (singletons) is to discard all the non-active (redundant) rules in the complete FRB that cannot be identified by the existing set of experimental data points. This is the main idea of our proposed concept of *grid-type fuzzy models*, explained in the next section of the chapter.

Before performing the calculation procedure for identifying the singletons of all fuzzy rules in the GFM, the shapes of the membership functions and their locations should be defined. In the example from Fig. 5 and throughout this chapter we have adopted the commonly used symmetrical *Gaussian* membership functions that are evenly distributed in the input space. The membership functions within a given input are designed to be with *equal width* (spread) σ and a predefined o*verlapping level* $OL = 0.5$, as shown in Fig. 6.

The membership degree $y \in [0, 1]$ of each membership function for a given data point with input x is calculated according to the following formula:

$$y = \exp\left(-\frac{(x-c)^2}{2\sigma^2}\right) \tag{2}$$

Here c and σ denote the *center* (location) of the *Gaussian* membership function and its *width* (spread) respectively. The following empirical formulas have been derived for determining the value for the spread of the membership function for two most commonly used overlapping levels, namely $OL = 0.5$ and $OL = 0.4$.

$$\sigma = 0.5A/1.17 \text{ for } OL = 0.5 \tag{3}$$

and

$$\sigma = 0.5A/1.35 \text{ for } OL = 0.4. \tag{4}$$

The parameter A represents the *interval* between the centers of two neighboring membership functions. In the example from the above Fig. 3 we have the interval $A = C2 - C1 = 0.25$.

The next step in designing the fuzzy model is to determine the values of all singletons (consequents) of the fuzzy rules. This is usually done in iterative way by repeatedly using the available set of input-output experimental data from a given operation (mode) of the real process. In the most often case these are data obtained by performing "passive experiments" (observations) of the process, so we can only observe (but not control) their distribution in the input space. As a result, some areas in the input space become more densely populated with experimental data, compared to other areas that might be even empty in some particular cases.

Then, in the case of creating a *complete* GFM that includes all fuzzy rules with their respective singletons, we need to have sufficient number of experimental data that "populate" the entire input space, in order to identify all the singletons. Here the ideal (but rarely possible) case is the one, shown in Fig. 7 with evenly distributed (generated) data within the entire input space, i.e. data with uniform distribution in the 2-dimensional input space.

As it was mentioned above, in the practical cases of real experimental (training) data, they usually occupy only a part of the input space. This means some of the fuzzy rules (i.e. their singletons) cannot be calculated, because of insufficient or missing at all data.

It becomes clear that the identified (trained) GFM is highly dependent on the concrete distribution of the available training data set in the input space. Therefore each particular training data set will produce a specific subset of identified fuzzy rules with their respective singletons that will be obviously located within the area of the collected experimental data.

This idea is implemented and used in the concept of the partial fuzzy grid models, explained in the next chapter.

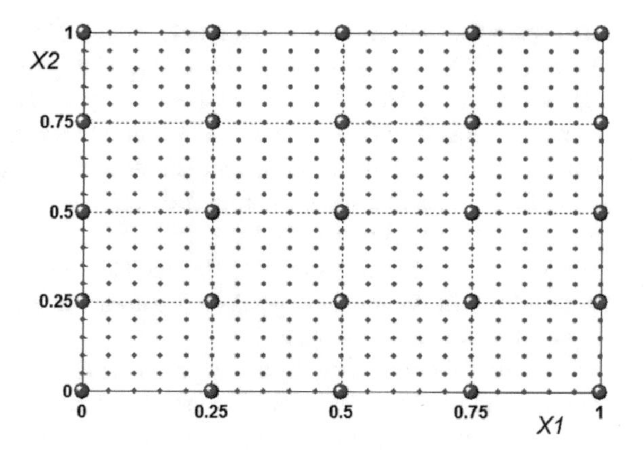

Fig. 7 Example of a data set consisting of *441* evenly distributed data in the 2-dimensional input space, used for learning the GFM

4 The Concept of Creating the Partial Grid-Type Fuzzy Models

There are certain types of real processes, such as the PV systems that normally operate under different conditions, depending on the performance requirements and the environmental changes. In this case, the experimental (observed) data taken from each operating condition form a kind of respective *data cloud* in the input space that has its specific shape and distribution. This data cloud is used as a training data set for defining the parameters (singletons) of the GFM. It is obvious that the model trained by this particular data cloud will approximate to some accuracy level only the area in the input space, where the data cloud is located. It will not be able to approximate reliably areas that are far away from the data cloud. Therefore we will call such models *partial* Grid-type Fuzzy Models (partial GFM).

The proposed *partial* GFM in this chapter play the role of *sub-models* in the overall *composite* model that describes the performance of the whole PV system. Each of the partial GFM is constructed and trained in an individual way, so that to represent closely the particular operating condition of the PV system.

The common point between all partial GFM is that they share the *same* predefined complete grid-type structure of the fuzzy rule base, but use only a *part* of it, depending on the data for the particular operation. Such concept of building the partial GFM, based on one common structure is done on purpose, in order to allow the models (and their respective operating conditions) to be compared between each other. In such way the partial GFM can be used for distinguishing one concrete operating condition from another one (or from all others). Such distinction can be easily performed by analyzing the prediction error of a given partial GFM, when it is used for modeling all existing operating conditions. Then it is obvious to expect that the GFM will show its best performance (i.e. the minimal prediction error) for data that are *within the data cloud*, from which it has been created.

In the further on introduced concept for creating the partial GFM, all of these models share the same predefined locations of the grid points (fuzzy rules) in the input space. However, in the general case, not all of the fuzzy rules in the complete grid can be identified by using the concrete training data (data cloud) that represent a given operation. The reason is that there will be insufficient (or even missing) data points for some of the fuzzy rules, which means that the singletons for these rules cannot be estimated. As a result, a *gap* in the complete grid appears.

All fuzzy rules that can be successfully identified by using the available training data (the data cloud) constitute the subset of the *active fuzzy rules* in the complete grid of all N fuzzy rules from (1). The actual number and locations of these active fuzzy rules will obviously depend on the concrete location and distribution of the training data in the input space.

Now the following definition of a partial grid-type fuzzy model can be given as follows: The *partial* GFM is such sub-model of the *complete* GFM that contains the set of *all active* fuzzy rules, obtained from the respective training data set (the data cloud).

From a fuzzy modeling point of view, the notion of *active* fuzzy rule in a partial GFM can be defined as such rule that has activation (firing) degree *above* or *equal* to a predetermined *activation level AL* (e.g. $AL = 0.25$) for *at least one* data point from the available training data set.

Then all data points that have activated this fuzzy rule with a degree *equal* or *above* the predefined *AV* are considered as *significant data* points for this rule. If the number of all data points available for constructing a given fuzzy model is denoted as M, then the number of all significant data points for a particular given fuzzy rule in this fuzzy model will be denoted as: $Ms \leq M$.

In the real practical situations, several data clouds (input-output operations) are usually collected separately, each of them representing different operations of the system. For example, in the case of the PV system, these could be four general data clouds that represent four typical seasonal operations of the system during *Winter, Spring, Summer* and *Autumn*.

Then each of the collected data sets will be used for creating a respective partial GFM that has its own subset of active fuzzy rules, i.e. its own partial Fuzzy Rule Base. It is clear that each of these partial FRB will be a subset of the general complete FRB with the predetermined complete grid structure.

A graphical illustration of the above concept for creating partial FGM is given in Fig. 8. Here simulated data are used to represent two different operations (*two* data clouds), named as *Operation1* and *Operation2*, each of them with *850* data distributed differently in the two-dimensional input space *X1-X2*. These two data clouds have been used to create two respective partial GFM with their subsets of active fuzzy rules.

The extraction of the subsets of the active fuzzy rules has been done according to the above definitions for active fuzzy rules and for activation level *AV*. Here *Gaussian* membership functions with uniform distribution, as in Fig. 8 have been used with assumed activation level of $AL = 0.25$ for this extraction. The result, as seen in Fig. 8 (the ball-type symbols) is: *13* active fuzzy rules extracted from *Operation1* and *11* active rules extracted from *Operation2*.

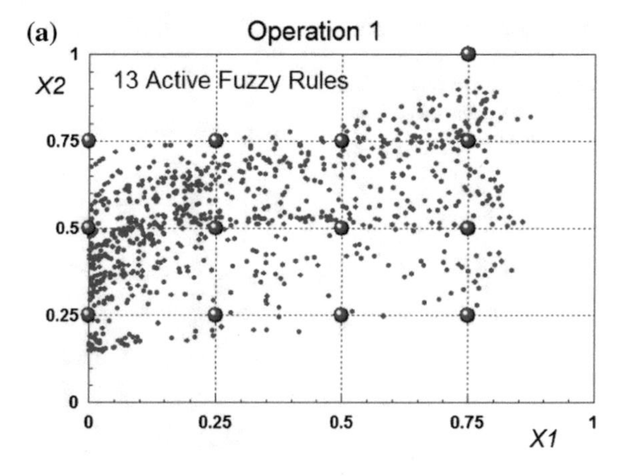

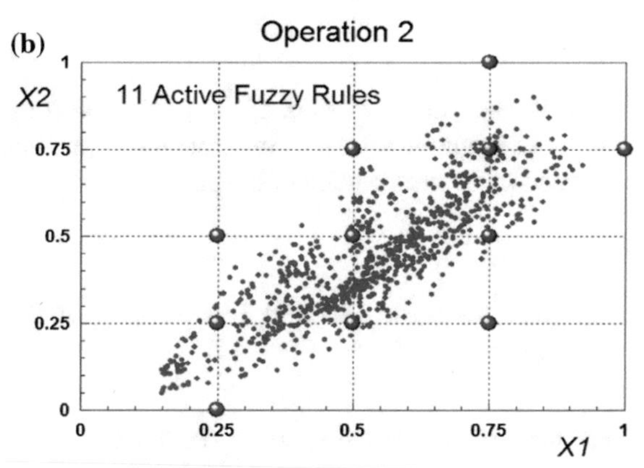

Fig. 8 Illustration of two different operations of a process that are used for creating two different partial fuzzy rule bases

It can be noticed from the plots in Fig. 8 that the two subsets of active fuzzy rules partially overlap with each other (i.e. they have common active fuzzy rules). This does not come at surprise, because it corresponds to the partial overlapping of the respective data sets in *Operation1* and *Operation2*.

It is also noticed that each of the active rules has in general different number of data points, associated to it. This also depends on the concrete data distribution in the input space. A visualization of the number of the data points for each active rules for both operations in given in the column (bar) plot in Fig. 9.

From Fig. 9 it is also easy to notice that some fuzzy rules (i.e. grid points) from the complete grid structure of *25* fuzzy rules are "empty". For example such are the rules: *5, 10, 11, 15, 16, 21, ...* This means that these rules do not have any singletons assigned to them. The reason is that there are no data points in this area of the input space that can activate the respective fuzzy rule at or above an assumed activation level of $AL = 0.25$.

Extraction of the set of all active fuzzy rules is the *first step* in the process of construction of the partial GFM, namely defining the *structure* of the GFM. The next

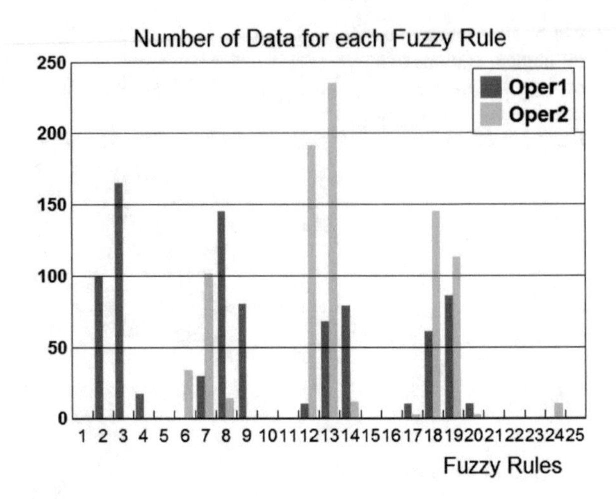

Fig. 9 Number of the data points belonging to the respective active fuzzy rules, for the data sets *Operation1* and *Operation2*

step is to *tune* the parameters of the partial GFM by use of the available training data set of input-output data. An original computation algorithm for such tuning is presented in the next section of the paper.

5 Algorithm for Iterative Learning of the Singletons of the Partial GFM

Once the structure of the fuzzy model **is** fixed, the next step is to tune (learn) the parameters of the model. The proposed grid-type fuzzy models have a special structure with predefined fixed locations and widths of the membership functions. Therefore the only remaining parameters for tuning are the *singletons* of the fuzzy rules. Basically these parameters can be obtained in a pure algebraic way, by using the least squares algorithm (LSA) [1, 2]. However such algorithmic approach is sensitive to the data quality and measurement noise, which means the algorithm may stop working in the cases with contradictory (e.g. same inputs, but different outputs) and highly noised data.

Therefore there is a large number of *numerical optimization* and *learning* algorithms that have been proposed in the literature to cope with the problem of parameter tuning, based on real experimental data. All of them have their specific features and are based on using specific heuristic rules with some assumptions for the parameters that have to be predefined by the user prior to running the algorithm. This makes such learning algorithms problem-dependent, which means they could perform well on some types of data, while performing unsatisfactory with other types of experimental data.

From a theoretical viewpoint, the learning algorithms that are based on different kinds of heuristics cannot guarantee the "best tuning", in a sense of finding the global optimum and rather tend to find a kind of local optimum solution.

In this section we propose another iterative algorithm, which is suitable for learning the singletons of both types of GFM, namely *complete* and *partial* GFM. This is relatively simple algorithm, which in practice doesn't need predefining of any parameters before its running.

The proposed learning algorithm works by recursively updating the singletons of all *active* fuzzy rules at every iteration. In order to increase the diversity in the search and thus the probability of finding the global optimum, the indexes (numbers) of fuzzy rules are *randomized* before each iteration. This is accomplished by a special function for random generation of *permutations* of length N (the number of all fuzzy rules in the GFM).

The following is the statement of the problem for running the proposed learning algorithm. We are given a *training* data set of M data points in the input-output space and we have already extracted the subset of all $N_a \le N$ *active* fuzzy rules, according to the assumed activation level ($AL = 0.5$ in this paper). The objective is to determine the values of all singletons: $u_i,\ i = 1, 2, \ldots, N_a$ such as to minimize the performance criterion for the error *RMSE*:

$$RMSE = \sqrt{\sum_{i=1}^{M} (y_i - y_{mi})^2 \Big/ M} \tag{5}$$

Here y_i represents the real (*measured*) output and y_{mi} is the *modeled* output for the given data point i, $i = 1, 2, \ldots, M$.

For calculating the fuzzy model that produces the modelled output, we have to choose the type of the *fuzzy inference* operation (*minimum* in this paper, but the *product* operation can also be applied), and the *defuzzyfication method*, which is chosen here as the most popular *weighted mean average* procedure:

$$y_{mi} = \sum_{j=1}^{N} u_j \mu_j \Big/ \sum_{j=1}^{N} \mu_j, i = 1, 2, \ldots, M \tag{6}$$

Here $\mu_j, j = 1, 2, \ldots, N$ are the *activation* (firing) degrees for the respective fuzzy rules. Then, it is obvious that when the modeled outputs get closer to the original (measured) outputs, i.e. $y_{mi} \approx y_i, i = 1, 2, \ldots, M$, then the *RMSE* approaches *zero* ($RMSE \to 0$).

The learning algorithm starts with the following *initialization* step:

Step 0. *Initialize* the values of all N_a singletons. Note that each fuzzy rule $FR_i, i = 1, 2, \ldots, N_a$ has its own subset M_s $(1 \le M_s \le M, i = 1, 2, \ldots, N_a)$ of so called *significant data* points for this fuzzy rule. Here we recall the definition that every data point that activates a given fuzzy rule at a level *equal* or *greater* than the predetermined activation level *AL* is called *significant* data point.

The approximate *initial values* of the singletons can be estimated in simple way as the *mean value* of the real outputs for all M_s significant data points for each singleton, i.e.

$$u_{i0} = \sum_{j=1}^{M_s} y_j \Big/ M_s, i = 1, 2, \ldots, N_a \tag{7}$$

Once the initial values of all singletons are known, they are used to calculate the *initial fuzzy model* by performing the *standard* fuzzy operations (*fuzzyfication*, *fuzzy inference* and *defuzzyfication*). The predicted (modeled) outputs $y_{mi}, i = 1, 2, \ldots, M$ are used to estimate the initial *RMSE*, denoted as $RMSE_O$.

During the whole iterative calculation process, the singletons obtained in the previous iteration are denoted by $u_{i\,old}, i = 1, 2, \ldots, N$, while the new (*updated*) singletons in the new iteration are denoted by $u_{i\,new}, i = 1, 2, \ldots, N$.

At every iteration of the learning process we *update* all N_a singletons, by analyzing the currently obtained model accuracy *RMSE*. This is performed in the following steps *1, 2* and *3*. The updating procedure of all N_a singletons during a given iteration is made in a sequence of "one by one", where the indexes of the singletons (the fuzzy rules) are randomized, in order to increase the diversity of the search for a global solution. The following are the computational details of these three steps:

Step 1. *Recursive step* for updating the value of the current randomly selected singleton denoted by the index $s, s \in [1, 2, \ldots, N]$. This singleton has a certain subset of $M_S \leq M$ *significant* data points associated to it. Then, by using the currently obtained (learned) fuzzy model we can calculate the modeled outputs of all significant data points, as follows:

$$y_{mj} = \left[\left(\sum_{i=1}^{N_a-1} u_{i\,old}\mu_i + u_{s\,old}\mu_s\right) \Big/ \sum_{i=1}^{N_a} \mu_i\right], j = 1, 2, \ldots, M_S \tag{8}$$

Simple transformations of (8) yield the following equation for calculating all M_S different (individual) values for the s-th singleton:

$$u_{sj} = \left(y_{mj} \sum_{i=1}^{N} \mu_i - \sum_{i=1}^{N-1} u_{i\,old}\mu_i\right) \Big/ \mu_s, j = 1, 2, \ldots, M_S \tag{9}$$

Finally, the updated value of the s-th singleton is updated as average of all M_S individual values from above, namely:

$$u_{s\,new} = \sum_{j=1}^{M_S} u_{sj} \Big/ M_S \tag{10}$$

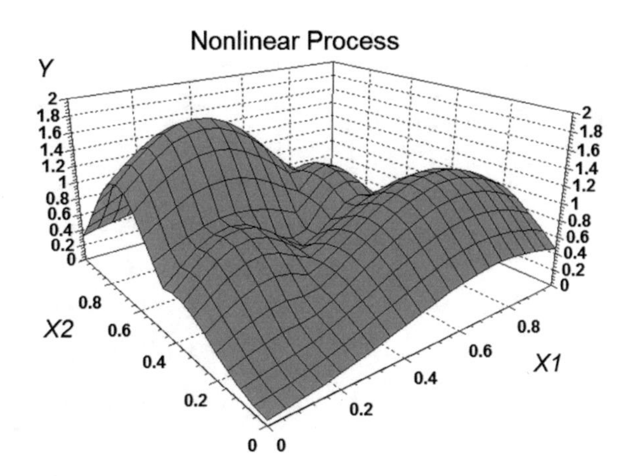

Fig. 10 Example of the 2-dimensional test nonlinear process, used to evaluate the performance of the learning algorithm

Step 2. Select the next singleton to be updated from the randomized list of singletons and go to Step 1. If the list of the singletons to be updated is empty, go to Step 3.

Step 3. Calculate the fuzzy model with all currently updated singletons and estimate the model accuracy by calculating the *RMSE* (5).

Step 4. Check the stopping criterion, which takes into account the current *RMSE* and the *trend of changing* the error with iterations. If the stopping criterion is *not* yet satisfied, go to the next iteration in Step 1. Otherwise, *save* the final obtained parameters (singletons) of the fuzzy model and *exit* the algorithm.

Illustration of the performance of the above proposed learning algorithm for creating grid-type fuzzy models is given in the sequel on the following simulated nonlinear test process with two inputs *X1* and *X2* and one output *Y*, as shown in Fig. 10.

Here the training data set from Fig. 7 with *441* uniformly scanned data in the *2*-dimensional input space is used to learn the *singletons* of the *complete* grid fuzzy model with $N = 5 \times 5 = 25$ fuzzy rules. The learning algorithm has converged in a few (less than *3*) iterations and the modeling error of the produced complete GFM was: $RMSE = 0.0835$. The fast convergence process can be seen from the plots in Fig. 11 of all *25 singletons* in the complete GFM of the test nonlinear process.

Many experiments by using different nonlinear test examples have been performed and they have shown that the above described learning algorithm has very fast and reliable convergence in creating both types, complete and partial GFM. The convergence was achieved practically within a few iterations and sometimes with small, rapidly decreasing oscillations. Therefore in checking the stopping criterion in Step 4, we have introduced a simple *heuristics* to reliably stop the algorithm, when a smooth and practically *not changing* trend of the singletons values is observed.

It is natural to expect that by increasing the number of the fuzzy rules in the *complete grid* of the FGM, we will obtain a model with a higher accuracy. This is illustrated in Fig. 12 for the case of a complete GFM with $7 \times 7 = 49$ fuzzy rules. Then a smaller learning error of $RMSE = 0.0404$ was obtained. From Fig. 12 it is

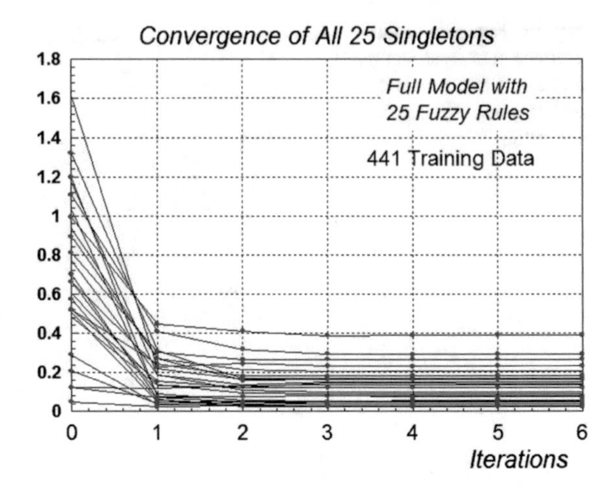

Fig. 11 Convergence of the values of all *25* singletons during iterative learning of the complete grid-type fuzzy model

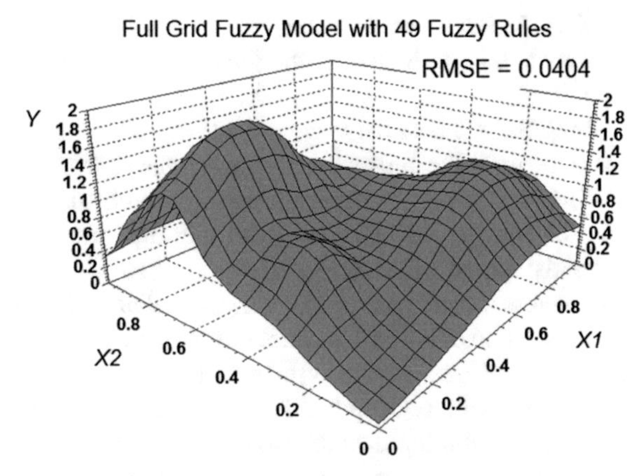

Fig. 12 The response surface of the complete GFM with $7 \times 7 = 49$ fuzzy rules for the test nonlinear process

easily noticed that respective response surface from the new created complete GFM is very close to the response surface of the original nonlinear process from Fig. 10.

The above learning algorithm was also used for creating two *partial* GFM of the nonlinear process from Fig. 10 by using the simulated input-output data from the two operations shown in Fig. 8. The following results were obtained: a partial GFM with *13* active fuzzy rules and $RMSE = 0.1117$ was created for *Operation1* and another partial GFM with *11* active rules and $RMSE = 0.0988$ was created for *Operation2*.

6 The Use of the Partial GFM for Recognition of Different Operating Modes of the PV System

As it became clear from the previous Sections, evaluating the level of performance of a real PV system under certain operating (atmospheric) conditions is a difficult task, because of the large and frequent dynamical changes of the inputs of the system. This makes difficult to judge whether the current operation of the system can still be considered as *normal* (under these atmospheric conditions), or there is a kind of *anomaly* in the operation of the systems, due to a malfunction in some of its components (a hardware reason). For example, the PV system is operating in the middle of the *Spring* season, but its performance is much closer to the typical *Winter* operation, i.e. an operation that has much less power production. Here it would be interesting and very practical to develop a *model-based* method that is able to *recognize* the current operation as such that belongs to one of the operations from the given list of *predetermined* typical operating conditions.

Such task is further on called *recognition of different operating modes* of the PV system. Theoretically, it can be considered as a classification task in which the class of the new (*unknown*) operation is determined by using the information from the existing typical operating modes—the given classes, represented by their respective partial grid-type fuzzy models. Then a relatively simple procedure of calculating and examining the *similarity level* between the new operation and all the existing partial GFM will eventually give answer of the classification task.

Different ways for estimating the similarity level between two operating modes of the PV system can be considered. For example, *one* relatively simple and reasonable way could be to compare *two* different performance errors of the same partial GFM. One is the performance error *RMSEo* of the partial GFM on the concrete *typical* operation mode, for which it has been created, and the other is the performance (prediction) error R*MSEp* of the same model on the data set from the *new* operation. Then the *absolute difference* |*RMSEo*–*RMSEp*| can be assumed as the *similarity level* between these two operations.

Another way to estimate the similarity level between two operating modes could be to calculate the percentage *of* the *overlapping area* (a value from 0% to 100%) between the area, covered by the grid points (fuzzy rules) of the respective partial *GFM* and the area, covered by the data points in the *new* operating mode. As an example, an overlapping area of *12%* means a *low* similarity level between the two operating modes, while an overlapping area of *87%* would suggest a *high* level of similarity between the two modes. Here we don't specify the way, in which these areas will be calculated, because it can be done in different ways.

One simple, but reliable method for finding the overlapping area is to calculate the number of all *non-confident* data points from the new observed operating mode. By definition, the non-confident data points are such data points that cannot be calculated reliably by the respective partial GFM. This means they cannot activate any of the fuzzy rules in the GFM with a level, greater or equal to the predefined activation level AL (e.g. $AL = 0.25$).

Then the *similarity level* can be given directly in percentage of the *overlapping area*. This is the *ratio* of the number of the non-confident points to *all data* points for the new observed operating mode and represents a value between *0.0* and *1.0*.

A *third way* to estimate the similarity level between two operating modes is to combine the two above calculation methods, namely the *difference* of the prediction errors and the *ratio* between the non-confident and all data points in one *fuzzy decision* making procedure. The computational details about this procedure can be performed in different ways and are beyond the scope of this chapter.

In the followings we will illustrate the general approach to recognition of different operating modes in a PV system on the example of the simplified 2-dimensional structure of the PV system from Sect. 2. The selected two inputs are: *X1*—solar radiation and *X3*—solar panel temperature. The output *Y* is the output electric power of the PV system.

We have collected large real data sets from all *4* seasons of the year: *Winter, Spring, Summer* and *Autumn* in order to create *4* respective partial GFM that represent *4* typical operating modes of the PV system.

All data sets have been divided into two subsets: one for *training* the respective partial GFM by using the learning algorithm in Sect. 5, and another for *testing* the performance (accuracy) of the created GFM. In creating the partial grid-type fuzzy models, the following predefined parameters have been used: *9* evenly distributed symmetrical *Gaussian membership functions* per input, which leads to a number of $N = 9 \times 9 = 81$ complete fuzzy rules; *overlapping ratio* between the neighboring membership functions: $OL = 0.50$, according to (3) and *activation level* of $AL = 0.25$.

The following Fig. 13 illustrates the dynamic behavior of the two inputs *X1* and *X2* and the output *Y* of the PV system for the first *1000* samplings during the season *Autumn*. In a similar way, Fig. 14 depicts the dynamic behavior of the PV system during the season *Winter*. The difference in the behaviors as magnitude of their parameters is obvious and it is due to the different atmospheric conditions in the two seasons.

The partial GFM for all *4* operating modes (the *4* seasons) have been created by using the iterative learning algorithm from Sect. 5. Here training data sets of *850* samplings have been used for the seasons *Spring, Summer* and *Autumn*, and *800* samplings have been used for the season *Winter*.

From the 2-dimensional plots *X1*-*X3* in Fig. 15 it can be noticed that the distribution of the training data in the input space *X1*-*X3* differs significantly from one to another season, but there are also some overlapping areas between the seasons, especially between *Spring, Summer* and *Autumn*.

The created partial GFM for all *4* seasons (the operating modes) have different number N_a of active fuzzy rules that represent the structure and location of the respective training data in the input *X1*-*X3* space. The sets of grid points for all *4* partial GFM are given in Fig. 16 and their similarity with the original data sets from Fig. 14 in the sense of covered areas from the input space can be easily noticed. The following ranges for the two input parameters have been predefined: [*Xmin1, Xmax1*] = [*0.0, 1200.0*] for Input *X1* and [*Xmin3, Xmax3*] = [*−5.0, 55.0*] for Input *X3*.

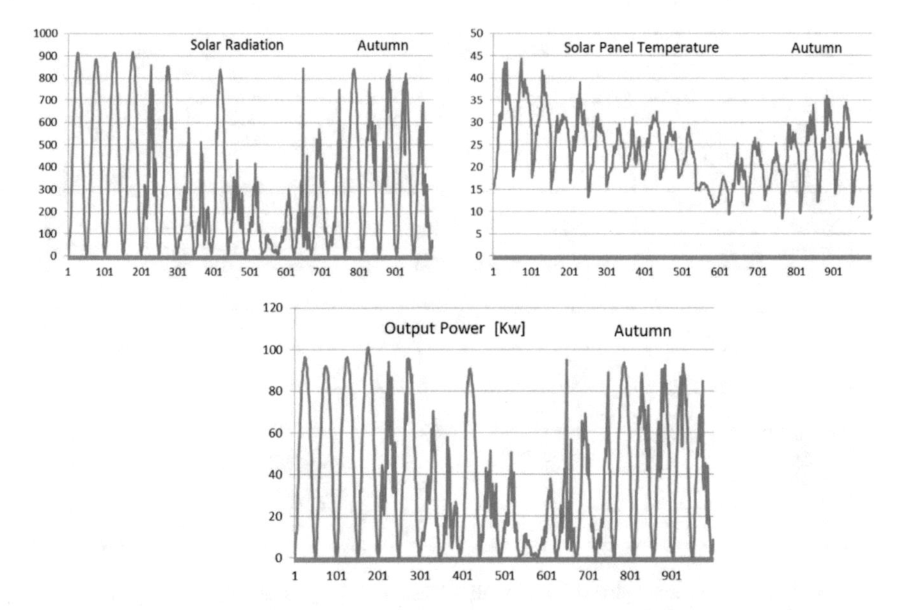

Fig. 13 The dynamic behavior of the two inputs *X1* and *X3* and the Output *Y* of the PV system during the *Autumn* season

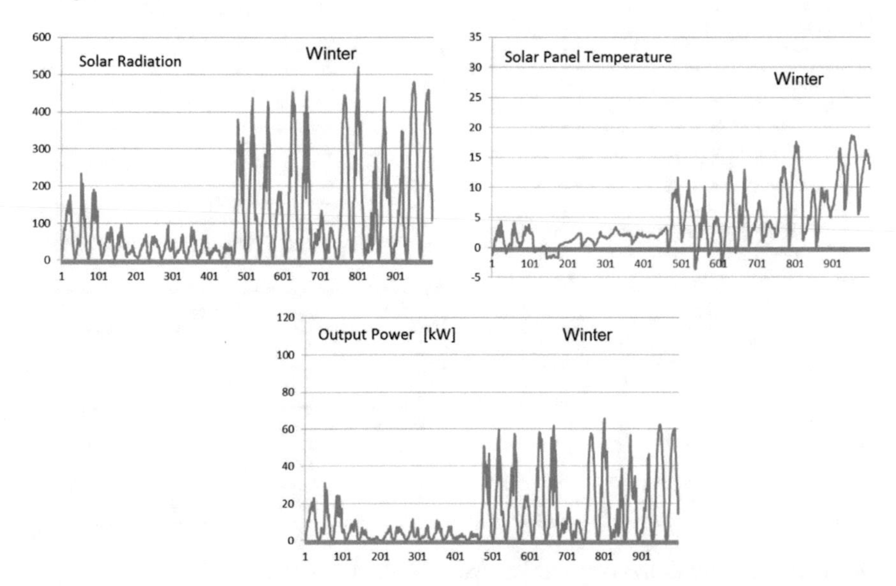

Fig. 14 The dynamic behavior of the two inputs *X1* and *X3* and the Output *Y* of the PV system during the *Winter* season

Table 1 shows the results from creating and testing the partial grid-type fuzzy modes for all *4* seasons. These results include the obtained *RMSE* from the *training*

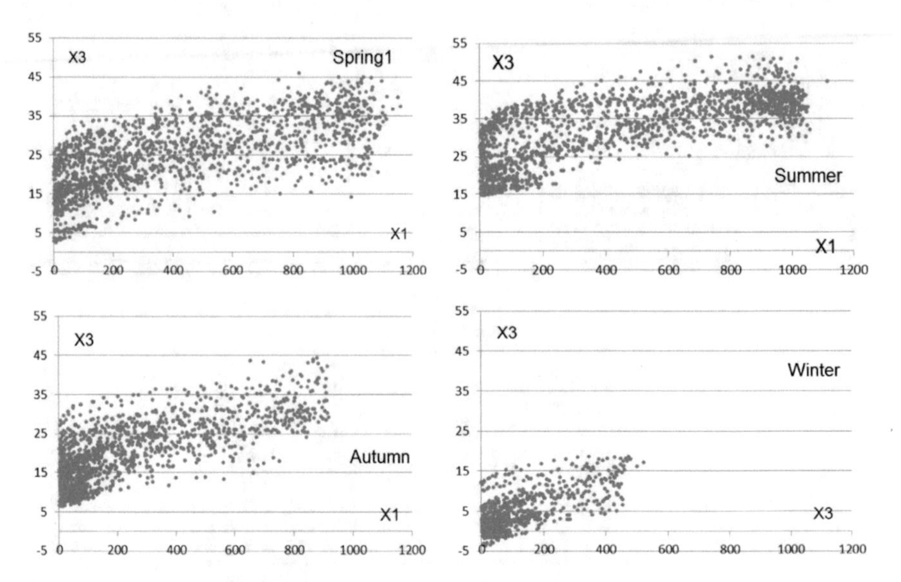

Fig. 15 The 2-dimensional plots *X1-X3* for all *4* seasons of the year that show the different distribution of these two parameters in the input space

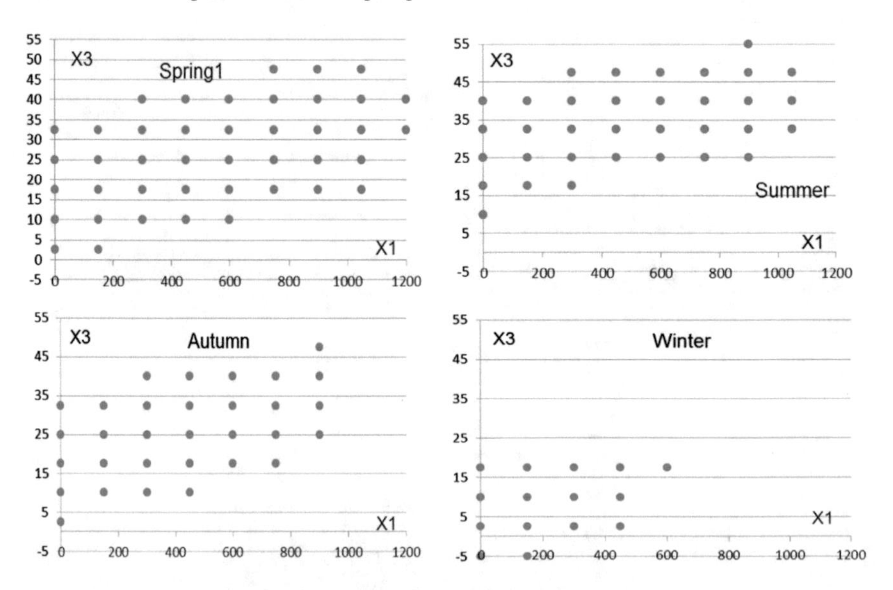

Fig. 16 Grid points structure of the obtained partial GFM for all *4* seasons

data sets and from the respective *test* data sets. This is performed for verification of the models. Then the last row in the table gives the number of the active fizzy rules N_a (number of the grid points) for each partial GFM. It is seen that the number of

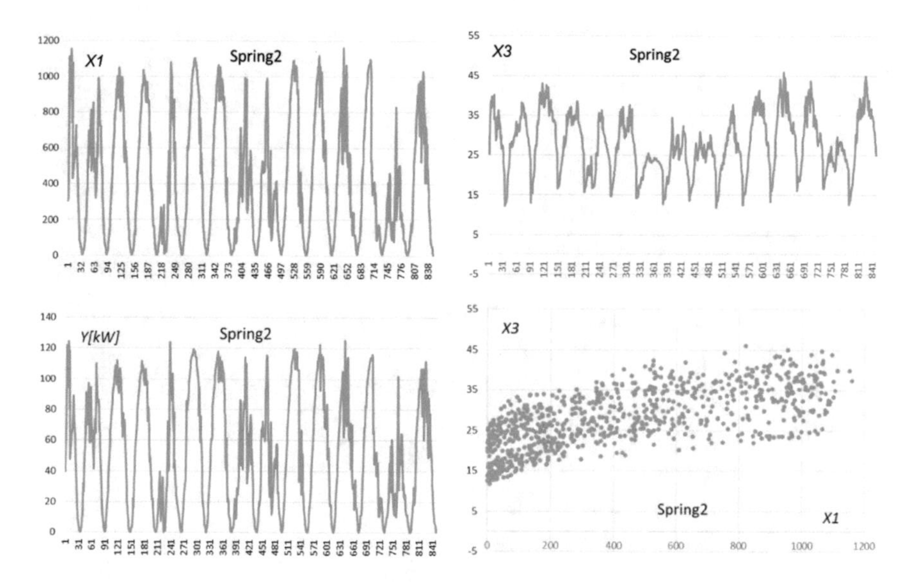

Fig. 17 New *unknown* operating mode, called *Spring2* with 850 experimental data

the active fuzzy rules for all models is much less than the complete number of $9 \times 9 = 81$ fuzzy rules.

It is seen from the table that the obtained training errors are sufficiently small. They are within the range: $1.7 \leq RMSE \leq 2.6$, which constitutes up to 2.1% error from the nominal full power of *120* kW of the PV system. The respective *RMSE* obtained from the test data sets for all *4* models are also small enough, up to *3.09*, which means about *2.5%* from the nominal power of the PV system. All this means that the created partial GFM have sufficient accuracy to be used for the purpose of recognition of different operating modes of the PV system.

Now the above described *mode recognition* procedure will be illustrated for *two new* data sets that have been taken from the spring and summer season and will be considered as two *unknown* operating modes to be recognized. Each of the data sets consists of *850* data samplings. They are named *Spring2* and *Summer2* since they actually belong to these two seasons and are depicted in Fig. 17 and Fig. 18 respectively. It is important to note that these are different, *new data sets* that have *not* been included in the training and test data sets used for creating the partial GFM for *Spring* and *Summer* in Table 1.

The comparison results for each of the above two new (unknown) operation modes with the partial GFM for all *4* seasonal modes *Spring, Summer, Autumn* and *Winter* are given in Table 2. Here two types of estimation of the *similarity level* between the new (unknown) operating mode and each of the respective seasonal modes (the partial GFM) have been used. The first type estimates the similarity level directly by taking into account the value of the prediction error *RMSEp* of the respective seasonal model on the new data set. Then the smallest prediction error will be con-

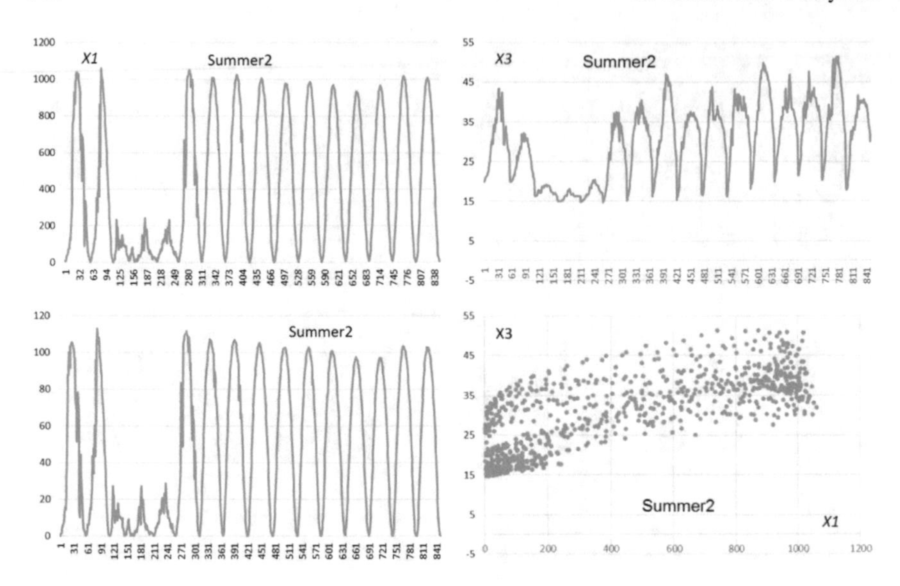

Fig. 18 New *unknown* operating mode, called *Summer2* with 850 experimental data

Table 1 Main characteristics of all *4* seasonal partial GFM

Partial GFM	Spring	Summer	Autumn	Winter
Training data sets	*850*	*850*	*850*	*800*
RMSEo	*2.4435*	*1.8667*	*1.7527*	*2.5118*
Test data sets	*850*	*794*	*531*	*236*
RMSEt	*2.7568*	*3.0975*	*1.9909*	*2.4029*
Active fuzzy rules N_a	*42*	*34*	*31*	*15*

sidered as the biggest similarity level between the two operations. The second type of estimation uses the *Absolute Difference* between the model error *RMSEo* and the prediction error *RMSEp* of a given seasonal model on the unknown operating mode, i.e. |*RMSEo–RMSEp*|. Then again the smallest difference will be considered as the biggest similarity level.

The results in Table 2 show clearly that both types of similarity level estimations produce the *same correct* recognition results, namely: the new operation *Spring2* belongs to the typical *Spring* operating mode and the new operation *Summer2* belongs to the typical *Summer* operating mode. These decisions are shown in *bold* in the Table.

The *same correct* recognition results have been also extracted by taking into account another parameter, namely the percentage of the *Overlapping Area* for each of the *4* partial GFM with the respective new (unknown) operating mode. Here a maximal level of *100%* overlapping area is observed in the cases of using the *Spring* and the *Summer* models. These results coincide with the correct answers.

Table 2 Results from recognition of the two new operating modes: *Summer2* and *Spring2*

New operating modes	Comparison with the seasonal partial GFM	Partial GFM for spring	Partial GFM for summer	Partial GFM for autumn	Partial GFM for winter
Spring2 850 data	*RMSEp:*	***2.7569***	*4.0833*	*5.6578*	*4.8404*
	Abs Difference:	**—**	—	—	—
	|RMSEo-RMSEp|	***0.3134***	*2.2166*	*3.9051*	*2.3286*
	Overlapping Area:	***100%***	*96.82%*	*93.88%*	*8.71%*
Summer2 850 data	*RMSEp:*	*4.4970*	***3.0589***	*3.5182*	*6.0137*
	Abs Difference:	—	**—**	—	—
	|RMSEo-RMSEp|	*2.0535*	***1.1922***	*1.7655*	*3.5019*
	Overlapping Area:	*95.65%*	***100%***	*96%*	*4.71%*

7 The Use of the Partial GFM for Condition Monitoring of Photovoltaic Systems

Once the partial fuzzy grid models have been created by using training data from several respective typical operations of a the PV system, they can be applied successfully to solving the problem of condition monitoring of the PV system.

Condition monitoring is a specific activity that is usually performed *periodically* (in *batch-mode*) by a human operator or in *real-time mode* (fully automatically) over the continuous operation of the system.

The main objective of the condition monitoring is to determine the "health condition" of the system and to discover its possible deviation from any (or all) of the *normal* operating conditions. If such deviation is detected and it is considered to be at a significant level, it is defined as *anomaly detection* and a respective *alert* is generated that requires appropriate actions, additional analysis and possible corrections in the system operation.

For analysis and comparison of the different operating conditions of the PV system, an approximate *performance evaluation* of the system should be done by using the input-output data from a given (short or longer) operation period.

As it was discussed in the first section of this chapter, there is no need of using extremely precise models for evaluating the performance of the system, but rather we need models with good *generalization* capabilities, being able to capture the *general process behavior.* Then these models will be used for comparing different operating modes of the PV system by distinguishing the different behaviors of the PV system, that might be because of different atmospheric conditions, or because of some malfunction in the system.

In the process of comparison, we have to estimate *how close* is the current behavior (the model) to any of the previous known behaviors. This is done by calculating and comparing the respective *similarity degrees*. Here two general cases may arise. The *first case* is when the current behavior of the system is *most similar* to a given previous (normal or abnormal) behavior. Then a respective decision for normality or abnormality in the operation of the system is produced.

The *other possible case* arises, when the current behavior of the system is *not similar to any* of the previously saved behaviors. Then a decision that the system is operating in a *new* (*unseen*) condition (operating mode) is made. It might be a *new normal* or a *new abnormal* condition and this should be rectified by more detailed analysis, usually by human operator. As a result, the new operating condition will be saved and added to the *Model Base* of all the previous recorded conditions (behaviors) of the system and will be used in the future condition monitoring process.

In order to properly distinguish between the different behaviors (operating modes) of the PV system, we have to define the notion of *similarity* between any pair of operating modes in a numerical way. Additionally, we have to create appropriate models that are able to capture the most typical behaviors of the PV system for several typical conditions (for example: *Summer* behavior and *Autumn* behavior).

Now it becomes clear that there are certain characteristics of the partial GFM that make them an appropriate choice for condition monitoring, as follows. *First*, these models are easy to create, since all of them share a *common* and *predefined* structure of the grid. This means that the locations and the widths of the membership functions for all fuzzy rules are fixed in advance and do not need to be optimized. Obviously this fact affects the accuracy of the partial GFM, but as already mentioned, this is not the priority in the condition monitoring.

Second, although the proposed learning algorithm for the singletons of the complete and partial GFM does not guarantee finding the global minimum for the RMSE, but may stop at a local minimum, it has extremely fast convergence. This speeds up the process of creating all partial models.

Third, each partial GFM has a reasonably good approximation only within the area of the input-output space, covered by its active fuzzy rules and performs poorly outside this area. Therefore the partial GFM is only reliable for use within the input area of the specified operation, for which it is created. This is an essential feature that can be used in condition monitoring for the purpose of distinguishing one from another operating condition.

Fourth, the partial GFM have at least two important parameters that give a general description of the concrete operating condition of the process, namely the number of the *active fuzzy rules* (corresponding to the area of operation) and the obtained *RMSE* that describes the accuracy of this model. These parameters are useful for definition and estimation of the similarity between different pairs of operations.

Now, the entire process of *periodical* (*batch-type*) condition monitoring can be summarized in the following general steps:

1. *Create* partial GFM for all available operating conditions (modes) of the PV system and keep them in a *Model Base*. This is a one-time (off-line) computation step that includes the learning of all models in the Mode Base;
2. *Collect* an input-output data set (*batch* data set) of sufficient size for the newly observed (still unknown) operating mode of the system;
3. Calculate the *predictive performance* of all partial GFM from the Model Base on the data from the concrete observed operating mode. This is easily done by calculating the respective *RMSE* from all models.
4. Find the *minimal* error *RMSEmin* that is obtained from a certain partial GFM in the Model Base;
5. *Check* the obtained *RMSEmin* against a predetermined *threshold Th.* The amount of this threshold depends on the value of the modeling error *RMSEo* obtained by the partial GFM, when it was trained (once only) by a respective training data set. It is quite reasonable to predefine the threshold as a value, slightly above the modeling error, e.g. $Th = 1.3 \times RMSEo$.

    (a) If $RMSEmin \leq Th$, then a decision is made that the current observed operation is *most similar* to the operation belonging to the respective GFM, which produced that *RMSE*;
    (b) If $RMSEmin > Th$, then the current observed operation is marked as a *different new operation* and it is saved as a *new member* of the Model Base for the future condition monitoring.

The above explained batch-type condition monitoring is a kind of *discrete* type of monitoring that needs a data set of sufficient size (e.g. at least *50–100* samplings) for reliably estimating the new operation.

In many real cases we are interested in running the condition monitoring procedure *continuously*, i.e. in a *real time* (at each sampling time). Then we have to specify the way, in which the data set for each new observed operating mode is accumulated. This is done in our experiments in this chapter by using the *moving window* approach, with a predefined *length Lw* of the window. Here *Lw* is the assumed number of the data points (samplings) that constitute the current portion of the observed (unknown) operating mode.

For a reliable and plausible estimation of this operating mode, it is advisable that the window length *Lw* is at least *20–30* samplings. At each sampling period the observed operating mode is updated, by adding the new sampling into the window and discarding the oldest (last) sampling. Thus the new data set for the operating mode contains only one new data point, compared to the data set taken in the previous window. In such way a gradual and smooth change of the type of the operating mode between the steps of the monitoring process can be expected.

The above steps and considerations are taken into account for the following simulations that illustrate the work of the real-time condition monitoring system with a fixed window length of $Lw = 30$ samplings. We have selected all the experimental data from the same PV system, as in the previous examples.

First, we have trained *4* partial grid-type models for each of the seasons: *Summer, Autumn, Winter* and *Spring* by using *300* data samples from each of these seasons.

Then, we produced a relatively long series of *2400* experimental data, by concatenating *600* data samplings from each of the four seasons in the following sequence: *Summer* → *Autumn* → *Winter* → *Spring*.

The iterative learning algorithm from Sect. 5 has been used to create the *4* seasonal models and the following modeling errors *RMSE*o for all models have been obtained: *Summer* → *RMSEo* = *1.6413*; *Autum* → *RMSEo* = *1.8874*; *Winter* → *RMSEo* = *0.9885*; *Spring* → *RMSEo* = *2.4959*.

The threshold *Th* for each of the seasonal operating mode has been defined as a value that is *30%* higher than the respective error *RMSEo* of the model, i.e. $Th = 1.3 \times RMSEo$.

The real-time condition monitoring system was run for the whole sequence of *2400* samplings that contain the sequence of *600* samplings from each of the *4* seasons. As a result, four parallel plots have been produced that represent the *prediction error RMSE* of each of the seasonal models for the concrete data window of *30* samplings. For easy evaluation of the monitoring results produced by each of the models, the respective threshold *Th* is also plotted as a horizontal line over the time.

The graphical results from the conditional monitoring process are presented in Fig. 19 in the form of *4* parallel plots—one for each of the Hypotheses: *Summer, Autumn, Winter* and *Spring*.

There could be some particular cases in which all data within the current window cannot be estimated reliably, because they cannot active any of the fuzzy rules in the respective partial GFM with activation level, equal or more than the predetermined $AL = 0.25$. Such data are called *non-confident* data. Obviously, such non-confident data are outside the operational range of the respective partial GFM. If this is the case for all data within the current data window, the result of the estimation by the current partial GFM is set to a large constant value, i.e. the prediction error is artificially set to $RMSE = 10.0$ in the simulations of Fig. 19.

For proper understanding of the evaluation results from the real-time condition monitoring in Fig. 19, some explanations are necessary. First, for this simulation example we know in advance what could be the *ideal result* from the monitoring. There are *4* equal portions of *600* samplings that are connected in the following seasonal sequence: *Summer* → *Autumn* → *Winter* → *Spring*. Therefore we expect ideally the monitoring system to detect the respective *4* seasons, each of them with *600* samplings, in the same sequence. This means that the prediction errors *RMSE* for the hypothesis *Summer* (channel *1*) should be *below* the respective threshold for the first *600* samplings and then should go higher (above the bar) for all other samplings.

In the case of hypothesis *Autumn* (channel 2), the prediction errors should go *below* the respective threshold *Th only* between the sampling *601* and sampling *1200* and should be higher (above the bar) in all other areas of the samplings.

In a similar way, for the hypothesis *Winter* (channel *3*), the prediction errors should be below the respective threshold *Th only* between the samplings *1201* and *1800*. Finally, for the hypothesis Spring (channel 4), the prediction errors should be below the respective threshold *Th only* between the samplings *1801* and 2400.

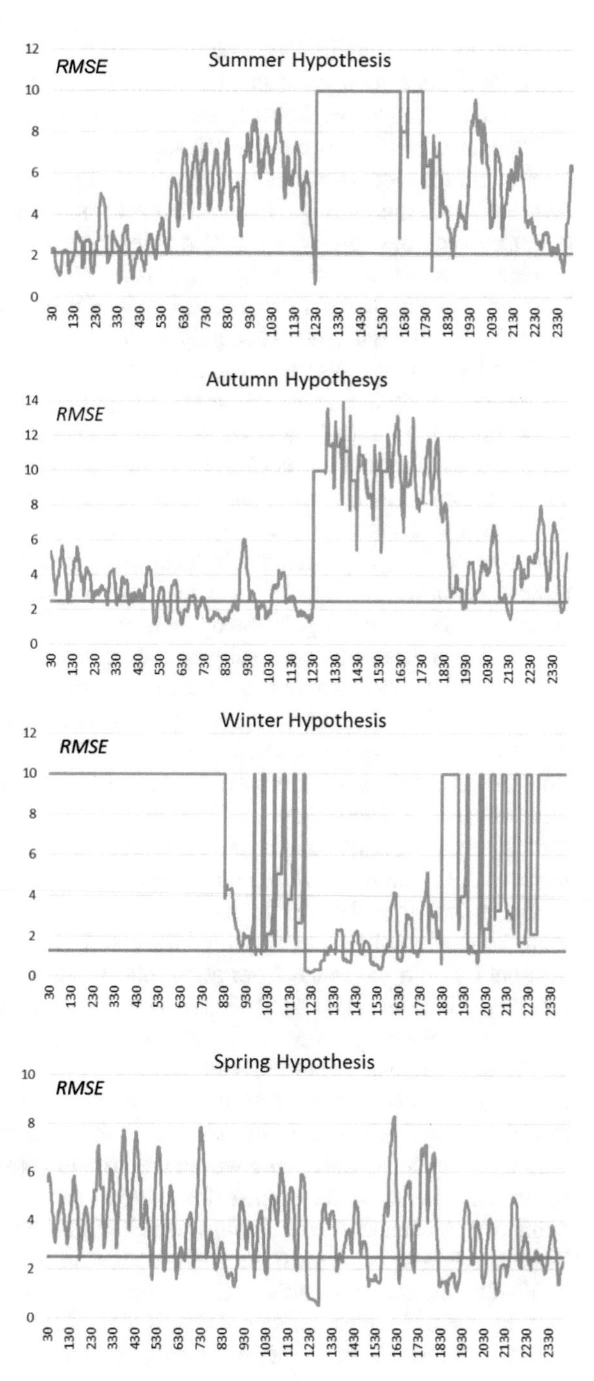

Fig. 19 Evaluation results from the *4*-channel real-time condition monitoring system, based on the grid-type GFM for *4* seasons

From the results in Fig. 19 it is obvious that we have not obtained such ideal results and instead there are a lot of fluctuations of the prediction errors *RMSE* below and above the respective thresholds (the bars) for all the channels.

There are several reasons for such approximate results, which will be explained. *First* of all, the model accuracy and the level of generalization play the most important role in the monitoring. If a good balance between these two properties cannot be achieved during training, then the monitoring results will deteriorate. *Second*, the window length *Lw* also plays significant role, mainly in affecting the oscillations of the predicted *RMSE* for all channels. Obviously, a longer window length (e.g. *Lw* = *50, 60, …, 100*) will lead to damping the oscillations, because of the effect of *averaging*. However, at the same time, long window length will make it more difficult to detect correctly the moment (the sampling time), at which the PV systems enters a new operating mode. The *third* and also very important reason for the approximate and possibly not correct results from the condition monitoring could be that the behavior of the system in the further time sampling *deviates* from the *original* operating mode behavior that has been used for learning of the seasonal models. This could be partially amended by generating (learning) *more* typical models that represent a larger number of typical behaviors (operating modes) of the system, thus covering a wider area of possible operations.

Looking at the evaluation results from the condition monitoring system in Fig. 19, we are still able to detect correctly in approximate way the four different seasonal behaviors of the PV system. For example, channel *1* (*Summer*) is the only one of all *4* channels that have the prediction error *RMSE below* the respective threshold *Th* (in average) for the approximate range between *1* to about 550–600 samplings. The other channels *2, 3* and *4* have a *RMSE* that is clearly above the respective thresholds for the same range of samplings. This suggests that the most probable operating mode of the PV systems within this range is the *Summer* mode.

Going further on through the samplings, we notice that there is a similar situation with the channel *2* (*Autumn*), where the *RMSE* keeps being below the respective threshold within the range from about *600* to *1200*. All the other channels (*1*, *3* and *4*) produce errors that are clearly above their respective thresholds. This is an indication that the operating mode within this range is *Autumn*.

The next seasonal operation—the *Winter*, shown by channel *3* is a bit more tricky to recognize. It is very easy to distinguish *Winter* from the operating modes (the hypothesis) of *Summer* and *Autumn*, but at the same time there is some similarity with the behavior in the *Spring* within the sampling range of *1200–1800*. Here still the channel *3* (the *Winter*) shows more often a lower *RMSE*—below the threshold, compared to channel 4 (the *Spring*), which highly fluctuates around the threshold. Therefore *Winter* is the most likely choice for the correct operating mode.

Finally, the analysis of the last range of samplings: from *1800* till *2400* shows that the channels (channel 4) are clearly out of contention, because of their high level RMSE, above the respective thresholds. The remaining channel *4* (the *Spring* hypothesis) is the only one that shows reasonable (although fluctuating) errors around the threshold. Then the conclusion is that *Spring* is the most probable operating mode within this sampling range.

8 Discussions and Conclusions

The proposed in this chapter partial GFM have a simplified structure with some predefined parameters, such as the size of the grid, shape of the membership functions and their overlapping level. The remaining are the singletons of the fuzzy rules that can be easily learned from the available input-output data by the presented iterative learning algorithm with fast convergence.

The partial GFM are suitable type of models that are able to capture in a satisfactory way the behavior of the PV system under different operating conditions (modes). Therefore they have been used in this research as members of the so called Model Base that accumulates the models of all currently known operating modes of the PV system. Then these partial GFM are used for solving the tasks of performance evaluation, recognition of different operating modes and condition monitoring of the PV system. The algorithmic steps of all these *3* tasks have been explained in the chapter and their performance has been shown by using respective experimental data. The common thing here is that in all tasks calculating the *similarity level* between a given pair of operating modes is important element of the whole calculation procedure. It is suggested in the chapter that such similarity can be estimated in several different ways, namely: by using the prediction error *RMSE* of each model, by using the percentage of the overlapping area between the real data set and the operating range of the respective model, or by combining them in an integrated fuzzy logic based decision making procedure.

The last option was not yet developed in this chapter and is set as one of the future directions for further research. Another task that waits for its reasonable solution is how to automate and improve the fidelity of the decision making in the real-time monitoring system, because in its current version this process is still a kind of a human-based decision.

Finally we feel full confidence that the described methodology in this chapter for performance evaluation and condition monitoring of PV systems, with some small improvements would become a true realistic and applicable approach.

References

1. Wang, L.X., Mendel, J.: Generating fuzzy rules by learning from examples. IEEE Trans. Syst. Man Cybern. **22**, 1414–1427 (1992)
2. Jin, Y., von Seelen, W., Sendhoff, B.: On generating flexible, complete, consistent and compact fuzzy rule systems from data using evolution strategies. IEEE Trans. Syst. Man Cybern.—Part B **29**, 829–845 (1999)
3. Yaochu, Jin: Fuzzy modeling of high-dimensional systems: complexity reduction and interpretability improvement. IEEE Trans. Fuzzy Syst. **8**(2), 212–221 (2000)
4. Abdulkadir, M., Samosir, A.S., Yatim, A.H.: Modeling and simulation based approach of photovoltaic system in simulink model. ARPN J. Eng. Appl. Sci. **7**(5), 616–623 (2012)
5. Aoun, N., Chenni, R., Nahman, B., Bouchouicha, K.: Evaluation and validation of equivalent five-parameter model performance for photovoltaic panels using only reference data. Energy Power Eng. **14**, 235–245 (2014)

6. Stoyanov, V., Vachkov, G.: Performance evaluation of photovoltaic systems by use of RBF network models. In: Proceedings of the 16th IFAC Conference TECIS 2015, Sozopol, Bulgaria, 24–27 Sept 2015, pp. 28–33 (2015)
7. Stoyanov, V., Vachkov, G.: Fuzzy Modeling of Photovoltaic Systems by Use of Incomplete Fuzzy Rule Base with RBF Activation Functions. In: Proceedings of the International Conference Automatics and Informatics'15, 4–7 Oct 2015, Sofia, Bulgaria, pp. 75–78 (2015)
8. Combs, W.E., Andrews, J.E.: Combinatorial rule explosion eliminated by a fuzzy rule configuration. IEEE Trans. Fuzzy Syst. **6**, 1–11 (1998)

Intelligent Control of Uncertain Switched Nonlinear Plants: NFG Optimum Control Synthesis via Switched Fuzzy Time-Delay Systems

Georgi M. Dimirovski, Jinming Luo and Huakuo Li

1 Introduction

Fuzzy systems are well known to provide for means to emulate intelligent decision and control algorithms [10, 24, 25, 30, 39, 40] for dynamic processes that are difficult or much too complex to model analytically [5, 8, 24–26] under various operating conditions. Since 1985, when it were first proposed and practically applied, Takagi-Sugeno (T-S) fuzzy system models have been proven as a rather effective means to represent nonlinear dynamic processes exhibiting various phenomena including nonlinearities, time delays, and uncertainties system (e.g., see [7, 10, 23, 25, 29, 37]). On the grounds of T-S class of fuzzy models a number of control solutions have been derived and many applications developed; for instance, see works [10, 13, 19, 27, 36] and references therein. Recent but rather instructive surveys are found in [41] by Zadeh and in [19] by Ojleska et al. In general, fuzzy systems and controls appear to be a category [8] of complex nonlinear systems [5, 26, 41].

On the other end of the complexity spectrum of general nonlinear systems [3, 5, 16, 34, 35, 43], during the last few decades switched systems and switching-based control [2, 6, 12, 15, 28, 31, 38, 43, 42] have been in the research focus of many scientists and engineers worldwide. A switched system is consisted of a family of

G. M. Dimirovski (✉)
Faculty of Electrical Engineering and Information Technologies, St. Cyril and St. Methodius University, Rugjer Boskovic 18, 1000 Skopje, Republic of Macedonia
e-mail: dimir@feit.ukim.edu.mk; gdimirovski@dogus.edu.tr

G. M. Dimirovski
Faculty of Engineering, Dogus University of Istanbul Acibadem, Zeamet Sokak 21, 34722 Istanbul, Turkey

J. Luo · H. Li
State Key Laboratory of Synthetic Automation, Northeastern University, Shenyang 110819, People's Republic of China
e-mail: Luojinming830@163.com.cn

V. Sgurev et al. (eds.), *Learning Systems: From Theory to Practice*, Studies in Computational Intelligence 756, https://doi.org/10.1007/978-3-319-75181-8_12

continuous-time or discrete-time subsystems and a switching rule that orchestrates the switching between them. These appear of considerable importance due to successes in practical applications and their importance in the theory developments. A considerable number of results for both the analysis and the synthesis of switched systems are available by now; for instance, see [3, 13, 15, 21, 20, 23, 36] and references therein.

Especially, if each subsystem of a switched system appears represented by a fuzzy system model, then the overall switched system is known as a switched fuzzy system. A switched fuzzy system evolves according to a sort of "hard switching" between the subsystems, which are fuzzy systems themselves, and a kind of "soft switching" among the fuzzy-rule binds of linear dynamic systems within the class of T-S fuzzy models [36, 41]. These two switching strategies and their interaction indeed may lead to some very complex behaviours of switched fuzzy systems. It is therefore that the results for such switched fuzzy systems in the literature are considerably limited as pointed above.

In the study of switched systems, most works are about the stability and stabilization issues [2, 6, 12, 15, 28, 31, 38, 43, 42]. The issues of optimization and optimal control of such nonlinear systems [11, 23, 35] by and large appear embedded into the solving stabilization control [4, 6, 11, 12, 16, 22–25, 30, 27, 43]. In addition, it is well understood that parametric uncertainties are among the principal factors responsible for the degraded system stability and performance. In the current study a specific type of T-S fuzzy system model is employed. Local dynamics in different regions of plant's state space, as typical for all T-S models, are represented by linear dynamics, while the overall plant representation model is emulated by means of a fuzzy blend through nonlinear fuzzy membership functions. Since time delays and uncertainties are inherent features of many physical processes and potential sources of instability or poor performance, these ought to be observed in fuzzy system models. Thus a T-S fuzzy system possessing time delays and uncertainties [10, 22, 23] is adopted as the representation model for the considered wide category of complexity nonlinear systems [8].

In the literature, typically, the uncertainty in plant system model is considered only. In practice, however, the controlling infrastructure may exhibit a certain degree of operating errors, which could also cause destabilization and/or performance degradation of the closed-loop system. A controller for a given plant is thus expected to be insensitive or non-fragile to such operating errors. Therefore, the robust stability against parametric uncertainties not only in the plant but also in the controller implementation is an important problem [22, 23, 28, 38] to handle at the design stage. It is furthermore desirable the synthesized system to be asymptotically stable and endowed with a certain level of performance index, which invoked the concept of non-fragile control. The non-fragile guaranteed cost (NGC) optimum control problems have been studied in both frameworks of fuzzy systems setting [30, 38] and of switched systems setting [2, 28]. To the best of authors' awareness, for the uncertain switched fuzzy systems with time delays, no results on the non-fragile guaranteed cost control problem have been reported up to now.

Motivated by the above considerations, for a class of switched fuzzy systems with time delays, this work studies the optimal guaranteed cost control problem which is robust not only for the system uncertainty and but also for the controller fragility. The innovative points in this contribution may be identified as follows. Firstly, it should be noted that none of subsystems are assumed to be asymptotically stable. Nonetheless a sufficient condition for asymptotical stability of the closed-loop system and an upper bound of the guaranteed cost function are established via the multiple Lyapunov functions approach by employing Lurie type of candidate Lyapunov functions [16, 18, 34]. In comparison with the existing results on switched fuzzy systems, the results of this paper have two distinct features. The optimum guaranteed-cost control problem for switched fuzzy systems with delays is believed to be studied in such a depth for the first time. Furthermore, unlike in the classical guaranteed-cost control problem, in the present study the uncertainty of fuzzy controller gain affects the system matrices as well as it enters the cost function criterion.

Contents which is presented in this contributed chapter, is organized as follows. In the next section the needed mathematical preliminaries are introduced, and the formal statements of the task problem is presented. Then in the subsequent section the main novel results, which comprise two theorems, are derived and proved. There after the section with computed numerical and simulation results for a typical illustrative examples are presented. Finally, a brief section summarizing the concluding remarks and some hint on future research is given, which is followed by references used.

2 Mathematical Preliminaries and Problem Statement

First a note on notation in this paper is given in here. Expression $X > 0\,(X \geq 0)$ denote the positive definite (positive semi-definite) matrix X, while $X < 0\,(X \leq 0)$ means that the matrix X is negative definite (negative semi-definite). Symbols I and O, respectively, represent the identity matrix and zero matrix with appropriate dimensions. Also, note further the space of fuzzy premises of the plant process state vector can be depicted as shown in Fig. 1, which is one alternative partitioning of the state space of a given plant process to be controlled. Further this work follows closely the original conference paper by Dimirovski et al. [7].

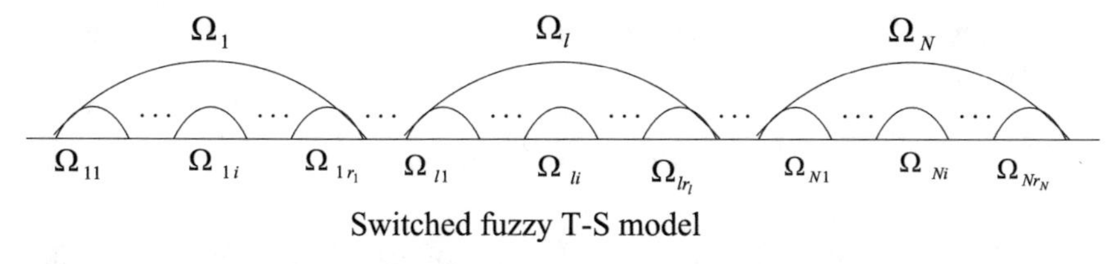

Fig. 1 An illustration of feasible partitioning of the space of fuzzy premises of the plant process state vector

A class of switched uncertain nonlinear dynamic processes possessing time delays, all subsystems of which are considered rather complex [5] and not amenable to analytical representation modelling, represents the investigated goal class of systems [39, 40]. However, such complex systems can be described [15] by an innovated class of switched Takagi-Sugeno (T-S) type of fuzzy system models [7]. This class of T-S fuzzy-rule based representation models of general nonlinear, time-delay, uncertainty dynamic plant processes are given below:

$$Plant\ Rule: If\ x_1\ is\ \Omega^i_{\sigma 1},\ x_2\ is\ \Omega^i_{\sigma 2}, \ldots, x_N\ is\ \Omega^i_{\sigma N},\ Then$$
$$\dot{x}(t) = (A_{\sigma i} + \Delta A_{\sigma i}(t))\, x(t) + A_{d\sigma i} x(t-\tau) + B_{\sigma i} u_\sigma(t) \tag{1}$$

the premises of which are assumed to partition the plant's state space as illustrated in Fig. 1 (see [19–21]). Moreover, the symbol σ in this T-S fuzzy model indicates all those quantities are under switching, that is to say (1) represents a class of switched T-S fuzzy model possessing both time delays and uncertainties. In particular, the piecewise-constant function

$$\sigma = \sigma(t)\colon\ [0, +\infty) \rightarrow M = \{1, 2 \ldots, N_{sw}\} \tag{2}$$

does represent the switching signal. Notice that this is a rather general mapping, which describe arbitrary switching signals.

The switching signal (2) is in direct one-to-one correspondence with the switching sequence

$$\sum = \{x_0; (l_0, t_0), (l_1, t_1), \ldots (l_k, t_k) \ldots, | l_k \in M\}. \tag{3}$$

This set-theoretic representation of the switching sequence means the l_k-th subsystem is in active status with the initial state $x_0 = x(t_0)$ and initial time instant t_0 for any interval $\forall t \in [t_k, t_{k+1})\,; i = 1, 2 \ldots, r_l$, and r_l is the number of inference rules. Furthermore, R^k_l denotes the i-th fuzzy inference rule; Ω^i_{ln} represents the fuzzy subset; $x(t) \in R^n$ is the system state vector; $u_l(t)$ is the control input vector; A_{li} and B_{li} are constant matrices with appropriate dimensions; τ denotes the bounded constant time delay, and it is assumed a nonnegative real-valued. Quantity $\Delta A_{li}(t)$ is a time-varying matrix function representing the uncertainty, which is assumed to satisfy the Assumption 1, presented below.

Assumption 1 [37] The uncertainty considered here is the norm bounded one having the form

$$\Delta A_{li}(t) = D_{li} M_{li}(t) E_{li}, \tag{4}$$

where D_{li} and E_{li} are known constant matrices of appropriate dimensions while $M_{li}(t)$ is an unknown time-varying matrix function satisfying $M^T_{li}(t) M_{li}(t) \leq I$.

It is already well known [24, 25] by means of the singleton fuzzification, product inference engine and average-centre defuzzification, the overall model of the system (1) can be inferred as follows:

$$\dot{x}(t) = \sum_{i-1}^{r_\sigma} h_{\sigma i}(x(t))[(A_{\sigma i} + \Delta A_{\sigma i}(t))x(t) + A_{d\sigma i}x(t-\tau) + B_{\sigma i}u_\sigma(t)] \tag{5}$$

In this representation model (5), the quantity $h_{\sigma i}\,(x\,(t))$ appears as

$$h_{\sigma i}\,(x\,(t)) = \frac{\prod_{p=1}^{n} \mu_{\sigma p}^{i}(x_p)}{\sum_{i=1}^{\tau_\sigma} \prod_{p=1}^{n} \mu_{\sigma p}^{i}(x_p)}, \tag{6}$$

where $\mu_{\sigma p}^{i}(x_p)$ denotes the membership function of x_p in the fuzzy set $\Omega_{\sigma p}^{i}$ for $p = 1, 2 \ldots n$. Though, for simplicity, the symbol $h_{\sigma i}(t)$ is used to denote $h_{\sigma i}(x(t))$ and also note that $0 \leq h_{\sigma i}(t) \leq 1$, $\sum_{i=1}^{r_\sigma} h_{\sigma i}(t) = 1$.

On the grounds of the technology of parallel distributed compensation architecture [30], the PDC, the state feedback controller is considered next. Given the fuzzy representation model (1) adopted to describe the complex, nonlinear, time-delay, uncertainty plant process to be controlled, it is naturally consistent to adopt the following fuzzy representation model to describe the fuzzy controller as follow:

$$\begin{aligned} &Controller\ Rule\ R_\sigma^i\!: \text{ If } x_1 \text{ is } \Omega_{\sigma 1}^{i}, x_2 \text{ is } \Omega_{\sigma 2}^{i}, \ldots, x_N \text{ is } \Omega_{\sigma N}^{i}, \\ &\qquad\qquad \text{Then } u_{\sigma i}\,(t) = (K_{\sigma i} + \Delta K_{\sigma i}\,(t))\,x\,(t)\,. \end{aligned} \tag{7}$$

In here, quantity $K_{\sigma i}$ is the controller gain to be derived in the course of design synthesis, while $\Delta K_{\sigma i}(t)$ is aimed at capturing the possible uncertainty in controller's gain [4, 7, 38]. The latter is also assumed to satisfy an adequate and relevant assumption that corresponds to Assumption 1 for the uncertainty of plant process to be controlled.

Assumption 2 [4, 14, 38] The possibly existing controller uncertainty is assumed to be norm-bounded of the form

$$\Delta K_{li}(t) = D_{ali} M_{ali}(t) E_{ali}, \tag{8}$$

where D_{ali} and E_{ali} are known constant matrices of appropriate dimensions while $M_{ali}(t)$ is an unknown time-varying matrix function satisfying $M_{ali}^{T}(t)M_{ali}(t) \leq I$.

Finally, following the same fuzzy inference procedure [25], the overall controller is derived as

$$u_\sigma(t) = \sum_{i-1}^{r_\sigma} h_{\sigma i}(t)[K_{\sigma i} + \Delta K_{\sigma i}(t)]x(t). \tag{9}$$

Remark 1 It should be noted, usually, the controller gain uncertainty includes both additive and multiplicative norm-bounded forms. In this paper, only an additive form of (8) is taken into consideration in order to solve the optimum non-fragility fragility problem.

Definition 1 [23, 35] The guaranteed-cost functional of the system (5), for some positive definite matrix Q, may be defined by means of the following linear-quadratic index of operating performance

$$J = \int_{0}^{\infty} [x^T(t)Qx(t) + u_\sigma^T(t)Iu_\sigma(t)]dt. \quad (10)$$

Notice that this linear quadratic functional expression is also called cost function of the control system to be designed in due course.

Definition 2 [28, 31] Consider the system (5). If there exist the state feedback controller $u_l(t)(l \in M)$ for each subsystem, an appropriate switching law $\sigma(t)$, and a positive scalar J^* such that the closed-loop system is asymptotically stable and the value of the function (8) satisfies $J \leq J^*$ for all the admissible uncertainties, then value J^* is called a non-fragile guaranteed-cost (NGC) optimum and $u_l(t)$ is called a guaranteed-cost control law.

Remark 2 [14, 15] It should be noted, the NGC control problem is different from the classical guaranteed-cost control, because $\Delta K_{li}(t)$ not only affects system matrices but also enters the cost function J^* through the switched control.

3 Time New Results Based on a Novel View on Plant Complexity

The main objective of this paper is to design a guaranteed cost control law and a switching law such that under their combined action the system (5) is asymptotically stable and the guaranteed-cost function (8) remains less or equal to the NGC value. For this purpose, recent works [13–15] have been revisited, re-studied and further improved in the here reported research. A solvability condition for the above defined NGC problem is derived via the approach of multiple candidate Lyapunov functions [2, 14] of Lurie type [16, 18]. Thereafter, an optimal non-fragile guaranteed-cost control problem is investigated to the full. For this purpose, the following three lemmas are necessary.

Lemma 1 [29] *Let D, E, K and $M(t)$ be real-valued matrices of appropriate dimensions, also with $M^T(t)M(T) \leq I$. Then following relationships hold true:*

(a) *For any real-valued scalar $\mu > 0$, it holds valid $DM(t)E + (DM(t)E)^T \leq \mu^{-1}DD^T + \mu E^T E$;*

(*b*) *For any real-valued symmetric matrix* $W = W^T > 0$ *and any real-valued scalar* $\mu > 0$, *there hold valid* $W - \mu DD^T > 0$ *and* $(K + DM(t)E)^T W^{-1}(K + DM(t)E) \leq K^T(W - \mu DD^T)K + \mu^{-1}E^T E$.

Lemma 2 [23, 30] *For any vectors* x_1, x_2 *and a matrix* R *of appropriate dimensions, there holds valid* $x_1^T Rx_2 + x_2^T Rx_1 \leq x_1^T RQ^{-1}R^T x_1 + x_2^T Qx_2$ *for any symmetric positive definite matrix* Q.

Lemma 3 [4, 31] *For any two matrices* X *and* Y, *and any positive real-valued constant* $\varepsilon > 0$, *there holds true* $X^T Y + Y^T X \leq \varepsilon X^T X + \varepsilon^{-1} Y^T Y$, *where* $X \in R^{m*n}$, $Y \in R^{m*n}$.

Then the following first main result is derived in the present research study.

Theorem 1 *Suppose for the system (5) there exist a set of non-positive scalars* β_{lv} *and sets of symmetric positive-definite matrices* P_l *and of gain matrices* K_{li}, $l, v \in M$, $v \neq l$ *satisfying the following matrix inequalities*

$$\prod\nolimits_{lij} + T_{lij} + P_l S_{lij} P_l + R_{lij} + \sum_{v=1, v\neq l}^{N} \beta_{lv}(P_l - P_v) < 0, \tag{11}$$

where the respective matrix quantities are:

$$\prod\nolimits_{lij} = P_l A_{li} + A_{li}^T P_l + P_l B_{li} K_{lj} + K_{lj}^T B_{li}^T P_l + P_l A_{lj} + A_{lj}^T P_l + P_l B_{lj} K_{li} + K_{li}^T B_{lj}^T P_l,$$

$$S_{lij} = D_{li} D_{li}^T + B_{li} D_{alj} D_{alj}^T B_{li}^T + D_{lj} D_{lj}^T + B_{lj} D_{ali} D_{ali}^T B_{lj}^T + A_{dli} Q^{-1} A_{dli}^T + A_{dlj} Q^{-1} A_{dlj}^T,$$

$$T_{lij} = 2E_{ali} E_{ali}^T + E_{li} E_{li}^T + 2E_{alj} E_{alj}^T + E_{lj} E_{lj}^T + 2I + 2Q,$$

$$R_{lij} = K_{lj}^T (I - D_{alj} D_{alj}^T)^{-1} K_{lj} + K_{li}^T (I - D_{ali} D_{ali}^T)^{-1} K_{li}.$$

Then there exists the guaranteed-cost control law (9) such that the system (5) is asymptotically stable under combined action with the switching law

$$\sigma = \arg min \left\{ x^T(t) P_l x(t), \quad \forall l \in M \right\}, \tag{12}$$

and the cost function (8) possesses the NGC value $J^* = min \left\{ x_0^T P_l x_0, l \in M \right\}$ *for any non-zero initial state* $x_0 = x(t_0)$, *i.e. for arbitrary but constant initial disturbance impulse may occur acting on the plant process.*

Proof In the first step, choose a candidate Lyapunov-Lurie functions for the system (5) as the following linear quadratic functional of optimal control theory [35]:

$$V_l(x(t)) = x^T(t) P_l x(t) + \int_{t-\tau}^{t} x^T(s) Q x(s) ds. \tag{13}$$

Next, calculate the time derivative along the state vector trajectory of the system (5) to obtain:

$$
\begin{aligned}
\dot{V}_l(x(t)) &= \dot{x}^T(t)P_l x(t) + x^T(t)P_l\dot{x}(t) + x^T(t)Qx(t) - x^T(t-\tau)Qx(t-\tau) \\
&= \sum_{i=1}^{r_l}\sum_{j=1}^{r_l} h_{li}(t)h_{lj}(t)x^T(t)\{[A_{li} + \Delta A_{li}(t) + B_{li}(K_{lj} + \Delta K_{lj}(t))]P_l \\
&\quad + P_l[A_{li} + \Delta A_{li}(t) + B_{li}(K_{lj} + \Delta K_{lj}(t))]\}x(t) \\
&\quad + \sum_{i=1}^{r_l}\sum_{j=1}^{r_l} h_{li}(t)h_{lj}(t)\{[x^T(t)Qx(t) - x^T(t-\tau)] \\
&\quad + x^T(t-\tau)A_{dli}^T P_l x(t) + x^T(t)P_l A_{dli}x(t-\tau)]\} \\
&\le \sum_{i=1}^{r_l}\sum_{j=1}^{r_l} h_{li}(t)h_{lj}(t)x^T(t)\{[A_{li} + \Delta A_{li}(t) + B_{li}(K_{lj} + \Delta K_{lj}(t))]\cdot \\
&\quad \cdot P_l + P_l[A_{li} + \Delta A_{li}(t) + B_{li}(K_{lj} + \Delta K_{lj}(t))]\}x(t) \\
&\quad + \sum_{i=1}^{r_l}\sum_{j=1}^{r_l} h_{li}(t)h_{lj}(t)\{[x^T(t)Qx(t) + x^T(t)P_l A_{dli}Q^{-1}A_{dli}^T P_l x(t)]\} \\
&= \sum_{i=1}^{r_l} h_{li}^2(t)\{x^T(t)[I + (K_{li} + \Delta K_{li}(t))^T(K_{li} + \Delta K_{li}(t)) \\
&\quad + P_l A_{li} + A_{li}^T P_l + P_l B_{li}K_{li} + K_{li}^T B_{li}^T P_l + P_l\Delta A_{li} + \Delta A_{li}^T P_l \\
&\quad + P_l B_{li}\Delta K_{li} + \Delta K_{li}^T B_{li}^T P_l + Q + P_l A_{dli}Q^{-1}A_{dli}^T P_l]x(t) \\
&\quad - x^T(t)x(t) - x^T(t)(K_{li} + \Delta K_{li}(t))^T(K_{li} + \Delta K_{li}(t))x(t)\} \\
&\quad + \sum_{i=1}^{r_l}\sum_{j=1}^{r_l} h_{li}(t)h_{lj}(t)\{x^T(t)[2I + (K_{li} + \Delta K_{li}(t))^T\cdot \\
&\quad \cdot(K_{lj} + \Delta K_{lj}(t)) + (K_{lj} + \Delta K_{lj}(t))^T(K_{li} + \Delta K_{li}(t)) \\
&\quad + P_l A_{li} + A_{li}^T P_l + P_l B_{li}K_{lj} + K_{lj}^T B_{li}^T P_l \\
&\quad + P_l\Delta A_{li} + \Delta A_{li}^T P_l + P_l B_{li}\Delta K_{lj} + \Delta K_{lj}^T B_{li}^T P_l \\
&\quad + P_l A_{lj} + A_{lj}^T P_l + P_l B_{lj}K_{li} + K_{li}^T B_{lj}^T P_l \\
&\quad + P_l\Delta A_{lj} + \Delta A_{lj}^T P_l + P_l B_{lj}\Delta K_{li} + \Delta K_{li}^T B_{lj}^T P_l \\
&\quad + 2Q + P_l A_{dli}Q^{-1}A_{dli}^T P_l + P_l A_{dlj}Q^{-1}A_{dlj}^T P_l]x(t) \\
&\quad - 2x^T(t)x(t) - x^T(t)(K_{li} + \Delta K_{li}(t))^T(K_{lj} + \Delta K_{lj}(t))x(t) \\
&\quad - x^T(t)(K_{lj} + \Delta K_{lj}(t))^T(K_{li} + \Delta K_{li}(t))x(t) \\
&\le \sum_{i=1}^{r_l} h_{li}^2(t)\{x^T(t)[I + K_{li}^T(I - D_{ali}D_{ali}^T)^{-1}K_{li} \\
&\quad + E_{ali}E_{ali}^T + P_l A_{li} + A_{li}^T P_l + P_l B_{li}K_{li} + K_{li}^T B_{li}^T P_l \\
&\quad + P_l D_{li}D_{li}^T P_l + P_l B_{li}D_{ali}D_{ali}^T B_{li}^T P_l + E_{ali}E_{ali}^T + E_{li}E_{+li}^T + P_l A_{dli}Q^{-1}A_{dli}^T P_l + Q]x(t) \\
&\quad - x^T(t)x(t) - x^T(t)(K_{li} + \Delta K_{li}(t))^T(K_{li} + \Delta K_{li}(t))x(t)\} \\
&\quad + \sum_{i=1}^{r_l}\sum_{j=1}^{r_l} h_{li}(t)h_{lj}(t)\{x^T(t)[2I + K_{lj}^T(I - D_{alj}D_{alj}^T)^{-1}K_{lj} + K_{li}^T(I - D_{ali}D_{ali}^T)^{-1}K_{li} \\
&\quad + E_{ali}E_{ali}^T + E_{alj}E_{alj}^T + P_l A_{li} + A_{li}^T P_l + P_l B_{li}K_{li} + K_{li}^T B_{li}^T P_l + P_l D_{li}D_{li}^T P_l
\end{aligned}
$$

$$
\begin{aligned}
&+ P_l B_{li} D_{alj} D_{alj}^T B_{li}^T P_l + E_{alj} E_{alj}^T + E_{li} E_{+li}^T + P_l A_{lj} + A_{lj}^T P_l + P_l B_{lj} K_{li} + K_{li}^T B_{lj}^T P \\
&+ P_l D_{lj} D_{lj}^T P_l + P_l B_{lj} D_{ali} D_{ali}^T B_{lj}^T P + \ E_{ali} E_{ali}^T + E_{lj} E_{lj}^T \\
&+ P_l A_{dli} Q^{-1} A_{dli}^T P_l + P_l A_{dlj} Q^{-1} A_{dlj}^T P_l + 2Q]x(t) \\
&- 2x^T(t)x(t) - x^T(t)(K_{li} + \Delta K_{li}(t))^T (K_{lj} + \Delta K_{lj}(t)x(t) \\
&- x^T(t)(K_{lj} + \Delta K_{lj}(t))^T (K_{li} + \Delta K_{li}(t)x(t)\}.
\end{aligned}
$$

By virtue of the inequality (11) and by using the LMI technique [9] on the grounds of applying the celebrated S-procedure for nonlinear control [34] and the linear matrix inequality solutions technique [32, 33] due to V. A. Yakubovich, it can be found that:

$$
\begin{aligned}
\dot{V}_l(x(t)) &< \sum_{i=1}^{r_l} h_{li}^2(t)[-x^T(t)x(t) - x^T(t)(K_{li} + \Delta K_{li}(t))^T (K_{li} + \Delta K_{li}(t)x(t)] \\
&+ \sum_{i<j}^{r_l} \sum_{j=1}^{r_l} h_{li}(t)h_{lj}(t)[-2x^T(t)x(t) - x^T(t)(K_{li} + \Delta K_{li}(t))^T (K_{lj} + \Delta K_{lj}(t)x(t) \\
&- x^T(t)(K_{lj} + \Delta K_{lj}(t))^T (K_{li} + \Delta K_{li}(t)x(t)] = -[x^T(t)x(t) + u_l^T(t)u_l(t)] < 0.
\end{aligned}
$$

whenever $\beta_{lv} \leq 0$. Furthermore, by means of integrating both sides of this expression it follows:

$$
\begin{aligned}
J < -\int_0^{+\infty} \dot{V}_l(x(t))dt &= -\sum_{k=0}^{\infty} \int_{t_k}^{t_{k+1}} \dot{V}_{lk}(x(t))dt \\
&= -[V_{l_0}(x(t_1)) - V_{l_0}(x(t_0)) + V_{l_1}(x(t_2)) - V_{l_1}(x(t_1)) + \ldots] \\
&= V_{l_0}(x(t_0)) = x_0^T P_{l_0} x_0 \ = \min\{x_0^T P_l x_0, l \in M\} = J^*
\end{aligned}
$$

Therefore the system (5) is asymptotically stable in the closed loop under combined control (2.9) and switching law (12), while the cost function (13) possesses a NGC optimum value $J^* = \min\left\{x_0^T P_l x_0,\ l \in M\right\}$, which completes the proof. □

Remark 3 For non-switched fuzzy systems, work [31] gave a stabilization result which is a special case of the above Theorem 1.

Remark 4 It should be noted, according to optimal control theory [35], different feasible solutions may result in different guaranteed-cost upper bounds of Theorem 1.

It is therefore that the task of optimizing the matrix P_l in order to achieve the minimal NGC for the stabilized closed-loop system yields the next Theorem 2.

Theorem 2 *An optimal non-fragile guaranteed cost J_{opt}^* can be achieved via solving the optimization problem*

$$
\min_{P_l} \gamma \ s.t. \tag{14}
$$

has relevant solution that holds with the inequality

$$\begin{pmatrix} \gamma & x_0^T \\ x_0 & P_l \end{pmatrix} > 0, \quad \forall l \in M. \tag{15}$$

Proof By means of Schur Complement [9], it is reasonably easy now to find a minimal guaranteed-cost upper bound as $J_{opt}^* = \min\{x_0^T P_l x_0, l \in M\}$, where matrices $\tilde{P}_l$ are solutions of the optimization problem (14). □

4 Computing Numerical and Simulation Results for a Case Study

In this section, an illustrative case study example from the literature [23, 38] is fully elaborated on by using the above proven new theorems. The considered example is an uncertain switched nonlinear dynamic process represented by means of a fuzzy system Takagi-Sugeno model with time delays and uncertainties that is composed of two subsystems. The fuzzy-rule representation model [7] is as follows

$$\begin{aligned}
R_1^1 &: \text{If } x_1 \text{ is } \Omega_{11}, \text{Then } \dot{x}(t) = (A_{11} + \Delta A_{11}(t))x(t) + A_{d11}x(t-\tau) + B_{11}u_{11}(t),\\
R_1^2 &: \text{If } x_1 \text{ is } \Omega_{12}, \text{Then } \dot{x}(t) = (A_{12} + \Delta A_{12}(t))x(t) + A_{d12}x(t-\tau) + B_{12}u_{12}(t),\\
R_2^1 &: \text{If } x_1 \text{ is } \Omega_{21}, \text{Then } \dot{x}(t) = (A_{21} + \Delta A_{21}(T))x(t) + A_{d21}x(t-\tau) + B_{21}u_{21}(t),\\
R_2^2 &: \text{If } x_1 \text{is } \Omega_{22}, \text{Then } \dot{x}(t) = (A_{22} + \Delta A_{22}(t))x(t) + A_{d22}x(t-\tau) + B_{22}u_{22}(t).
\end{aligned}$$

In there, the respective system's matrices are given as presented further below:

$$A_{11} = \begin{bmatrix} -22 & 10 \\ -230 & -10 \end{bmatrix}, A_{12} = \begin{bmatrix} -25 & 10 \\ -300 & -10 \end{bmatrix}, A_{21} = \begin{bmatrix} -22 & 10 \\ 200 & -10 \end{bmatrix},$$

$$A_{22} = \begin{bmatrix} -24 & 10 \\ -300 & -10 \end{bmatrix}, A_{d11} = A_{d12} = A_{d21} = \begin{bmatrix} 1 & 0 \\ 0 & 1 \end{bmatrix}, A_{d22} = \begin{bmatrix} -1 & 0 \\ 0 & 1 \end{bmatrix},$$

$$B_{11} = B_{12} = B_{21} = \begin{bmatrix} 1 & 0 \\ 0 & 1 \end{bmatrix}, B_{22} = \begin{bmatrix} -1 & 0 \\ 0 & 1 \end{bmatrix},$$

$$D_{11} = \begin{bmatrix} 1 & 0 \\ 1 & 0 \end{bmatrix}, D_{12} = \begin{bmatrix} 2 & 0 \\ 1 & -5 \end{bmatrix}, D_{21} = D_{22} = \begin{bmatrix} 2 & 0 \\ 1 & -4 \end{bmatrix},$$

$$E_{11} = \begin{bmatrix} -10 & 0 \\ 0 & 0 \end{bmatrix}, E_{12} = \begin{bmatrix} -1 & 0.2 \\ -6 & 0 \end{bmatrix}, E_{21} = E_{22} = \begin{bmatrix} 1 & -0.2 \\ -5 & 0 \end{bmatrix}$$

$$D_{a11} = D_{a12} = D_{a21} = D_{a22} = \begin{bmatrix} 0.01 & -0.1 \\ 0 & 0.1 \end{bmatrix},$$

$$E_{a11} = \begin{bmatrix} 1 & -1 \\ -1 & 1 \end{bmatrix}, E_{a21} = E_{a22} = E_{a12} = \begin{bmatrix} 1 & 0 \\ 0 & 1 \end{bmatrix},$$

$$M_{11}(t) = M_{12}(t) = M_{21}(t) = M_{22}(t) = \begin{bmatrix} \sin t & 0 \\ 0 & \cos t \end{bmatrix},$$

$$M_{a11}(t) = M_{a12}(t) = M_{a21}(t) = M_{a22}(t) = \begin{bmatrix} \sin t & 0 \\ 0 & \cos t \end{bmatrix}.$$

The membership functions, the selection of which is a considerable task on its own, have been empirically selected to be as follows:

$$h_{11}(x_1(t)) = 1 - \frac{1}{1+e^{-2x_1(t)}}, \quad h_1(x_1(t)) = \frac{1}{1+e^{-2x_1(t)}},$$

$$h_{21}(x_1(t)) = 1 - \frac{1}{1+e^{-2(x_1(t)-0.3)}}, \quad h_{22}(x_1(t)) = \frac{1}{1+e^{-2(x_1(t)-0.3)}}.$$

By making use of MATLAB LMI toolbox [1, 17] (also see [32–35]) the optimum positive definite matrices P_1, P_2 and feedback gain matrices have been computed as follows:

$$Q = \begin{bmatrix} 1.71660.1450 \\ 0.14500.0758 \end{bmatrix}, P_1 = \begin{bmatrix} 5.2136 - 0.1310 \\ -0.13100.2015 \end{bmatrix}, P_2 = \begin{bmatrix} 4.2762 - 0.1057 \\ -0.1057\ 0.1812 \end{bmatrix}$$

$$K_{11} = \begin{bmatrix} -5.1475 & 0.1267 \\ -0.0572 & -0.1939 \end{bmatrix}, \quad K_{12} = \begin{bmatrix} -5.1419\,0.1288 \\ 0.1090 - 0.1889 \end{bmatrix}$$

$$K_{21} = \begin{bmatrix} -0.5601 & -0.5228 \\ -0.0818 & -0.1806 \end{bmatrix}, \quad K_{22} = \begin{bmatrix} 1.0824 - 0.5972 \\ 0.1429 & -0.1783 \end{bmatrix}$$

The guaranteed-cost function then has been found to possess an optimum NGC value $J^*_{opt} = 4.6686$.

The simulation results are computed for the closed-loop switched fuzzy control systems under the assumption the initial disturbance of the plant state vector $x(0) = [-1 \ \ 1]^T$ equally strong in both controlled variables but in opposite directions. Under the designed fuzzy feedback control along with switching law that governs the switching among fuzzy time delay subsystems the state responses of controlled plant (5) converge rather quickly back to the equilibrium steady-state as demonstrated by Fig. 2. On the other hand, Fig. 3 depicts the PDC-fuzzy controller's generated input signals to the switched fuzzy time-delay plant sytems (5). Figure 2 depicts plant's state responses in closed loop under combined fuzzy feedback control (9) and switching signal (12); these are depicted in Figs. 3 and 4, respectively.

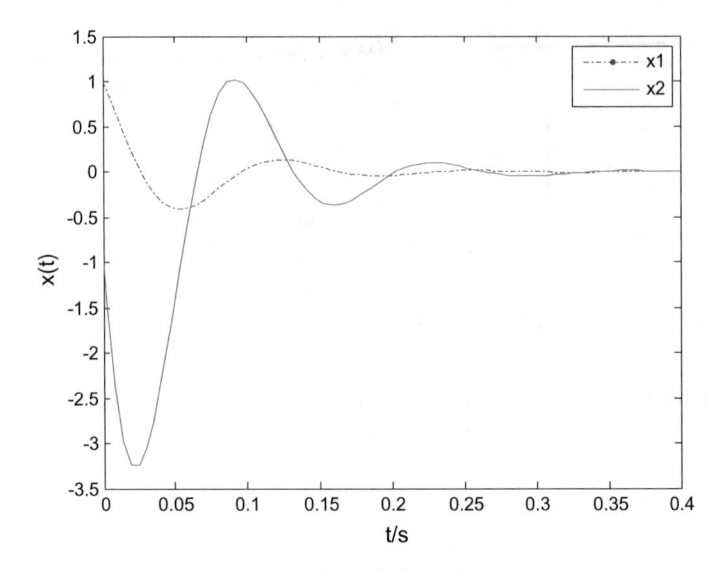

Fig. 2 Closed-loop switched fuzzy control systems: time-domain responses of plant state variables representing the system state trajectories in the time domain of the switched fuzzy system with time delays and uncertinties

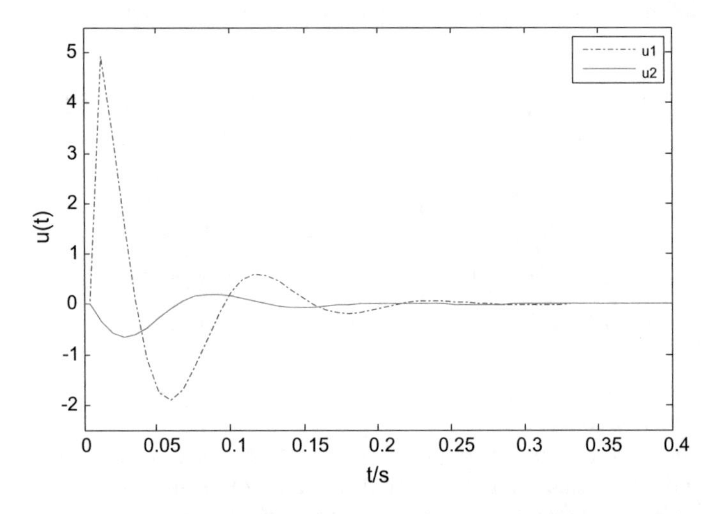

Fig. 3 Closed-loop switched fuzzy control systems: time-domain responses of plant control variables which are generated by the designed PDC fuzzy controller, which restore quickly the initially disturbed steady-state equilibrium of the plant thus reinforcing stable operation

Under the designed combined control strategy comprising the PDC-fuzzy controller (9) and the switching law (12) the state responses of the controlled plant process not only asymptotically but rather quickly do converge the operating equilibrium state. This closed-loop control systems reinforcement of stable plant operation,

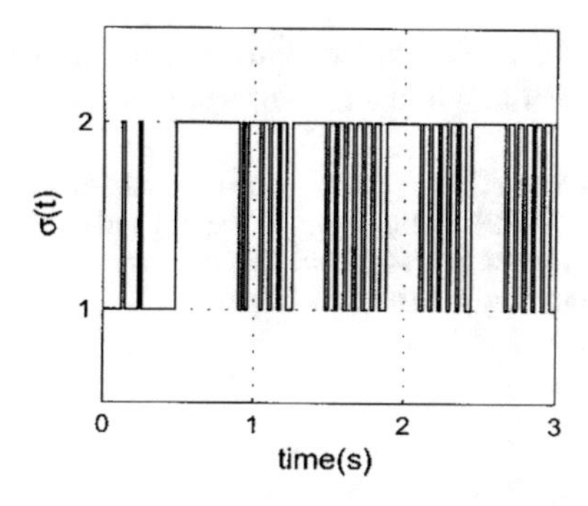

Fig. 4 Closed-loop switched fuzzy control systems: the switching signal that governs the switching among fuzzy subsystems of the switched fuzzy time-delay model of plant process within the closed loop switched fuzzy control system; it becomes a stationary one once the steady-state equilibriums is being restored

however, requires considerable control efforts during the first 100 ms approximately. Note that Fig. 3 depicts input signals to the switched fuzzy time-delay uncertainty system (5) that are generated by the fuzzy PDC controller. On these grounds and following the fuzzy-logic inference theory with a reasonable accuracy, one can conclude that the original systems of class (1) shall function according to these synthesis design results.

5 Concluding Remarks and Future Research

The problem of intelligent control for complex time-delay uncertain plants, which are only amenable to Takagi-Sugeno fuzzy-rule based modelling, has been investigated and solved via synergy of fuzzy-logic control and non-fragile guaranteed cost control principles. A fuzzy-system based intelligent version of the NGC problem for a class of switched fuzzy uncertain system models with time delays has been solved via the approach of multiple Lyapunov-Lurie functions. Furthermore, an optimization problem which minimizes the NGC optimum cost of the plant has been solved at the same setting. In this way, there have been designed a non-fragile state-feedback controller and an appropriate switching law. While plant's complexity system dynamics was assumed not amenable to representation modelling other than by employing T-S rules, a fundamental assumption of physical importance is that plant state variables are online real-time detectable and measurable.

In the cases when plant's states are not measurable or much too complicated to detect and measure, then the next research task to be solved the construction of combined fuzzy state observer and fuzzy output feedback controller in addition to the switching rule for switched fuzzy systems. This is the topic of an ongoing research which is related to such fuzzy observed based output feedback control of the investigated class of complex plants. Furthermore, a topic of future research is to solve the same task problem in the case of actuators may fail or malfunction.

Acknowledgements This research has been funded by the National Science Foundation of P. R. China (grant No. 61174073). Also, in part it was supported by the Ministry of Education & Science of R. Macedonia (grant No. 14-3154/1). Authors are grateful to professor Jun Zhao for his co-operation and help on many occasions in the past.

References

1. Boyd, S.P., El Ghaoui, L., Feron, E., Balakrishnan, V.: Linear Matrix Inequalities in Systems and Control Theory, SIAM Studies in Applied Mathematics, vol. 15. The SIAM, Philadelphia, PA (1994)
2. Branicky, M.S.: Multiple Lyapunov functions and other analysis tools for switched and hybrid systems. IEEE Trans. Autom. Control **43**(4), 475–482 (1998)
3. Chen, J., Liu, F.: Robust fuzzy control for nonlinear system with disk pole constraints. Control Decis. **22**(9), 983–988 (2007)
4. Choi, D.J., Park, P.G.: Guaranteed cost controller design for discrete-time switching fuzzy systems. IEEE Trans. Syst. Man Cybern.: Part B Cybern. **34**(1), 110–119 (2004)
5. Dimirovski, G.M.: Fasciating ideas on complexity and complex systems control. In: Dimirovski, G.M. (ed.) Complex Systems—Relationships between Control, Communications and Computing, Studies in Systems, Decision and Control, vol. 55, vi–xxxix. Springer International Publishing AG Switzerland, Cham, CH (2016)
6. Dayawansa, W.P., Martin, C.F.: A converse Laypunov theorem for a class of dynamical systems which undergo switching. IEEE Trans. Autom. Control **44**(4), 751–760 (1999)
7. Dimirovski, G.M., Luo, J., Li, H., Zhao, J.: Intelligent control of uncertain switched nonlinear plants: NGF optimum control of switched time-delay fuzzy systems revisited. In: Proceedings ISO2016 of the IEEE International Conference on Intelligent Systems, Sofia, Bulgaria, 4–6 September 2016, pp. 510–515. Bulgarian IM/CS/SMC Society on behalf of the IEEE, Piscatawy, NJ, Sofia, BG (2016)
8. Dimirovski, G.M., Gough, N.E., Barnett, S.: Categories in systems and control. Int. J. Syst. Sci. **8**(9), 1081–1090 (1977)
9. Gahinet, P., Nemirovskii, A., Laub, A.J., Chilali, M.: The LMI Tool Box. The MathWorks, Inc., Natick, NJ (1995)
10. Huang, D., Nguang, S.K., Patel, N.: Fuzzy control of uncertain nonlinear networked control systems with random time-delays. In: Proceedings of the 27th American Control Conference, New York City, NY, 11–13 July 2007, pp. 299–304. IEEE on behalf of the AACC, Piscataway, NJ (2007)
11. Krassovskiy, N.N.: Towards the theory of optimal control (in Russian). Avtomatika i Telemekhanika **18**(11), 960–970 (1957)
12. Li, L.-L., Zhao, J., Dimirovski, G.M.: Multiple Lyapunov functions approach to observer-based H_{inf} control for switched systems. Int. J. Syst. Sci. **44**(5), 812–819 (2013)
13. Luo, J., Dimirovski, G.M.: Intelligent ontrol for switched fuzzy systems: synthesis via nonstandard Lyapunov functions. In: Sgurev, V., Hadjiski, M., Kacprzyk, J., Atanassov, K. (eds.) Studies in Computational Intelligence, vol. 657, pp. 115–141. Springer, Berlin, Heidelberg (2017)
14. Luo, J., Dimirovski, G.M., Zhao, J.: Non-fragile guaranteed-cost control for a class of switched systems. In: Proceedings of the 31st Chinese Control Conference, Hefei, CN, 25–27 July 2012, pp. 2112–2116. Technical Committee on Control Theory, the CAA, Beijing, CN (2012)

15. Luo, J., Dimirovski, G.M., Zhao, J.: Non-fragile guaranteed-cost control for a class of switched systems. In: Proceedings of the 32nd Chinese Control Conference, Xi'an, CN, 26–28 July 2013, pp. 2174–2179. Technical Committee on Control Theory, the CAA, Beijing, CN (2013)
16. Lurie, A.I.: Some Nonlinear Problems of the Automatic Control Theory (in Russian). Gostehizdat, Moskva – Lningrad (1951)
17. MathWorks: MATLAB. The MathWorks, Inc., Natick, NJ (1994)
18. Meyer, K.R.: Liapunov function for the problem of Lurie. Proc. Natl. Acad. Sci. USA **53**, 501–503 (1965)
19. Ojleska, V., Kolemisevska-Gugulovska, T., Dimirovsky, G.: Recent advances in analysis and design for switched fuzzy systems: a review. In: Proceedings of the 6th IEEE International Conference on Intelligent Systems IS2012, Sofia, BG, 6–8 September 2012, pp. 248–257. Bulgarian IM/CS/SMC Society on behalf of the IEEE, Piscataway, NJ, Sofia, BG (2012)
20. Ojleska, V.M., Kolemisevska-Gugulovska, T.D., Dimirovski, G.M.: Influence of the state space partitioning into regions when designing switched fuzzy controllers. I Facta Univ. Ser. Autom. Control Robot. **9**(1), 131–139 (2010)
21. Ojleska, V.M., Kolemisevska-Gugulovska, T.D., Dimirovski, G.M.: Switched fuzzy control systems: exploring the performance in applications. Int. J. Simul.: Syst. Sci. Technol. **12**(2), 19–29 (2011)
22. Park, J.H., Jung, H.Y.: On the design of non-fragile guaranteed cost controller for a class of uncertain dynamic systems with state delays. Appl. Math. Comput. **150**(1), 245–257 (2004)
23. Ren, J.S.: Non-fragile LQ fuzzy control for a class of non-linear descriptor system with time delays. In: Proceedings of the 4th International Conference on Machine Learning and Cybernetics, Guangzhou, 18–21 August 2005, pp. 1121–1126. The Institute of Automation, CAS, Beijing, CN (2005)
24. Takagi, T., Sugeno, M.: Fuzzy identification of systems and its applications to modelling and control. IEEE Trans. Syst. Man Cybern. **15**(1), 116–132 (1985)
25. Tanaka, K., Wang, H.O.: Fuzzy Control System Design and Analysis: A Linear Matrix Inequality Approach. Wiley, New York, NY (2001)
26. Wang, F.-Y.: Computational experiments for behaviour analysis and decision evaluation of complex systems. J. Syst. Simul. **16**(5), 893–897 (2004)
27. Wang, J.L., Shu, Z.X., Chen, L., Wang, Z.X.: Non-fragile fuzzy guaranteed cost control of uncertain nonlinear discrete-time systems. In: Proceedings of the 26th Chinese Control Conference, Zhangjiajie, Hunan, CN, 26–31 July 2007, pp. 2111–2116. The CAA, Technical Committee on Control Theory, CAS, Beijing (2007)
28. Wang, R., Zhao, J.: Non-fragile hybrid guaranteed cost control for a class of uncertain switched linear systems. J. Control Theory Appl. **1**, 32–37 (2006)
29. Wang, W.J., Kao, C.C., Chen, C.S.: Stabilization, estimation and robustness for large-scale time-delay systems. Control Theory Adv. Technol. **7**, 569–585 (1991)
30. Wang, H.O., Tanaka, K., Griffin, M.: An approach to fuzzy control of nonlinear systems: stability and design issues. IEEE Trans. Fuzzy Syst. **4**(1), 14–23 (2006)
31. Wang, R., Feng, J.X., Zhao, J.: Design methods of non-fragile controllers for a class of uncertain switched linear systems. Control Decis. **21**(7), 321–326 (2006)
32. Yakubovich, V.A.: Solution to some matrix inequalities occuring in the theory of automatic control (in Russian). Dokladi akademii nauk SSSR **143**(6), 1304–1307 (1962)
33. Yakubovich, V.A.: Solution to some matrix inequalities occuring in the nonlinear theory of control (in Russian). Dokladi akademii nauk SSSR **156**(2), 271–281 (1964)
34. Yakubovich, V.A.: S-procedure in the nonlinar control theory (in Russian). Vestnik Leningradskogo Universiteta, Seriyua Mathematika, mekhanika, astronomiya **1**, 62–77 (1971)
35. Yakubovich, V.A.: Maximum principle in the optimization problem of controllers (in Russin). Vestnik Leningradskogo Universiteta, Seriyua Mathematika, mekhanika, astronomiya **7**, 55–68 (1980)
36. Yang, H., Dimirovski, G.M., Zhao, J.: Switched fuzzy systems: representation modelling, stability analysis, and control design. In: Kacprzyk, J. (ed.) Studies in Computational Intelligence—Intelligent Techniques and Tool for Novel Systems Architectures, vol. 109, pp. 169–184. Springer, Berlin, Heidelberg (2008)

37. Yee, J.S., Yang, G.H., Wang, J.L.: Non-fragile guaranteed cost for discrete-time uncertain linear systems. Int. J. Syst. Sci. **32**(7), 845–853 (2001)
38. Yin, P.Q., Yu, L., Zheng, K.: T-S model based non-fragile guaranteed cost fuzzy control for nonlinear time-delay systems. Control Theory Appl. **25**(1), 38–44 (2008)
39. Zadeh, L.A.: A rationale for fuzzy control. J. Dyn. Syst. Meas. Control Trans. ASME Ser. G **94**(1), 3–4 (1972)
40. Zadeh, L.A.: Outline of a new approach to the analysis of complex systems and decision processes. IEEE Trans. Syst. Man Cybern. **3**(1), 28–44 (1973)
41. Zadeh, L.A.: Is there a need for fuzzy logic? Inf. Sci. **178**, 2751–2779 (2008)
42. Zhao, J., Dimirovski, G.M.: Quadratic stability of a class of switched nonlinear systems. IEEE Trans. Autom. Control **49**(4), 574–578 (2004)
43. Zhao, J., Hill, D.J.: On stability, L_2-gain and H_{inf}-control for switched systems. Automatica **44**, 1220–1232 (2008)

Multidimensional Intuitionistic Fuzzy Quantifiers and Level Operators

Krassimir Atanassov, Ivan Georgiev, Eulalia Szmidt and Janusz Kacprzyk

1 Short Remarks on Intuitionistic Fuzzy Propositional and Predicate Logics

This paper is an extension of our communication, based on [13].

In classical logic (e.g., [24]), to each proposition (sentence) we juxtapose its truth value: truth—denoted by 1, or falsity—denoted by 0. In the case of fuzzy logic [28], this truth value is a real number in the interval [0, 1] and it is called "truth degree". In the intuitionistic fuzzy case (see [3, 5, 7, 10]) we add one more value—"falsity degree"—which is again in the interval [0, 1]. Thus, to the proposition p, two real numbers, $\mu(p)$ and $\nu(p)$, are assigned with the following constraints:

$$\mu(p), \nu(p) \in [0, 1] \text{ and } \mu(p) + \nu(p) \leq 1.$$

K. Atanassov (✉)
Department of Bioinformatics and Mathematical Modelling, Institute of Biophysics and Biomedical Engineering Bulgarian Academy of Sciences, 105 Acad. G. Bonchev Str., 1113 Sofia, Bulgaria
e-mail: krat@bas.bg

K. Atanassov · I. Georgiev
Prof. Asen Zlatarov University, 8010 Bourgas, Bulgaria
e-mail: ivandg@btu.bg

E. Szmidt · J. Kacprzyk
Systems Research Institute Polish Academy of Sciences, ul. Newelska 6, 01-447 Warsaw, Poland
e-mail: szmidt@ibspan.waw.pl

E. Szmidt · J. Kacprzyk
Warsaw School of Information Technology, ul. Newelska 6, 01-447 Warsaw, Poland
e-mail: kacprzyk@ibspan.waw.pl

V. Sgurev et al. (eds.), *Learning Systems: From Theory to Practice*, Studies in Computational Intelligence 756, https://doi.org/10.1007/978-3-319-75181-8_13

Here, we define only the operations "disjunction", "conjunction" and "implication", originally introduced in [3], which have classical logic analogues, as follows:

$$V(p \vee q) = \langle \max(\mu(p), \mu(q)), \min(\nu(p), \nu(q)) \rangle,$$

$$V(p \wedge q) = \langle \min(\mu(p), \mu(q)), \max(\nu(p), \nu(q)) \rangle,$$

$$V(p \rightarrow q) = \langle \max(\nu(p), \mu(q)), \min(\mu(p), \nu(q)) \rangle.$$

The classical negation is defined by

$$V(\neg p) = \langle \nu(p), \mu(p) \rangle.$$

We extend V to all propositional formulas in the usual way.

A formula A is said to be an *intuitionistic fuzzy tautology* (*IFT*, for short) if for all evaluations V we have that $V(A) = \langle a, b \rangle$ implies $a \geq b$. Two formulas A and B are *equivalent*, denoted $A \equiv B$, if $A \rightarrow B$ and $B \rightarrow A$ are IFTs.

The idea for evaluation of the propositions was extended for predicates (see [18–21, 24, 27]) as follows (see, e.g., [5, 7, 12, 22]).

Let x be a variable, obtaining values in a set E and let $P(x)$ be a predicate of the variable x. Let

$$V(P(x)) = \langle \mu(P(x)), \nu(P(x)) \rangle.$$

The IF-interpretations of the quantifiers *for all* ($\forall$) and *there exists* ($\exists$) are introduced in [5, 12, 22] by

$$V(\exists x P(x)) = \langle \sup_{y \in E} \mu(P(y)), \inf_{y \in E} \nu(P(y)) \rangle,$$

$$V(\forall x P(x)) = \langle \inf_{y \in E} \mu(P(y)), \sup_{y \in E} \nu(P(y)) \rangle.$$

In addition, we assume that E is a finite set, so we can use the following denotations:

$$V(\exists x P(x)) = \langle \max_{y \in E} \mu(P(y)), \min_{y \in E} \nu(P(y)) \rangle,$$

$$V(\forall x P(x)) = \langle \min_{y \in E} \mu(P(y)), \max_{y \in E} \nu(P(y)) \rangle.$$

The geometrical interpretation of the quantifiers is illustrated in Figs. 1 and 2, where $x_1, ..., x_5$ are the possible values of the variable x and $V(P(x_1)), ..., V(P(x_5))$ are the corresponding IF-estimations.

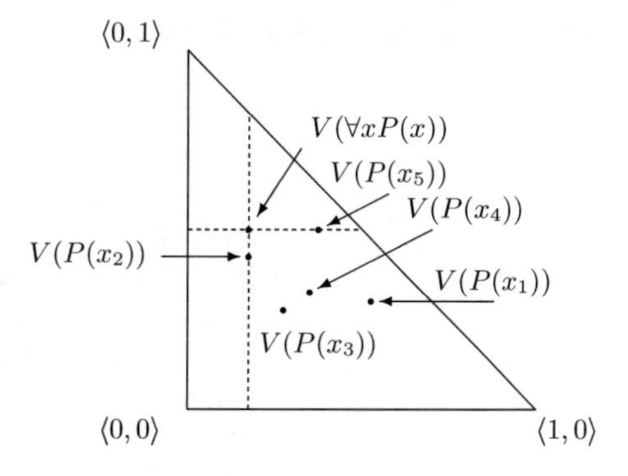

Fig. 1 Quantifier $\forall$ in the intuitionistic fuzzy interpretation triangle

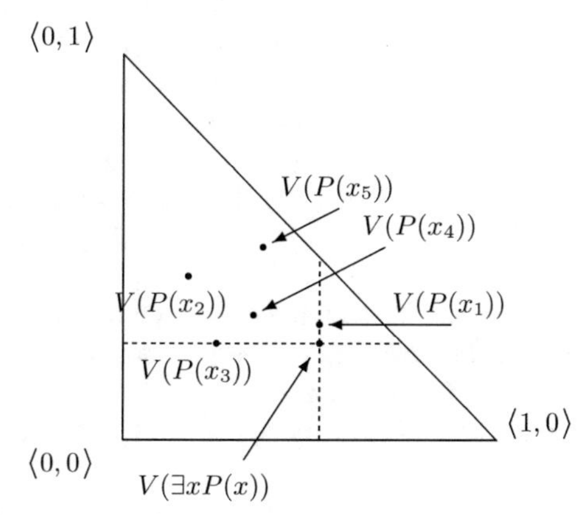

Fig. 2 Quantifier $\exists$ in the intuitionistic fuzzy interpretation triangle

The most important property of the two quantifiers is that each of them juxtaposes to every unary predicate P a point (exactly one for each quantifier) in the IF-interpretational triangle.

In [7] the following assertions are formulated and proved.

Theorem 1 *The logical axioms of the theory $\mathcal{K}$ (see [24]):*

(a) $\forall x A(x) \to A(t)$,
for the fixed variable t,
(b) $\forall x(A \to B) \to (A \to \forall x B)$,
where the variable x *is not free in* A,

are IFTs.

Theorem 2 *The following formulae (see, e.g. [24]) are IFTs:*

$$\begin{aligned}
&(a)\ (\forall x A(x) \to B) \equiv \exists x(A(x) \to B),\\
&\quad \textit{where the variable } x \textit{ is not free in } B,\\
&(b)\ \exists x A(x) \to B \equiv \forall x(A(x) \to B),\\
&\quad \textit{where the variable } x \textit{ is not free in } B,\\
&(c)\ B \to \forall x A(x) \equiv \forall x(B \to A(x)),\\
&\quad \textit{where the variable } x \textit{is not free in } B,\\
&(d)\ B \to \exists x A(x) \equiv \exists x(B \to A(x)),\\
&\quad \textit{where the variable } x \textit{ is not free in } B,\\
&(e)\ (\forall x A \wedge \forall x B) \equiv \forall x(A \wedge B),\\
&(f)\ (\forall x A \vee \forall x B) \to \forall x(A \vee B),\\
&(g)\ \neg\forall x A \equiv \exists x \neg A,\\
&(h)\ \neg\exists x A \equiv \forall x \neg A,\\
&(i)\ \forall x \forall y A \equiv \forall y \forall x A,\\
&(j)\ \exists x \exists y A \equiv \exists y \exists x A,\\
&(k)\ \exists x \forall y A \to \forall y \exists x A,\\
&(l)\ \forall x(A \to B) \to (\forall x A \to \forall x B).
\end{aligned}$$

The link between the interpretations of the quantifiers and the topological operators C (closure) and I (interior), defined over IFSs (see [5]), is obvious.

Here, for the first time, we introduce a intuitionistic fuzzy level operator over predicate P. For every $\alpha, \beta \in [0, 1]$, such that $\alpha + \beta \leq 1$:

$$N_{\alpha,\beta}(P) = \{x|\mu(P(x)) \geq \alpha \ \&\ \nu(P(x)) \leq \beta\} \subseteq E.$$

2 Main Results

2.1. Let $E, Z_1, Z_2, ..., Z_n$ be fixed finite linearly ordered sets.

By analogy with intuitionistic fuzzy multi-dimensional sets, introduced in [14–17], here for the first time, for a predicate P of the variables $x, z_1, z_2, ..., z_n$, ordered in the present form, we define an intuitionistic fuzzy evaluation function V for P of the form

$$V(P(x, z_1, z_2, ..., z_n)) = \langle \mu_P(x, z_1, z_2, ..., z_n), \nu_P(x, z_1, z_2, ..., z_n)\rangle,$$

where $x \in E$ is a (basic) variable, $z_1 \in Z_1$, $z_2 \in Z_2$, ..., $z_n \in Z_n$ are additional variables, $\mu_P(x, z_1, z_2, ..., z_n) \in [0, 1]$, $\nu_P(x, z_1, z_2, ..., z_n) \in [0, 1]$ and

$$\mu_P(x, z_1, z_2, ..., z_n) + \nu_P(x, z_1, z_2, ..., z_n) \leq 1.$$

Here, $\mu_P(x, z_1, z_2, ..., z_n)$ and $\nu_P(x, z_1, z_2, ..., z_n)$ are the degrees of validity and non-validity of $P(x, z_1, z_2, ..., z_n)$, respectively.

In the particular case, when $n = 1$, we obtain the case of temporal IFL (see [4]).

Having in mind the results from [14], we can define the following $(n+1)$-dimensional quantifiers:

(a) (partial) standard quantifier

$$V(\exists(x, z_1, z_2, ..., z_n)P(x, z_1, z_2, ..., z_n))$$

$$= \langle \max_{y \in E} \mu_P(y, z_1, z_2, ..., z_n), \min_{y \in E} \nu_P(y, z_1, z_2, ..., z_n)\rangle,$$

$$V(\forall(x, z_1, z_2, ..., z_n)P(x, z_1, z_2, ..., z_n))$$

$$= \langle \min_{y \in E} \mu_P(y, z_1, z_2, ..., z_n), \max_{y \in E} \nu_P(y, z_1, z_2, ..., z_n)\rangle$$

(b) (partial) i-quantifiers

$$V(\exists^i(x, z_1, z_2, ..., z_n)P(x, z_1, z_2, ..., z_n))$$

$$= \langle \max_{t_i \in Z_i} \mu_P(x, z_1, z_2, ..., z_{i-1}, t_i, z_{i+1}, ..., z_n),$$

$$\min_{t_i \in Z_i} \nu_P(x, z_1, z_2, ..., z_{i-1}, t_i, z_{i+1}, ..., z_n)\rangle,$$

$$V(\forall^i(x, z_1, z_2, ..., z_n)P(x, z_1, z_2, ..., z_n))$$

$$= \langle \min_{t_i \in Z_i} \mu_P(x, z_1, z_2, ..., z_{i-1}, t_i, z_{i+1}, ..., z_n),$$

$$\max_{t_i \in Z_i} \nu_P(x, z_1, z_2, ..., z_{i-1}, t_i, z_{i+1}, ..., z_n)\rangle$$

(c) general additional (shortly, a-) quantifier

$$V(\exists^a(x, z_1, z_2, ..., z_n)P(x, z_1, z_2, ..., z_n))$$

$$= \langle \max_{t_1 \in Z_1} ... \max_{t_n \in Z_n} \mu_P(x, t_1, t_2, ..., t_n),$$

$$\min_{t_1 \in Z_1} ... \min_{t_n \in Z_n} \nu_P(x, t_1, t_2, ..., t_n)\rangle,$$

$$V(\forall^a(x, z_1, z_2, ..., z_n)P(x, z_1, z_2, ..., z_n)$$

$$= \langle \min_{t_1 \in Z_1} ... \min_{t_n \in Z_n} \mu_P(x, t_1, t_2, ..., t_n),$$

$$\max_{t_1 \in Z_1} ... \max_{t_n \in Z_n} \nu_P(x, t_1, t_2, ..., t_n)\rangle$$

(d) general quantifier

$$V(\exists^g(x, z_1, z_2, ..., z_n)P(x, z_1, z_2, ..., z_n))$$

$$= \langle \max_{y\in E} \max_{t_1\in Z_1} ... \max_{t_n\in Z_n} \mu_P(y, t_1, t_2, ..., t_n),$$

$$\min_{y\in E} \min_{t_1\in Z_1} ... \min_{t_n\in Z_n} \nu_P(y, t_1, t_2, ..., t_n)\rangle,$$

$$V(\forall^g(x, z_1, z_2, ..., z_n)P(y, z_1, z_2, ..., z_n)$$

$$= \langle \min_{y\in E} \min_{t_1\in Z_1} ... \min_{t_n\in Z_n} \mu_P(y, t_1, t_2, ..., t_n),$$

$$\max_{y\in E} \max_{t_1\in Z_1} ... \max_{t_n\in Z_n} \nu_P(y, t_1, t_2, ..., t_n)\rangle$$

Theorem 3 *For each of the fifth pairs of quantifiers, the equalities*

$$\begin{aligned}
&V(\neg\exists(x, z_1, z_2, ..., z_n)\neg P(x, z_1, z_2, ..., z_n)) = \\
&= V(\forall(x, z_1, z_2, ..., z_n)P(x, z_1, z_2, ..., z_n)), \\
&V(\neg\forall(x, z_1, z_2, ..., z_n)\neg P(x, z_1, z_2, ..., z_n)) = \\
&= V(\exists(x, z_1, z_2, ..., z_n)P(x, z_1, z_2, ..., z_n)), \\
&V(\neg\exists^i(x, z_1, z_2, ..., z_n)\neg P(x, z_1, z_2, ..., z_n)) = \\
&= V(\forall^i(x, z_1, z_2, ..., z_n)P(x, z_1, z_2, ..., z_n)), \\
&V(\neg\forall^i(x, z_1, z_2, ..., z_n)\neg P(x, z_1, z_2, ..., z_n)) = \\
&= V(\exists^i(x, z_1, z_2, ..., z_n)P(x, z_1, z_2, ..., z_n)), \\
&V(\neg\exists^a(x, z_1, z_2, ..., z_n)\neg P(x, z_1, z_2, ..., z_n)) = \\
&= V(\forall^a(x, z_1, z_2, ..., z_n)P(x, z_1, z_2, ..., z_n), \\
&V(\neg\forall^a(x, z_1, z_2, ..., z_n)\neg P(x, z_1, z_2, ..., z_n)) = \\
&= V(\exists^a(x, z_1, z_2, ..., z_n)P(x, z_1, z_2, ..., z_n), \\
&V(\neg\exists^g(x, z_1, z_2, ..., z_n)\neg P(x, z_1, z_2, ..., z_n)) = \\
&= V(\forall^g(x, z_1, z_2, ..., z_n)P(x, z_1, z_2, ..., z_n), \\
&V(\neg\forall^g(x, z_1, z_2, ..., z_n)\neg P(x, z_1, z_2, ..., z_n)) = \\
&= V(\exists^g(x, z_1, z_2, ..., z_n)P(x, z_1, z_2, ..., z_n)
\end{aligned}$$

hold.

Proof Let us check the validity of the first equality.

$$V(\neg\exists(x, z_1, z_2, ..., z_n)\neg P(x, z_1, z_2, ..., z_n))$$

$$= \neg\exists(x, z_1, z_2, ..., z_n)\neg\langle\mu_P(x, z_1, z_2, ..., z_n),$$

$$\nu_P(x, z_1, z_2, ..., z_n)\rangle$$

$$= \neg\exists(x, z_1, z_2, ..., z_n)\langle\nu_P(x, z_1, z_2, ..., z_n),$$

$$\mu_P(x, z_1, z_2, ..., z_n)\rangle$$

$$= \neg\langle \max_{y \in E} \nu_P(y, z_1, z_2, ..., z_n), \min_{y \in E} \mu_P(y, z_1, z_2, ..., z_n)\rangle$$

$$= \langle \min_{y \in E} \mu_P(y, z_1, z_2, ..., z_n), \max_{y \in E} \nu_P(y, z_1, z_2, ..., z_n)\rangle$$

$$= V(\forall(x, z_1, z_2, ..., z_n)P(x, z_1, z_2, ..., z_n)).$$

The other equalities are proved in the same manner.

An important **Open Problem** is: Which other negations from the ones, defined in [11], also satisfy these equalities?

2.2. Here, we extend the defined above intuitionistic fuzzy level operator over predicate P to the following forms:

$$N^{w}_{\alpha,\beta}(P) = \{(z_1, z_2, ..., z_n) | \max_{y \in E} \mu_P(y, z_1, z_2, ..., z_n) \geq \alpha$$

$$\& \min_{y \in E} \nu_P(y, z_1, z_2, ..., z_n) \leq \beta\},$$

$$N^{s}_{\alpha,\beta}(P) = \{(z_1, z_2, ..., z_n) | \min_{y \in E} \mu_P(y, z_1, z_2, ..., z_n) \geq \alpha$$

$$\& \max_{y \in E} \nu_P(y, z_1, z_2, ..., z_n) \leq \beta\},$$

$$N^{w,i}_{\alpha,\beta}(P) = \{(x, z_1, z_2, ..., z_{i-1}, z_{i+1}, ..., z_n) |$$

$$| \max_{t_i \in Z_i} \mu_P(x, z_1, z_2, ..., z_{i-1}, t_i, z_{i+1}, ..., z_n) \geq \alpha$$

$$\& \min_{t_i \in Z_i} \nu_P(x, z_1, z_2, ..., z_{i-1}, t_i, z_{i+1}, ..., z_n) \leq \beta\},$$

$$N^{s,i}_{\alpha,\beta}(P) = \{(x, z_1, z_2, ..., z_{i-1}, z_{i+1}, ..., z_n) |$$

$$| \min_{t_i \in Z_i} \mu_P(x, z_1, z_2, ..., z_{i-1}, t_i, z_{i+1}, ..., z_n) \geq \alpha$$

$$\& \max_{t_i \in Z_i} \nu_P(x, z_1, z_2, ..., z_{i-1}, t_i, z_{i+1}, ..., z_n) \leq \beta\},$$

$$N^{w,a}_{\alpha,\beta}(P) = \{x | \max_{t_1 \in Z_1} ... \max_{t_n \in Z_n} \mu_P(x, t_1, t_2, ..., t_n) \geq \alpha$$

$$\& \min_{t_1 \in Z_1} ... \min_{t_n \in Z_n} \nu_P(x, t_1, t_2, ..., t_n) \leq \beta\},$$

$$N^{s,a}_{\alpha,\beta}(P) = \{x | \min_{t_1 \in Z_1} ... \min_{t_n \in Z_n} \mu_P(x, t_1, t_2, ..., t_n) \geq \alpha$$

$$\& \max_{t_1 \in Z_1} \dots \max_{t_n \in Z_n} \nu_P(x, t_1, t_2, \dots, t_n) \leq \beta\}.$$

As for the general quantifier, since its evaluation is just a point in the IF-triangle, the corresponding level operators will be propositions (with value 1 (true) or 0 (false)):

$$N_{\alpha,\beta}^{w,g}(P) = 1 \Leftrightarrow \max_{y \in E} \max_{t_1 \in Z_1} \dots \max_{t_n \in Z_n} \mu_P(y, t_1, t_2, \dots, t_n) \geq \alpha$$

$$\& \min_{y \in E} \min_{t_1 \in Z_1} \dots \min_{t_n \in Z_n} \nu_P(y, t_1, t_2, \dots, t_n) \leq \beta,$$

$$N_{\alpha,\beta}^{s,g}(P) = 1 \Leftrightarrow \min_{y \in E} \min_{t_1 \in Z_1} \dots \min_{t_n \in Z_n} \mu_P(y, t_1, t_2, \dots, t_n) \geq \alpha$$

$$\& \max_{y \in E} \max_{t_1 \in Z_1} \dots \max_{t_n \in Z_n} \nu_P(y, t_1, t_2, \dots, t_n) \leq \beta.$$

Theorem 4 *For every predicate P and for every $\alpha, \beta \in [0, 1]$, such that $\alpha + \beta \leq 1$, the inclusions:*

$$N_{\alpha,\beta}^{s}(P) \subseteq N_{\alpha,\beta}^{w}(P),$$
$$N_{\alpha,\beta}^{s,i}(P) \subseteq N_{\alpha,\beta}^{w,i}(P),$$
$$N_{\alpha,\beta}^{s,a}(P) \subseteq N_{\alpha,\beta}^{w,a}(P)$$

hold. Moreover,

$$N_{\alpha,\beta}^{s,g}(P) = 1 \Longrightarrow N_{\alpha,\beta}^{w,g}(P) = 1.$$

2.3. It is well known from classical logic that for each predicate P with argument x having a finite of number interpretations $a_1, a_2, \dots, a_n$:

$$V(\forall x P(x)) = V(P(a_1) \wedge P(a_2) \wedge \dots \wedge P(a_n)),$$

$$V(\exists x P(x)) = V(P(a_1) \vee P(a_2) \vee \dots \vee P(a_n)).$$

Now, having in mind the ideas from [8], we mention that from classical logic, it is well-known that for any two formulas A and B:

$$A \vee B = \neg A \rightarrow B, \tag{1}$$

$$A \wedge B = \neg(A \rightarrow \neg B). \tag{2}$$

Having the 185 intuitionistic fuzzy implications, described in [8] and the 53 intuitionistic fuzzy negations, generated by them and described in [11], Angelova and Stoenchev constructed in [1] 185 disjunctions and 185 conjunctions, using formulas (1) and (2).

On the other hand, as we can see in [6], the disjunction and the conjunction can have two forms and formulas (1) and (2) can be changed to the following new ones:

$$A \vee B = \neg A \to \neg\neg B, \tag{3}$$

$$A \wedge B = \neg(\neg\neg A \to \neg B). \tag{4}$$

In [2], Angelova and Stoenchev constructed 185 disjunctions and 185 conjunctions, using the formulas (3) and (4).

Therefore, formulas (1)–(4) must be rewritten to

$$A \vee_{i,1} B = \neg_{\varphi(i)} A \to_i B, \tag{5}$$

$$A \wedge_{i,1} B = \neg_{\varphi(i)}(A \to_i \neg_{\varphi(i)} B). \tag{6}$$

$$A \vee_{i,2} B = \neg_{\varphi(i)} A \to_i \neg_{\varphi(i)}\neg_{\varphi(i)} B, \tag{7}$$

$$A \wedge_{i,2} B = \neg_{\varphi(i)}(\neg_{\varphi(i)}\neg_{\varphi(i)} A \to_i \neg_{\varphi(i)} B), \tag{8}$$

where $\varphi(i)$ is the subsequent number of the negation that corresponds to the i-th implication (cf. Table 1.3 in [11]).

As it is discussed in [9], we see the possibility for constructing a third group of disjunctions and conjunctions. They will have the forms

$$A \vee_{i,3} B = \neg_1 A \to_i B, \tag{9}$$

$$A \wedge_{i,3} B = \neg_1(A \to_i \neg_1 B). \tag{10}$$

The first **Open Problem** that arises is: Construct all new disjunctions and conjunctions. It is interesting to check whether some disjunctions and conjunctions will coincide. The second **Open Problem** is to study the behaviour of the new disjunctions and conjunctions. For example, which of them will satisfy De Morgan Laws and in which form of these laws?

Another **Open Problem** is to study the properties of the disjunctions and conjunctions from (5)–(10). It is very important to check the validity of the separate axioms—of the intuitionistic logic, of Kolmogorov, of Lukasiewicz and Tarski, of Klir and Yuan, and the other ones.

In [9] an idea for lots of new quantifiers was discussed: For each new pair of conjunction and disjunction, we obtain a pair of quantifiers that have the forms

$$V(\forall_{i,j} x P(x)) = V(P(a_1) \wedge_{i,j} P(a_2) \wedge_{i,j} \ldots \wedge_{i,j} P(a_n)),$$

$$V(\exists_{i,j} x P(x)) = V(P(a_1) \vee_{i,j} P(a_2) \vee_{i,j} \ldots \vee_{i,j} P(a_n)),$$

where i $(1 \leq i \leq 185)$ and j $(1 \leq j \leq 3)$ are the indices of the respective pair of conjunction and disjunction that generates the new pair of quantifiers. One important point here is that for the definition to make sense, we need to explicitly define how to iterate the disjunctions and the conjunctions. We use the following left-associative conventions:

$$A_1 \wedge_{i,j} A_2 \wedge_{i,j} \dots \wedge_{i,j} A_n \wedge_{i,j} A_{n+1}$$

$$= (A_1 \wedge_{i,j} A_2 \wedge_{i,j} \dots \wedge_{i,j} A_n) \wedge_{i,j} A_{n+1},$$

$$A_1 \vee_{i,j} A_2 \vee_{i,j} \dots \vee_{i,j} A_n \vee_{i,j} A_{n+1}$$

$$= (A_1 \vee_{i,j} A_2 \vee_{i,j} \dots \vee_{i,j} A_n) \vee_{i,j} A_{n+1}.$$

In the case when $\wedge_{i,j}$ and $\vee_{i,j}$ are associative (and commutative), any other grouping (or rearranging) will give the same result.

Obviously, $\forall_{4,1}$ coincides with the standard quantifier $\forall$ and $\exists_{4,1}$ coincides with the standard quantifier $\exists$.

One special case is the following: using implication $\rightarrow_{139}$ and negation $\neg_1$ we obtain for $a, b, c, d \in [0, 1]$ and $a + b, c + d \leq 1$:

$$V(\langle a, b\rangle \vee_{139,3} \langle c, d\rangle) = \left\langle \frac{a+c}{2}, \frac{b+d}{2} \right\rangle = \langle a, b\rangle \wedge_{139,3} \langle c, d\rangle.$$

These operations are commutative, but not associative. If for each i: $V(P(x_i)) = \langle a_i, b_i\rangle$, then

$$V(\forall_{139,3} x P(x)) = \left\langle \sum_i p_i.a_i, \sum_i p_i.b_i \right\rangle = V(\exists_{139,3} x P(x)),$$

where $p_1 = \frac{1}{2^{n-1}}$ and $p_i = \frac{1}{2^{n-i+1}}$ for $2 \leq i \leq n$.

Hence, there exists a quantifier's interpretation for which both quantifiers "$\forall$" and "$\exists$" coincide. In this case, we check directly, that

$$\neg_1 \forall_{139,3} x \neg_1 P(x) = \forall_{139,3} x P(x).$$

It is very interesting that the quantifier $\forall_{139,3}$ can be regarded as an analogue of the weight center operator W (see, e.g. [6]).

For a finite linearly ordered set X, $i \in \{1, \dots, 185\}$, $j \in \{1, 2, 3\}$ and $(n + 1)$-dimensional predicate P we define

$$\min_{x \in X}^{i,j} P(x, y_1, \dots, y_n)$$

$$P(x_1, y_1, \dots, y_n) \wedge_{i,j} P(x_2, y_1, \dots, y_n) \wedge_{i,j} \dots \wedge_{i,j} P(x_m, y_1, \dots, y_n),$$

$$\max_{x \in X}^{i,j} P(x, y_1, \dots, y_n)$$

$$= P(x_1, y_1, \ldots, y_n) \vee_{i,j} P(x_2, y_1, \ldots, y_n) vee_{i,j} \ldots \vee_{i,j} P(x_m, y_1, \ldots, y_n),$$

where $x_1 < x_2 < \ldots < x_m$ are the elements of X, listed in increasing order. Both operations produce a predicate of $y_1, \ldots, y_n$.

Let us define

$$\overline{Q}_{i,j} = \begin{cases} \exists_{i,j}, & \text{if } Q_{i,j} \text{ is } \forall_{i,j} \\ \forall_{i,j}, & \text{if } Q_{i,j} \text{ is } \exists_{i,j} \end{cases}.$$

After these remarks, we continue with definitions of new quantifiers over the $(n+1)$-dimensional predicate P. Let for $1 \le i \le 185$ and for $1 \le j \le 3$: $Q_{i,j} \in \{\forall_{i,j}, \exists_{i,j}\}$. Let

$$\underset{x \in X}{\text{ext}}(Q_{i,j}) = \begin{cases} \max^{i,j}_{x \in X}, & \text{if } Q_{i,j} = \exists_{i,j} \\ \min^{i,j}_{x \in X}, & \text{if } Q_{i,j} = \forall_{i,j} \end{cases},$$

Then, for $i, i_1, \ldots, i_n \in \{1, 2, \ldots, 185\}$ and $j, j_1, \ldots, j_n \in \{1, 2, 3\}$, we define:

(e) general Q-additional quantifiers

$$V((Q^1_{i_1,j_1}, Q^2_{i_2,j_2}, \ldots, Q^n_{i_n,j_n})(x, z_1, z_2, \ldots, z_n)$$

$$P(x, z_1, z_2, \ldots, z_n))$$

$$= \langle \underset{t_1 \in Z_1}{\text{ext}}(Q^1_{i_1,j_1}) \ldots \underset{t_n \in Z_n}{\text{ext}}(Q^n_{i_n,j_n})\, \mu_P(x, t_1, t_2, \ldots, t_n),$$

$$\underset{t_1 \in Z_1}{\text{ext}}(\overline{Q}^1_{i_1,j_1}) \ldots \underset{t_n \in Z_n}{\text{ext}}(\overline{Q}^n_{i_n,j_n})\, \nu_P(x, t_1, t_2, \ldots, t_n)\rangle,$$

(f) general Q-quantifier

$$V((Q_{i,j}, Q^1_{i_1,j_1}, Q^2_{i_2,j_2}, \ldots, Q^n_{i_n,j_n})(x, z_1, z_2, \ldots, z_n)$$

$$P(x, z_1, z_2, \ldots, z_n))$$

$$= \langle \underset{y \in E}{\text{ext}}(Q_{i,j})\, \underset{t_1 \in Z_1}{\text{ext}}(Q^1_{i_1,j_1}) \ldots$$

$$\underset{t_n \in Z_n}{\text{ext}}(Q^n_{i_n, j_n})\, \mu_P(y, t_1, t_2, ..., t_n),$$

$$\underset{y \in E}{\text{ext}}(\overline{Q}_{i,j})\ \underset{t_1 \in Z_1}{\text{ext}}(\overline{Q}^1_{i_1, j_1}) \cdots$$

$$\underset{t_n \in Z_n}{\text{ext}}(\overline{Q}^n_{i_n, j_n})\, \nu_P(y, t_1, t_2, ..., t_n)\rangle,$$

Theorem 5 *For each of the third quantifiers, equalities*

$$V(\neg_1(Q^1_{i_1,j_1}, Q^2_{i_2,j_2}, ..., Q^n_{i_n,j_n})(x, z_1, z_2, ..., z_n)$$

$$\neg_1 P(x, z_1, z_2, ..., z_n))$$

$$= V((\overline{Q}^1_{i_1,j_1}, \overline{Q}^2_{i_2,j_2}, ..., \overline{Q}^n_{i_n,j_n})(x, z_1, z_2, ..., z_n)$$

$$P(x, z_1, z_2, ..., z_n))$$

$$V(\neg_1(Q_{i,j}, Q^1_{i_1,j_1}, Q^2_{i_2,j_2}, ..., Q^n_{i_n,j_n})(x, z_1, z_2, ..., z_n)$$

$$\neg_1 P(x, z_1, z_2, ..., z_n))$$

$$= V((\overline{Q}_{i,j}, \overline{Q}^1_{i_1,j_1}, \overline{Q}^2_{i_2,j_2}, ..., \overline{Q}^n_{i_n,j_n})(x, z_1, z_2, ..., z_n)$$

$$P(x, z_1, z_2, ..., z_n)).$$

Open Problem: Which of the equalities from Theorems 1 and 2 are valid for the new quantifiers?

3 Conclusion

The so defined multidimensional intuitionistic fuzzy quantifiers can obtain different applications in the area of artificial intelligence. For example, we can use them in procedures for decision making and for intercriteria analysis, in rules of intuitionistic fuzzy expert systems, and others.

All these multidimensional intuitionistic fuzzy quantifiers are first-order. In a next research, we will discuss possibilities for defining second and higher-order multidimensional intuitionistic fuzzy quantifiers. Some properties for standard predicates, discussed in [18–21, 23–27, 29–31] will be studied for the multidimensional intuitionistic fuzzy quantifiers.

In future, we will study the possibility to change the condition "Let E, Z_1, Z_2, ..., Z_n be fixed finite linearly ordered sets" with which Sect. 2.1 started. When the properties of the new intuitionistic fuzzy conjunctions and disjunctions are studied, probably, we will be able to change this condition with the condition "Let E, Z_1, Z_2, ..., Z_n be fixed finite partially ordered sets". So, the new constructions will give additional possibilities for application in some areas of the artificial intelligence.

Acknowledgements The first two authors are thankful for the support provided by the Bulgarian National Science Fund under Grant Ref. No. DFNI-I-02-5 "InterCriteria Analysis: A New Approach to Decision Making".

References

1. Angelova, N., Stoenchev, M.: Intuitionistic fuzzy conjunctions and disjunctions. Part 1. In: Annual of Section "Informatics" of the Union of Bulgarian Scientists, vol. 8 (2015) (in press)
2. Angelova, N., Stoenchev, M.: Intuitionistic fuzzy conjunctions and disjunctions. Part 2. In: Issues in Intuitionistic Fuzzy Sets and Generalized Nets, vol. 12 (2016) (in press)
3. Atanassov, K.: Two Variants of Intuitonistic Fuzzy Propositional Calculus. Preprint IM-MFAIS-5-88, Sofia (1988)
4. Atanassov K.: Remark on a temporal intuitionistic fuzzy logic. In: Second Scientific Session of the "Mathematical Foundation Artificial Intelligence" Seminar, Sofia, March 30: Preprint IM-MFAIS-1-90. Sofia vol. 1990, pp. 1–5 (1990)
5. Atanassov, K.: Intuitionistic Fuzzy Sets. Springer, Heidelberg (1999)
6. Atanassov, K.: On Intuitionistic Fuzzy Sets Theory. Springer, Berlin (2012)
7. Atanassov, K.: On Intuitionistic Fuzzy Logics: Results and Problems. In: Modern Approaches in Fuzzy Sets. In: Atanassov, K., Baczynski, M., Drewniak, J., Kacprzyk, J., Krawczak, M., Szmidt, E., Wygralak, M., Zadrozny, S. (eds.) Intuitionistic Fuzzy Sets, Generalized Nets and Related Topics, Volume 1: Foundations. SRI-PAS, Warsaw, pp. 23–49 (2014)
8. Atanassov, K.: On intuitionistic fuzzy implications. Issues Intuitionistic Fuzzy Sets Gen. Nets **12**, 1–19 (2016) (in press)
9. Atanassov, K.: On Intuitionistic fuzzy quantifiers. Notes on Intuitionistic Fuzzy Sets **22**(2), 1–8 (2016)
10. Atanassov, K.: Intuitionistic fuzzy logics as tools for evaluation of Data Mining processes. Knowl.-Based Syst. **80**, 122–130 (2015)
11. Atanassov, K., Angelova, N.: On intuitionistic fuzzy negations, law for excluded middle and De Morgan's Laws. Issues Intuitionistic Fuzzy Sets Gen. Nets **12** (2016) (in press)
12. Atanassov, K., Gargov, G.: Elements of intuitionistic fuzzy logic. I. Fuzzy Sets Syst. **95**(1), 39–52 (1998)
13. Atanassov, K., Georgiev, I., Szmidt, E., Kacprzyk, J.: Multidimensional intuitionistic fuzzy quantifiers. In: Proceedings of the 8th IEEE Conference Intelligent Systems, Sofia, 46 Sept 2016, pp. 530–534
14. Atanassov, K., Szmidt, E., Kacprzyk, J.: On intuitionistic fuzzy multi-dimensional sets. Issues Intuitionistic Fuzzy Sets Gen. Nets **7**, 1–6 (2008)

15. Atanassov, K., Szmidt, E., Kacprzyk, J., Rangasamy, P.: On intuitionistic fuzzy multidimensional sets. Part 2. In: Advances in Fuzzy Sets, Intuitionistic Fuzzy Sets, Generalized Nets and Related Topics. Vol. I: Foundations, Academic Publishing House EXIT, Warszawa, pp. 43–51 (2008)
16. Atanassov, K., Szmidt, E., Kacprzyk, J.: On intuitionistic fuzzy multi-dimensional sets. Part 3. In: Developments in Fuzzy Sets, Intuitionistic Fuzzy Sets, Generalized Nets and Related Topics, Vol. I: Foundations. Warsaw, SRI Polush Academy of Sciences, pp. 19–26 (2010)
17. Atanassov, K., Szmidt, E., Kacprzyk, J.: On intuitionistic fuzzy multi-dimensional sets. Part 4. Notes Intuitionistic Fuzzy Sets **17**(2), 1–7 (2011)
18. Barwise, J. (ed.): Studies in logic and the foundations of mathematics. In: Handbook of Mathematical Logic, North Holland (1989)
19. Crossley, J.N., Ash, C.J., Brickhill, C.J., Stillwell, J.C., Williams, N.H.: What is mathematical logic? London, Oxford University Press (1972)
20. van Dalen, D.: Logic and Structure. Springer, Berlin (2013)
21. Ebbinghaus, H.-D., Flum, J., Thomas, W.: Mathematical Logic 2nd edn. Springer, New York (1994)
22. Gargov, G., Atanassov, K.: Two results in intuitionistic fuzzy logic. Comptes Rendus de l'Academie bulgare des Sciences, Tome **45**(12), 29–31 (1992)
23. Lindstrm, P.: First-order predicate logic with generalized quantifiers. Theoria **32**, 186195 (1966)
24. Mendelson, E.: Introduction to Mathematical Logic. Princeton. D. Van Nostrand, NJ (1964)
25. Mostowski, A.: On a generalization of quantifiers. Fund. Math. **44**, 1236 (1957)
26. Mostowski, M.: Computational semantics for monadic quantifiers. J. Appl. Non-Class. Logics **8**, 107121 (1998)
27. Shoenfield, J.R.: Mathematical Logic, 2nd edn, Natick, MA, A. K. Peters (2001)
28. Zadeh, L.: Fuzzy logics. Computer 83–93 (1988)
29. http://www.cl.cam.ac.uk/~aac10/teaching-notes/gq.pdf
30. http://plato.stanford.edu/entries/generalized-quantifiers/
31. http://www.unige.ch/lettres/linguistique/files/5014/1831/0318/MMil_ch_14_GQs.pdf

Data Processing and Harmonization for Intelligent Transportation Systems: An Application Scenario on Highway Traffic Flows

Paulo Figueiras, Guilherme Guerreiro, Ricardo Silva, Ruben Costa and Ricardo Jardim-Gonçalves

1 Introduction

In the last few decades, technology changed the way people live, interact and work. The revolution made by smartphones, internet and sensors, lead to the collection of large volumes of data on a daily basis. Intelligent Transportation Systems (ITS) have to deal with the acquisition and processing of data coming from road sensors, sensing mobile devices, cameras, radio-frequency identification readers, microphones, social media feeds and other sources, to help daily commuters and transportation companies in the decision-making process. The efficient processing and storage of transportation data, can help mitigate many of the transportation challenges, such as excessive CO_2 emissions, traffic congestions, increased accident risks, and reduced quality of life.

All actors in ITS behave as data providers and consumers, leading to large volumes of available data to be processed. The growth in data production is being driven by individuals and their increased use of media (social networks), novel types of sensors and communication capabilities in vehicle and the traffic infrastructure, application

P. Figueiras (✉) · G. Guerreiro · R. Silva · R. Costa · R. Jardim-Gonçalves
CTS, UNINOVA, Dep. de Eng.ª Eletrotécnica, Faculdade de Ciências e Tecnologia, FCT, Universidade Nova de Lisboa, 2829-516 Caparica, Portugal
e-mail: paf@uninova.pt

G. Guerreiro
e-mail: g.guerreiro@campus.fct.unl.pt

R. Silva
e-mail: jpmas@dem.uminho.pt

R. Costa
e-mail: rddc@uninova.pt

R. Jardim-Gonçalves
e-mail: rg@uninova.pt

V. Sgurev et al. (eds.), *Learning Systems: From Theory to Practice*, Studies in Computational Intelligence 756, https://doi.org/10.1007/978-3-319-75181-8_14

of modern information and communication technologies (Cloud computing, Internet of Things, etc.) [1]. Thus, there is an emerging Big Data challenge in the ITS domain.

The big challenge in ITS is not on how to collect, but how to process and model large volumes of unstructured data for posterior analytics, which cannot be handled effectively by traditional approaches. There is a need for developing innovative services and applications capable of process and infer information in real-time for better support the decision making, but also to predict complex traffic related situations before they occur and take proactive actions.

Other aspects that must be considered are the challenges driven by a big data context, such as inconsistencies on the data itself (e.g. highway counters are susceptible to several anomalies), outdated data, the bandwidth of the connection. Such inconsistencies lead to the lack of quality on the data sets and, adding to that, the challenge in fusing and harmonizing large volumes of data from many sources at the same time (volume to variety ratio). These aspects bring particular importance to ETL (Extract-Transform-Load) processes, when traffic data needs to be loaded from sources into a harmonized data repository. ETL software houses have been extending their solutions to provide big data extraction, transformation and loading between big data platforms and traditional data management platforms, describing ETL now has "Big ETL" [2].

This work proposes and evaluates an ETL-based approach, able to process and model large volumes of raw traffic data efficiently. The proposed architecture should be able to: (i) take into account the data quality; (ii) the ability to cope with already existing data standards under the ITS domain, such as DATEX-II, to guarantee harmonization among data; (iii) provide a robust and scalable storage system.

The proposed architecture adopts "big data technologies", such as Apache Spark for data processing, and MongoDB for data storage. A more detailed overview of the proposed architecture, with evaluation of its performance, is presented in the next sections. This approach follows the principles of parallel, in-memory processing by Big Data technologies with all the advantages showed in studies [3, 4].

The approach presented here is being developed under the H2020 R&D OPTIMUM project [5], which operates in an environment of ubiquitous connectivity throughout the transportation network and its surroundings. Within this context one of the application scenarios of OPTIMUM project involves the development of a dynamic toll charging model, aiming at inducing behavioural changes in drivers by transferring heavy traffic from the urban and national roads to highways.

This paper is structured as follows: Sect. 2 presents relevant works related with the presented solution; Sect. 3 focuses on the application scenario; Sect. 4 presents the proposed architecture; Sect. 5 gives an overview on the implemented ETL process; Sect. 6 presents the results achieved so far; Sect. 7 encompasses the conclusion and paves the way for future work.

2 Related Work

The proposed work aims at developing a big-data platform for fusing and harmonize heterogeneous, dynamic streams of transportation data, provided by public authorities, transport operators as well as social media for a better management of transportation networks. The term "Big Data" is used for extremely large and complex data sets, which cannot be handled properly by traditional approaches and tools. Big data represents the assets characterized by high volume, velocity and variety requiring specific technology and analytical methods for its transformation into value [6].

The three aspects most considered to define big data in the literature are: (i) Volume, from the ever augmenting data collected; (ii) Velocity, from the growth on data acquisition; (iii) Variety, from heterogeneity of data formats and protocols used. Other aspects that must be considered are the issues in transporting data in a big data context, such as the many inconsistencies of the data itself (e.g. from gathering values from broken road sensors), outdated data, the speed variation of the internet connection or others. Proposing an ETL architecture that addresses the issues of big data, presents several challenges. Nevertheless, the literature shows some relevant works around the Big ETL concept, and possible approaches using big data technologies applied to the ITS domain. Next, is presented related work regarding the ITS frame, followed by the description of the most highlighted technologies' in them.

The authors in [3], present a technical architecture consisting of core Hadoop services and libraries for data ingest, processing and analytics, operating on an automotive domain processing a dataset of multiple terabytes and billions of rows. The authors in [4] propose a design scheme based on a distributed architecture, mainly using Kafka, Storm and Spark clusters, used to process smart city data, such as Map and POI data, GPS data, traffic data, video surveillance data, environment data, social activity data. Some results indicate that Spark performs better than other approaches. In [3, 4, 8, 9] is stated the many advantages of distributed processing in ETL and data analytics tasks, regarding structured and non-structured data, as well as done in [3, 4] a performance comparison between some technologies that implement the distributed approach. In [9], the authors enhance an existing mobility analytics framework to perform mobility analytics of mobile IoT gateway nodes (along with IoT end-devices), using Spark technology. The authors argue that Spark proved to be a suitable technology for infrastructure as well as ad hoc modes.

Considering the previous identified work, is relevant to portray a little bit these technologies. Hadoop [7] from Apache, is the most well-known framework for big data processes, was developed to process large data sets in a distributed manner, typically used for batch processing. Hadoop was designed for scalable applications and offers also his own type of storage. Apache Spark [8] is a high level and complete framework for Big Data processing, offering Spark SQL for working with structured data, Spark Streaming for stream processing, MLlib for machine learning libraries and GraphX for graphs and parallel graph computation. It runs on Hadoop but uses a different kind of working data sets, with Resilient Distributed Datasets (RDD), which are distributed through the cluster nodes memory when jobs are running. RDDs give efficient recovery after failure. Another great advantage of Spark is that runs in-memory, being more efficient in some operations such as iteration work.

Apache Storm [9] is a free and open source distributed real-time computation system. Storm is used for real-time analytics, online machine learning, continuous computation, distributed RPC, ETL, and more. Storm is fast: a benchmark clocked it at over a million tuples processed per second per node. It is scalable, fault-tolerant, guarantees your data will be processed, and is easy to set up and operate. For storage, the adoption of a NoSQL (Not Only SQL) technology is considered more appropriate. In the field of Big Data storage, there are technologies such as Hive, Cloudera [10], Cassandra [11] and MongoDB [12]. While the first two are based on Hadoop, the second two are based on NoSQL.

Data interoperability is also seen here as an important challenge to be tackled, since the proposed architecture needs to be compliant with a large diversity of transportation related data/services developed in different formats. In order to manage with several heterogeneous data sources, this work adopts an ITS standard based data model. The DATEX II [10] standard was first published in the end of 2006 and acknowledged in 2011 by the European Technical Specification Institute (ETSI) [11], for modelling and exchanging ITS related information, being an European standard for ITS since then.

From the beginning it has been developed to provide a way to standardize information covering the communication between traffic centres, service providers, traffic operators or media partners. Some of the main uses are: (i) Routing/rerouting using traffic management; (ii) Linking traffic management and traffic information systems; (iii) multi-modal information systems; (iv) information exchange between cars or between cars and traffic infrastructure systems.

Endeavours on how to use DATEX II so as to make ITS and Mobility-related frameworks interoperable are not new. In [13], a Cooperative ITS framework is devised in order to manage and optimize a Traffic Light Assistant. In this context, DATEX II is used as the messaging and communication means, being at the same time the interoperability provider, but also the messenger between road operators, users and infrastructure.

In [14] authors present a scalable Big Data multimodal framework able to manage both public and private road transport, by using DATEX II and other standards to represent data. In [15], a V2V (vehicle to vehicle), V2I (vehicle to infrastructure) and I2I (infrastructure to infrastructure) communication architecture is presented and DATEX II is innovatively used as the I2I messaging format. Finally, [16] presents the Norwegian case of a ITS framework for the Norwegian Public Roads Administration, in which several standards are used, such as DATEX II, while custom formats which would then be new local standards or extensions to the existing ones would be developed.

3 Application Scenario

The application scenario involves the development of a dynamic (toll) charging model, aiming to induce behavioural changes in drivers by transferring heavy traffic from the urban and national roads into highways. Such dynamic charging model will combine historical and real-time data, in order to calculate highway tolls' pricing with some hours in advance.

The design and development of a dynamic toll pricing model for highways implies that the design of a pricing model has to take into account traffic flows in real time and traffic flow prediction, resulting also in quality of service prediction for highways and national roads. The proposed dynamic toll pricing model changes tolls' prices, depending on several factors, such as real-time conditions of road networks, quality of service, road safety, environmental data, cost maintenance, toll revenues, congestion, traffic events and weather conditions.

In order to support the model, there is the need to accurately predict the status of road networks for real-time, short and medium term horizons, by using machine learning algorithms. Such algorithms will be used to feed the dynamic toll pricing model, reflecting the present and future traffic situations on the network. Since traffic data quantity and quality are crucial to the prediction of road networks' statuses, real-time and predictive analytics methods will use a panoply of data sources.

Therefore, the presented work addresses the development of an ETL platform based on Big Data technologies, able to capture, clean, harmonize and store traffic data from sensors, such as road-counters, telematics data, traffic incidents, GPS tracks and social media. The resulting data will be later used by machine learning and prediction components to fulfil the above prediction objectives.

Furthermore, the platform will be used on all use cases of the OPTIMUM project, which implies that the platform will be able to collect and process data from other countries, provided with different formats, increasing the volume, velocity and variety of the data sources handled by the proposed Big-ETL platform. Hence, the next section will emphasize on the nature of the project's data sources.

3.1 Data Sources

The testing pilot for the dynamic toll-charging model will be implemented in the North of Portugal, on six highways and their alternatives. In this context, data acquired during the pilot's execution are mainly provided by two partners: Infraestruturas de Portugal (IP), the Portuguese road infrastructure operator, for road network and traffic data, and Luis Simões (LS), a big logistics operator, for floating-car data.

Regarding the other pilots of the OPTIMUM project, data is provided by several entities, depending on the country. In the United Kingdom, traffic-related data sources are provided by UK's National Traffic Information System (NTIS) and the Birmingham City Council (BCC) provides traffic- and public transportation-related data; In Slovenia, the DARS Motorway Company provides several data sources, from weather- to traffic sensor-related data, and LPP, the Ljubljana Public Transport company provides public transportation data.

Data coming from such data sources has to be collected, cleaned and harmonized into predefined structures, depending on its nature and properties. This means that, for instance, traffic sensor data coming from any country of the described above will be stored into a generic structure for traffic sensor data, completely independent of the country of origin or the type of measurement the sensor performs. This work will focus primarily on the Portuguese pilot's data sources, namely sensor and traffic event data.

3.1.1 Sensor Data

Two different sensor-based data sources are used on the dynamic toll-charging pilot, and provided by IP: road counters and toll sensors. The first sensor-based data source is road counters. A road counter, or traffic counter is a device, often electronic in nature, used to count, classify, and/or measure the speed of vehicular traffic passing along a given roadway. The device is usually deployed in near proximity to the roadway and uses an on-road medium, such as pneumatic road tubes laid across the roadway, or piezo-electric sensors embedded in the roadway to detect the passing vehicles [17]. Other pilots also provide road counter data.

The second sensor-based source is toll sensors. Regarding toll sensors, electronic toll collection systems rely on these sensors to perform two of the four main stages of the electronic toll payment process: automated vehicle identification, automated vehicle classification. Both sensor types are handled by IP's SILEGO platform [18], which groups vehicle flow data from sensors throughout the road network. SILEGO is the receiver and integrator of all network data, ensuring the operational interface with IP counter sensors and toll sensors from concessionaires.

3.1.2 Traffic Event Data

IP maintains a Web portal, called Portal das Estradas,[1] which provides video and photography data on real-time traffic, coming from cameras scattered throughout the network, and information contained in message panels along the network.

To feed this portal IP developed a Web Service which contains traffic events captured in IP's traffic monitoring centre. To capture such data, workers at the centre analyse images and videos from the cameras on the road network, and receive detailed information from authorities on real-time events, such as accidents, road works, etc. The Web Service is accessible from the OPTIMUM project.

[1] http://www.estradas.pt

3.2 Methodology

The methodology adopted to implement the proposed ETL architecture was the CRISP-DM (Cross Industry Standard Process for Data Mining). It is considered a well-established methodology, providing a uniform framework and guideline for data miners. Although it was first published in 1999, CRISP-DM has been refined over the year, and with his six steps implementation assures that in the end of the process the knowledge and results are the expected for deployment. Summarily the six steps consist in: Business Understanding, Data Understanding, Data Preparation, Modelling, Evaluation and Deployment.

CRISP-DM is considered one of most widely methodologies for data mining projects, together with SEMMA and KDD [13, 14]. CRISP-DM was decided to be adopted here, mainly because is considered more complete and more practical to apply in real case scenarios with defined objectives.

3.3 The DATEX II Model

There are four main design principles involved in the creation of DATEX II: separation of concerns, in terms of their application domains, a rich domain model, which allows a comprehensive and well defined modelling of data, extensibility, allowing specific extensions depending on country or area, and data exchange (Fig. 1).

DATEX II supports and informs many ITS applications, in particular when cross-border trips are concerned. DATEX II provides the platform and technical specification for harmonized and standardized data modelling and data exchange in ITS

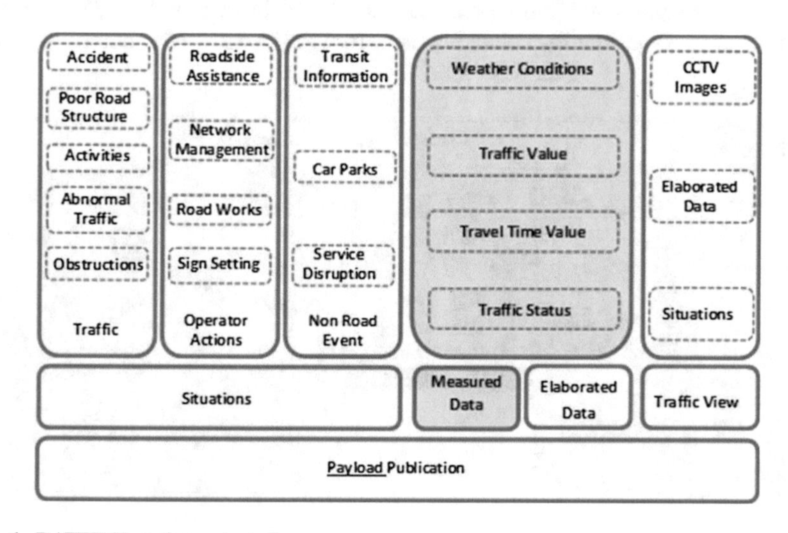

Fig. 1 DATEX II sections adopted

applications. For the sake of the work presented, the most important section is measured data, which provides a way of describing traffic sensor data. In the case of measured data, data sets are normally derived from direct inputs from outstations or equipment at specific measurement sites (e.g. loop detection sites or weather stations) which are received on a regular (normally frequent) basis.

4 Proposed Architecture

For the described application scenario, it is proposed an architecture to extract, transform and store efficiently (ETL stage) all the data previously described. The proposed architecture addresses the following technical requirements: (i) able to deal with raw data in many formats and sizes; (ii) assure data quality; (iii) efficient big data transformation and storage; (iv) being able to address interoperability at the data level, enabling the development of additional value added services for highways users; (v) and a robust and efficient distributed storage system, that is scalable in order to process data from additional traffic sensors.

To address the previous technical requirements, the main tools and standards used for developing the proposed architecture were (i) Apache Spark, used for large-scale data processing, which includes the tasks of data cleaning and transformation; (ii) DATEX-II, as data reference model which the data is stored; (iii) MongoDB, as a NoSQL approach, for storing and managing the traffic data.

Figure 2 depicts the conceptual architecture, where the data sources may arrive in different means (local or server documents, web services using SOAP or REST), and in different formats (txt, CSV, XLS, XML, JSON) and with all kinds of sizes. Spark supports the entire data collection, harmonization and cleaning processes.

All data adaptors and harmonizers are developed in Java or NodeJS. The data adaptors are responsible for collecting data from the various data sources and may

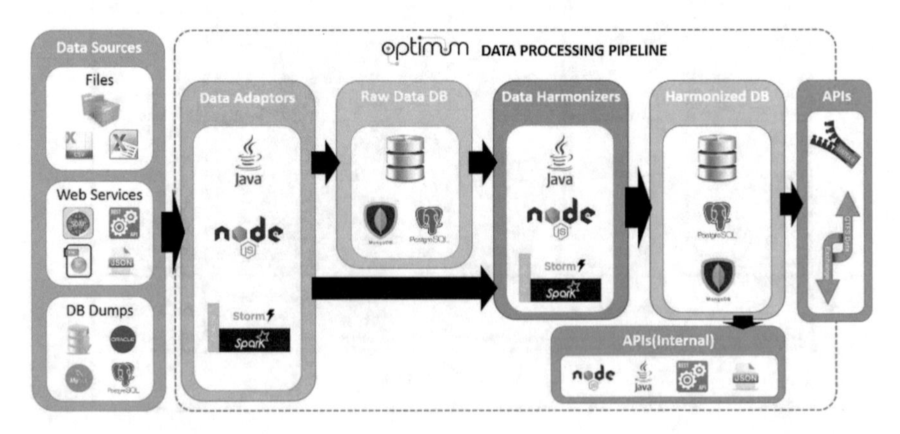

Fig. 2 Data processing pipeline

start the cleaning and harmonization process, depending on the veracity degree of data, i.e. its availability and quality. Data quality tests are made on all data sources, as explained in the following section.

In cases where the data needs little adaptations to reach the predefined structures, as in the case of public weather APIs, for instance, the adaptor stats the harmonization process; in other cases, such as data coming from Web Services, the adaptors convert data into JSON, and store the raw data in the raw data database.

The data harmonizers grab data from the adaptors or the raw data database and are responsible for cleaning and harmonizing data sources. There are specific harmonizers for specific data sources, depending on the nature of data, i.e. sensor data versus event data, national road sensor data versus highway sensor data, etc. Finally, harmonized data is stored in the harmonized database, which in turn feeds both internal (for use within the project) and external (through standard-based output adaptors) APIs.

5 Data Processing

The data processing procedure is presented in several steps. First, the data understanding step focused on finding the adequate harmonized data structures, with special attention to their compliancy with DATEX II, and on the data availability tests. Next, in the data preparation step, the adaptors and harmonizers were responsible for cleaning and harmonizing the data to be used or exported via APIs.

5.1 Data Understanding Step

5.1.1 Harmonized Structures

First, a conceptual analysis of the data was made, with objective of looking at all the traffic-related data sources and produce a list in which several data sources were aggregated into a common data structure, depending on their nature and context, and mapped into DATEX II concepts when possible.

The first step was to analyse all data sources from all pilots and decide which of these would have a common harmonized data structure For instance, all traffic sensor-related data sources should be held in a common sensor data structure; weather data coming from all pilots should be mapped into a common weather data structure, etc. First, each data source was assigned with an ID, as exemplified in Table 1 for some Portuguese data sources.

Second, for each data type, a list of the fields contained in the data sources was created and linked to each data source via its ID, as exemplified in Table 2 with just some of the fields. All the data sources represented by their IDs on the third column have the field presented on the second column in their structure.

Table 1 Data sources' ID assignment

Data source	ID
DTC_POR_ConcessionTollCrossingVolumes	1
DTC_POR_HighwayCounters	2
DTC_POR_RoadVolume	3
DTC_POR_TrafficEventsDB	5
DTC_POR_TrafficEventsWS	43

Table 2 Conceptual harmonization of data sources' fields

Structure name	Field	Data source ID
sensor_values	_id	1, 2, 3, 9, 10, 11, 12, 13, 14, 38, 44, 45
	sensor_id	1, 2, 3, 9, 10, 11, 12, 13, 14, 38, 44, 45
	date_time	1, 2, 3, 9, 10, 11, 12, 13, 14, 38, 44, 45
	a_vehicles/class_1_vehicles	1, 2, 3
	b_vehicles/class_2_vehicles	1, 2, 3
	c_vehicles/class_3_vehicles	1, 2, 3
	d_vehicles/class_4_vehicles	1, 2, 3
	class_5_vehicles	1, 2, 3
	light_vehicles	(calculated from the above flow values)
	heavy_vehicles	(calculated from the above flow values)
	total_vehicles	1, 2, 3, 38

The next step was to identify the data types for each field and which of these fields would map to a DATEX II concept, or XML tag, as shown in Table 3, for just some fields. This analysis exercise was repeated for several data sources, from sensor data and metadata to weather and environmental data, going through traffic event data, parking facility data, etc.

From the conceptual analysis, several data structures were created in order to house common data sources coming from all pilots. Another thing to bear in mind is that traffic sensor data sources coming from different pilots have different measurements or fields. For instance, the Portuguese sensor data has the occupancy percentage field, whereas the British sensor data has traffic headway field. Each of these fields only exist in their own country's sensor data.

The data structures are prepared to be dynamic in that sense, allowing for the presence of all the fields contained in all the data sources, related to the same data type. Some examples of the harmonized data structures are shown below as Mongo DB's BSON (Binary JSON) objects.

Figure 3 represents the harmonized structure for metadata of traffic sensors, whether they are loop, count or toll sensors. Figure 4 represents traffic sensor data. The "readings" array supports all types of measurements from the sensors, such as

Table 3 Mapping data fields to DATEX II

Field	Datex II concept	Data type
_id		ObjectId
sensor_id	datex:SiteMeasurements(measurementSiteReference)	ObjectId
date_time	datex:DateTimeValue(dateTime)*	Timestamp
a_vehicles/class_1_vehicles		Integer
b_vehicles/class_2_vehicles		Integer
c_vehicles/class_3_vehicles		Integer
d_vehicles/class_4_vehicles		Integer
class_5_vehicles		Integer
light_vehicles	datex:TrafficFlow(vehicleFlowValue) CarOrLightVehicle	Integer
heavy_vehicles		Integer
total_vehicles	datex:TrafficFlow(vehicleFlowValue)	Integer
gap_between_vehicles	datex:TrafficHeadway(averageDistanceHeadway)	Float

```
{
    "_id" : ObjectId("56ab5c569f6ed594781cc00f"),
    "concession_name" : "EP Grande Porto (ex AEDL)",
    "road_name" : "A1",
    "road_type" : "highway",
    "sensor_type" : "counter",
    "km_point" : 297.0,
    "sensor_id_holder" : "A1_297+975_CT3687_C",
    "section" : "Santo Ovideo - Coimbrões (A44) ",
    "state" : "active",
    "concession_holder" : "IP",
    "bearing" : "northbound",
    "country" : "pt",
    "location" : {
        "type" : "Point",
        "coordinates" : [
            -8.607517,
            41.11034
        ]
    }
}
```

Fig. 3 Example of the harmonized structure for traffic sensor metadata

average speeds, occupancy percentages, vehicle flows per class and per lane, etc. Finally, Fig. 5 shows the harmonized structure for traffic events. These structures take into account DATEX II enumerations, such as bearing or traffic event type.

5.1.2 Data Availability Tests

Several tests were made on all data sources, depending on the nature of the data. Some examples are shown below: In the case of traffic counters, available data spans

```
{
    "_id" : ObjectId("57015e7c60a5dee6439d5133"),
    "sensor_id" : "A1_297+975_CT3687_C",
    "date_time" : ISODate("2015-01-01T00:10:00.000+0000"),
    "volume" : 178,
    "average_speed" : 76,
    "occupancy" : 12,
    "flow" : 2136,
    "readings" : [
    {
      "type" : "flow",
      "value" : 0,
      "vehicle_class" : "optimum_pt_category_a"
    },
    {
      "type" : "flow",
      "value" : 2088,
      "vehicle_class" : "optimum_pt_category_b"
    },
(...)
    {
      "type" : "average_speed",
      "value" : 80,
    },
]
}
```


Fig. 4 Example of the harmonized structure for traffic sensor data


```
{
    "_id" : ObjectId("5759a6f9244a4217e4dff374"),
    "id" : "30ff66d3-25aa-9f54-e050-0010a46021411",
    "type" : "MaintenanceWorks",
    "road" : "M202",
    "bearing" : "bothWays",
    "description" : "Road works for road pavement",
    "start_date_time" : ISODate("2016-05-05T07:00:00.000+0000"),
    "end_date_time" : ISODate("2016-10-28T16:00:00.000+0000"),
    "location" : {
        "type" : "Point",
        "coordinates" : [
            41.7063924033925,
            -8.80627789033159
        ]
    },
    "source" : "IP",
    "version" : "1",
    "severity" : "unknown"
}
```


Fig. 5 Example of the harmonized structure for traffic event data

from January 1st, 2014 at 00:00 to November 31st, 2016 at 23:55 (still collecting), and it has a sample rate of five minutes, which means 12 readings per hour. Per year, the number of readings for one counter is 105120 readings.

Multiplying by the total number of sensors we get the number of total readings per year, 28382400. Of course some counters have time spans with no data, due to inactivity or mal-functioning periods. Data quality tests were made on sensor data,

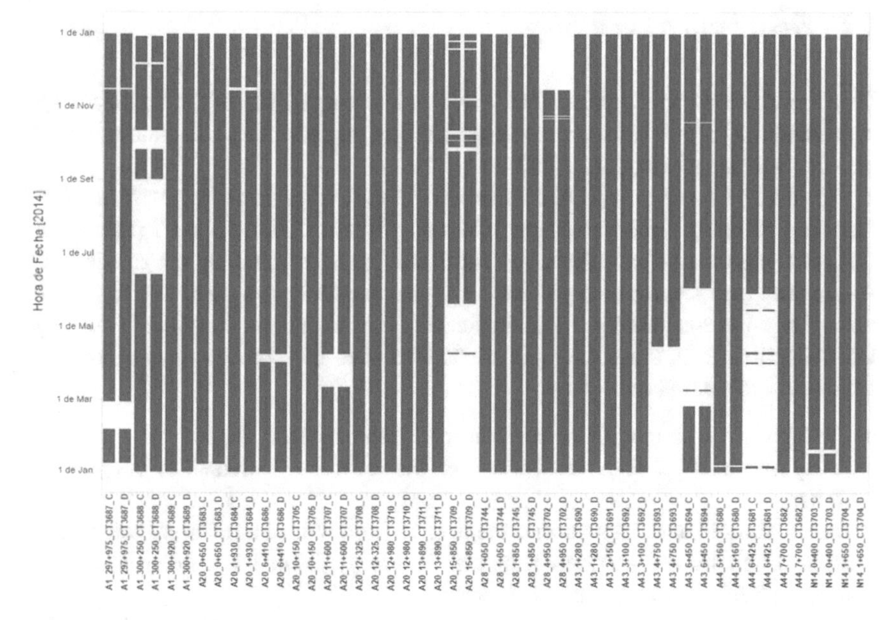

Fig. 6 Data availability test sample

in order to check the overall quality of the readings. A sample of the data availability test is shown in Fig. 6.

The sensors are organized in pairs, each for a specific direction of the highway. It is visible the time spans of inactivity for each sensor, represented by the white spaces. Figure 6 represents an example test made on a sample of 50 counters for the year of 2014 in the horizontal axis, while the vertical axis corresponds to the total timespan of the collected data.

From this test, a quality factor was introduced, η_{sensor}^{year}, which represents the percentage of data completeness for a sensor per year:

$$\eta_{sensor}^{year} = \frac{\# of\ stored\ readings\ per\ year}{\# of\ expected\ readings\ per\ year} * 100 \tag{1}$$

For instance, regarding the sensor with ID A20_0 + 650_CT3683_C, corresponding to the 7th column of Fig. 6, for the year 2014, the quality factor is:

$$\eta_{\text{A20\_0+650\_}CT3683\_C}^{2014} = \frac{103.206}{105.120} * 100 = 98,1792\% \tag{2}$$

5.2 Data Preparation Step

Data Preparation corresponds to the selection and preparation of the final data sets. This phase includes tasks, such as records, tables and attributes selection as well as cleaning and transformation of data.

Apache Spark, using its RDD abstraction to handle large datasets, provides several functions that allow multiple transformations of these RDDs. Figure 7 presents the ETL process, using Spark functions and RDD operations, taking advantage of its in—memory efficiency in repetitive tasks and parallel processing. In Fig. 7, the extraction source is a folder containing several files to process and store in a Mongo DB collection. Still, the same logic applied for such ETL process may reproduced for other data sources and types.

- The files in the source folder are converted into one RDD Object;
- Using Spark's Flat Map transformation, the records are divided in the previous RDD;
- Data is filtered and mapped into the desired harmonized structures, using Spark Map to Pair transformations. Data cleaning is performed to correct issues, like missing fields or wrong variable types;
- The RDD block produced in 3 is stored to a Mongo DB collection using MongoDB Hadoop Connector.

Some examples on the cleaning and harmonization processes are presented below. Sensor metadata and data for the Portuguese pilot are collected in two distinct ways. First, an historical dump in the form of CSV files is stored with the appropriate harmonized format. Second, a Web Service containing an updated stream of sensor data is continuously accessed in order to get (almost) real-time sensor data, and also needs to be transformed to comply with the harmonized format for sensor metadata and data.

In this case, the data coming from the Web Service has to be complemented with an Excel spreadsheet containing the locations, kilometre points and bearings of all the sensors in the Web Service list. The following example represents sensor metadata

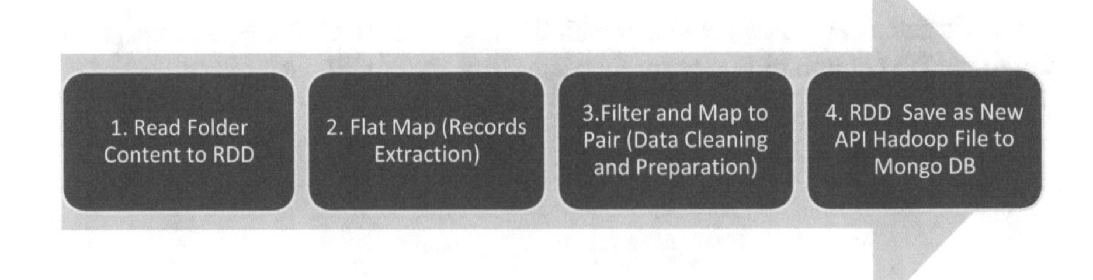

Fig. 7 ETL process

```
{
"_id" : ObjectId("57d297eeea968c19a8c6f605"),
"Concessionary" : "Brisa",
"ConcessionaryID" : "6",
"Concession" : "Auto-Estradas do Baixo Tejo, S.A.",
"ConcessionID" : "5",
"Road" : "A33",
"RoadID" : "20",
"Segment" : "A12/IC32-Alcochete",
"SegmentID" : "310",
"MeasurePoint" : "IC3 1+650 CT1869_C",
"MeasurePointID" : "1033"
}
```


Fig. 8 Raw BSON sensor metadata collected via Web Service


```
{
    "_id" : ObjectId("56ab5d6a9f6ed594781ddf01"),
    "sensor_id" : ObjectId("56ab5c569f6ed594781cc00f"),
    "a_vehicles" : 0,
    "b_vehicles" : 1836,
    "c_vehicles" : 180,
    "d_vehicles" : 0,
    "date_time" : ISODate("2014-01-08T10:55:00.000+0000"),
    "total_vehicles" : 2016,
    "light_vehicles" : 1836,
    "heavy_vehicles" : 180,
    "occupancy" : 9,
    "average_speed" : 98,
    "volume" : 168
}
```


Fig. 9 Raw BSON sensor data collected from CSV files

coming from the Web Service and stored in the raw data database with the structure, shown in Fig. 8. Table 4 shows the transformation processes applied on the fields of the sensor metadata coming from the Web Service.

Raw sensor data coming from CSV dumps is stored in MongoDB with the structure shown in Fig. 9. Table 5 shows the transformation processes applied on the fields of the sensor data coming from the CSV files.

Raw traffic event data coming from IP's Web Service is stored in MongoDB with the structure shown in Fig. 10. Table 6 shows the transformation processes applied on the fields of the traffic event data.

Finally, harmonized data can be exported to be used within the project, via internal JSON APIs, and also for consumers outside the project, via exporters to DATEX II.

6 Validation and Results

For validation and testing, CSV files containing raw traffic flow data were used as input. The proposed approach grows upon a typical ETL approach, by adopting big data technologies, therefore the metrics used for validation will measure the

Table 4 Transformations made on sensor metadata collected via Web Service

Original field	Harmonized field	Transformation
"Concession": "Auto-Estradas do Baixo Tejo, S.A."	"concession_name": "Auto-Estradas do Baixo Tejo, S.A."	The Concession field changes its name to concession_name
"Concessionary": "Brisa"	"concession_holder": "Brisa"	The Concessionary field changes its name to concession_holder
"Road": "A33"	"road_name": "A33"	The Road field changes its name to road_name
"Segment": "A12/IC32-Alcochete"	"section": "A12/IC32-Alcochete"	The Segment field changes its name to section
N/A	"sensor_type": "counter"	The sensor_type field is added depending on the type of road (highway, national, etc.)
"MeasurePoint": "IC3 1 + 650 CT1869_C"	"sensor_id_holder": "IC3 1 + 650 CT1869_C"	The MeasurePoint field changes its name to sensor_id_holder
N/A	"location": { "type": "Point", "coordinates": [41.7063924033925, −8.80627789033159]}	The location field is added from the Excel spreadsheet containing all the locations for the sensors and is converted to the GeoJSON format
N/A	"state": "active"	The state field is added, as all the sensors in the Web Service are active
N/A	"bearing": "eastbound"	The bearing field is added from the Excel spreadsheet containing all the bearings for the sensors and is converted to the DATEX II enumeration for bearing (northbound, southbound, etc.)
N/A	"km_point" : 1.65	The km_point field is added from the Excel spreadsheet containing all the kilometre points for the sensors

performance of the proposed approach versus a typical ETL approach without big data technologies.

The tests were performed using historical traffic flows from 2010 till 2016 in several highways, in a total of 36 CSV, resulting in 30 million records to be processed. Each record is composed by the concession's ID, toll's ID, date and flow per vehicle category. The data is cleaned, transformed and stored directly in the MongoDB historic database. Figure 11 highlights the performances using a traditional approach

Table 5 Transformations made on sensor data collected from CSV files

Original field	Harmonized field	Transformation
"sensor_id": ObjectId ("56ab5c569f6ed594781cc00f")	"sensor_id": "A1_297 + 975_CT3687_C"	The sensor_id field is changed from the _id field in the sensor metadata to the sensor_id_holder field (avoiding duplicate IDs when migrating to other instances of MongoDB)

"a_vehicles" : 0,
"b_vehicles" : 1836,
"c_vehicles" : 180,
"d_vehicles" : 180,
"light_vehicles" : 1836,
"heavy_vehicles" : 180,
"occupancy" : 9,
"volume" : 168,
"average_speed" : 98
	"readings": [{
	"type": "flow",
	"value": 0,
	"vehicle_class":
	"optimum_pt_category_a"},{
	"type": "flow",
	"value": 2088,
	"vehicle_class":
	"optimum_pt_category_b"},
	(…)
	{
	"type": "average_speed",
	"value" : 80,}]
		Readings for different vehicle categories are transformed to an array containing the type of reading, the value and the vehicle_class it applies to. Another field might be the lane number (not existent in the Portuguese data)
		Several vehicle classes were added in order to encompass all the vehicle categorizations. The readings array has a special field to map these new classes, called vehicle_class.

"total_vehicles" : 2016	"flow" : 2016	The names for the calculated flow values change. The word flow replaces the word vehicles

```
<Ocorrencia>
   <Id>3792681f-4877-fda3-e050-0010a46025e4f</Id>
   <Data>2016-08-25T00:00:00</Data>
   <DataInicio>2016-08-25T00:00:00</DataInicio>
   <DataFim>2016-08-25T08:00:00</DataFim>
   <CodigoPais>pt</CodigoPais>
   <Tipo>Accident</Tipo>
   <Estado>active</Estado>
   <Descricao>Accident in right lane involving two vehicles</Descricao>
   <DistritoInicial>Porto</DistritoInicial>
   <DistritoFinal>Porto</DistritoFinal>
   <ConcelhoInicial>Póvoa De Varzim</ConcelhoInicial>
   <ConcelhoFinal>Póvoa De Varzim</ConcelhoFinal>
   <Estrada>N205</Estrada>
   <Direccao>Ambos</Direccao>
   <Km>13.6</Km>
   <PontoGeometrico>
      <SRID>8320</SRID>
      <Ponto>
         <X>-8.72549342584446</X>
         <Y>41.43049481842990</Y>
      </Ponto>
   </PontoGeometrico>
</Ocorrencia>
```


Fig. 10 Raw traffic event data from IP's Web Service in XML format

without any big data technology, an approach using Spark configured to run locally with 2 threads, and using Spark configured to run locally with 4 threads.

Figure 12 depicts the performances of the proposed architecture, using the classical method, the method running Spark locally with 4 threads and using Spark on a distributed cluster environment (in standalone mode and 1 additional worker).

For the current validation process, it was decided to adopt Spark in standalone mode, due to its simplicity in setting the cluster and because it is already included in Spark. It is worth to mention that Spark was running locally on an i5-4200U, with 8 GB RAM machine, and the worker node was working on an i7-4790, with 8 GB RAM machine.

The results indicate that a "big data" approach, as the one presented here, is more suitable in terms of performance with respect to traditional approaches for ETL tasks. The classic method performance was compared with a local installation of Spark running with 2 threads and 4 threads. There is a clear increase of performance when using Spark, around 50% of decrease in the processing time. When comparing 2 thread and 4 thread modes, the improvement is not so obvious. For that reason, it was decided to test it on a cluster environment running Spark in standalone mode with an additional remote worker machine; the results showed a 25% gain when compared to the local mode with 4 threads.

Other Spark configurations were also tested, namely increasing the number of local threads and also increasing the number of remote workers. Such configurations did not clearly show substantial gains in terms of performance. By reading other related works, we have found out that Spark presents some issues related with garbage collection increase and file I/O time. Therefore, as future work, it is expected to test the approach on a larger data set.

Table 6 Transformations made on traffic event data collected from IP's Web Service

Original field	Harmonized field	Transformation
<Id>	"id"	The <Id> tag is changed to id
<DataInicio>	"start_date_time"	The <DataInicio> tag is changed to start_date_time
<DataFim>	"end_date_time"	The <DataFim> tag is changed to end_date_time
<CodigoPais>	"country"	The <CodigoPais> tag is changed to country
<Tipo>	"type"	The <Tipo> tag is changed to type
<Descricao>	"description"	The <Descricao> tag is changed to description
<Estrada>	"road"	The <Estrada> tag is changed to road
<Direccao>	"bearing": "bothWays"	The bearing field is converted to the DATEX II enumeration for bearing (northbound, southbound, etc.)
<Km>	"km_point"	The <Km> tag is changed to km_point
<PontoGeometrico> <SRID>8320</SRID> <Ponto> <X>-8.72549342584446</X> <Y>41.43049481842990</Y> </Ponto> </PontoGeometrico>	"location": { "type": "Point", "coordinates": [-8.72549342584446, 41.43049481842990]}	The location field is converted to the GeoJSON format
N/A	"version"	The version of the event is added, and updated anytime the same event is updated
N/A	"severity"	The severity is added; if there is no information, the value is unknown
N/A	"source"	The source provider is added, in this case, IP

7 Conclusions and Future Works

This work describes an ETL approach applied to the ITS domain, that is responsible for preparing and harmonizing traffic flow data, collected from a network of multiple sensors deployed highways. The main motivation that drives this work, is to create

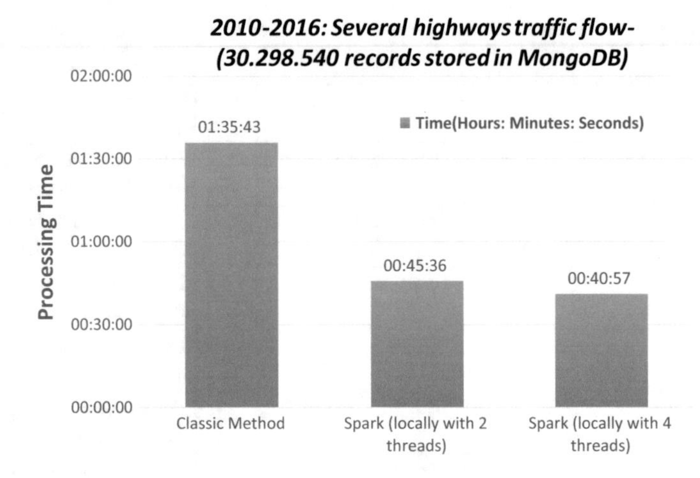

Fig. 11 Processing time "classical method" versus spark instance running locally

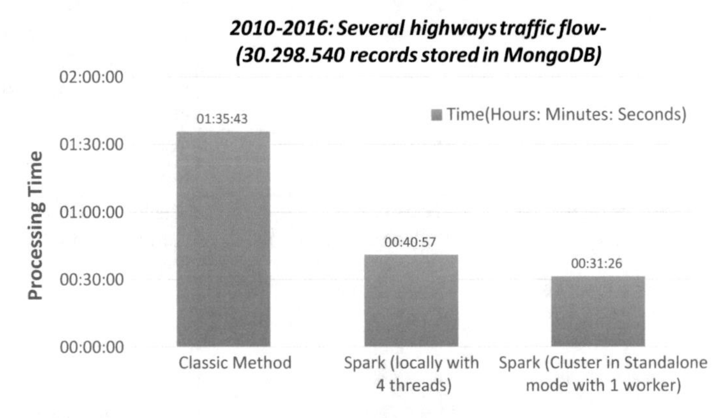

Fig. 12 Processing time "classical method" versus spark instance running locally versus spark instances running on a distributed cluster environment

data cleaning and transformation tasks into a huge volume of traffic data, available in heterogeneous formats.

The work presented here, is still part of ongoing work currently addressed under the EU H2020 OPTIMUM project. Preliminary results achieved so far enabled us to demonstrate considerable gains in performance, when compared to other traditional ETL approaches, and also form the basis for pointing out and discuss future work directions and opportunities in the area of the development of big data processing and mining methods under the ITS domain.

The authors have shown the Data Understanding and Preparation stages, presenting all the processes involved in the creation of the Big-ETL platform. Data availability tests were performed in order to check the integrity and quality of available data sources. Then, a data processing pipeline was developed in order to collect and harmonize data coming from heterogeneous sources. Finally, the results of this pipeline were shown, so as to validate the adopted Big Data approach based on Spark. These results show that the adoption of Spark as support for the Big Data harmonization procedure got really good results, in terms of performance, when compared with classical approaches.

Future work may include the integration of additional traffic data from other traffic sensors in order to cover a bigger geographical area, the implementation of a similar pipeline for real-time streams of data, using Apache Storm, for instance, and the application of the solution on a full-fledged server. Finally, the development of traffic forecasting services based on regression techniques are required in order to feed the dynamic toll charging model, on the other hand, CEP mechanisms are also required in order to identify relevant events (traffic accidents, traffic jams) in real-time which can heavily influence the status of the transportation network.

References

1. Fiosina, J., Fiosins, M., Müller, J.P.: Big data processing and mining for next generation intelligent transportation systems. Jurnal Teknologi **63**(3), 21–38 (2013)
2. Caserta, J., Cordo, E.: Big ETL: The Next 'Big' Thing, 9 Feb 2015 [Online]. http://data-informed.com/big-etl-next-big-thing/
3. Luckow, A., Kennedy, K., Manhardt, F., Djerekarov, E., Vorster, B., Apon, A.: Automotive big data: applications, workloads and infrastructures. In: IEEE International Conference on Big Data, Santa Clara, CA (2015)
4. Ma, S., Liang, Z.: Design and implementation of smart city big data processing platform based on distributed architecture. In: 10th International Conference on Intelligent Systems and Knowledge Engineering (ISKE), Taipei (2015)
5. OPTIMUM consortium, OPTIMUM Project, 1 Oct 2015 [Online]. http://optimumproject.eu/. [Acedido em 20 April 2016]
6. De Mauro, A., Greco, M., Grimaldi, M.: What is big data? A consensual definition and a review of key research topics. In: proceedings of the 4th International Conference on Integrated Information, Madrid (2015)
7. The Apache Software Foundation, Hadoop (2014) [Online]. [Acedido 2016]
8. The Apache Software Foundation, Apache Spark (2014) [Online]. https://spark.apache.org
9. The Apache Software Foundation, Apache Storm (2014) [Online]. https://storm.apache.org/
10. Cloudera, Inc., Cloudera Impala (2015) [Online]. http://www.cloudera.com/content/cloudera/en/products-and-services/cdh/impala.html
11. The Apache Software Foundation, Apache Cassandra Project (2015) [Online]. http://cassandra.apache.org/
12. MongoDB, Inc., MongoDB (2015) [Online]. https://www.mongodb.org/
13. Andersen, O., Krogh, B.B., Torp, K.: An open-source based ITS platform. In: Proceedings of IEEE International Conference on Mobile Data Management, vol. 2, pp. 27–32 (2013)
14. Taneja, M.: A mobility analytics framework for internet of things, pp. 113–118 (2015)
15. EIP/EIP + Project, Datex Easyway (2014) [Online]. http://www.datex2.eu/
16. European Telecommunications Standards Institute (2015) [Online]. http://www.etsi.org/
17. Freudenstein, J., Cornwell, I.: Tailoring a reference model for C-ITS architectures and using a DATEX II profile to communicate traffic signal information. In: Transport Research Arena, Paris, France (2014)
18. Duperet, A., Damas, C., Scemama, G.: Public authorities support for a large scale data infrastructure for mobility (LaSDIM). In: 22nd ITS World Congress, Bordeaux, France (2015)

Printed in the United States
By Bookmasters